CASS LIBRARY OF SCIENCE CLASSICS

No. 8

General Editor: Dr. L. L. LAUDAN, University College London

W0234986

PHILOSOPHICAL
EXPERIMENTS AND OBSERVATIONS

PHILOSOPHICAL

EXPERIMENTS

AND

OBSERVATIONS

ROBERT HOOKE

EDITED BY W. DERHAM

LONDON AND NEW YORK

First published by
FRANK CASS AND COMPANY LIMITED

First published   1726
New impression, with index, of the First edition   1967

Published 2005 by Routledge
2 Park Square, Milton Park, Abingdon, Oxfordshire OX14 4RN
711 Third Avenue, New York, NY 10017

First issued in paperback 2014

Routledge is an imprint of the Taylor and Francis Group, an informa business

ISBN 13: 978-0-714-61115-0 (hbk)
ISBN 13: 978-0-415-76033-1 (pbk)

Publisher's Note
The publisher has gone to great lengths to ensure the quality
of this reprint but points out that some imperfections
in the original may be apparent

Publisher's Note to the 1967 Edition

This is an exact facsimile reproduction of Derham's edition of the *Philosophical Experiments and Observations of the late Eminent Dr. Robert Hooke* (1726) except that an analytical table of contents, prepared by the General Editor, has been added.

Editor's Note to the 1967 Edition

Shortly after Hooke died in 1703, his miscellaneous papers and unpublished manuscripts were entrusted to Richard Waller, who edited and published some of them in a volume titled *The Posthumous Works of Robert Hooke* (1705; reprinted, Frank Cass, 1968). Waller himself died however before he was able to complete the task of republishing Hooke's papers and they were eventually handed on to William Derham. After delaying for what some of Hooke's followers thought to be a scandalously long time, Derham finally published this volume in 1726. It contains numerous papers and notes by Hooke as well as a number of important papers and letters written by Hooke's contemporaries and found, evidently, among Hooke's literary remains. For a discussion of several of the items contained in this work the reader is referred to Margaret 'Espinasse's *Robert Hooke* (London, and Berkeley, Calif., 1956) and Geoffrey Keynes' *Bibliography of Dr. Robert Hooke* (Oxford, 1960).

Analytic Table of Contents

PAGE

TABLE OF CONTENTS

TABLE OF CONTENTS

TABLE OF CONTENTS

Philoſophical

EXPERIMENTS

AND

OBSERVATIONS

Of the late Eminent

Dr. *ROBERT HOOKE,*

S. R. S.

And Geom. Prof. *Greſh.*

AND

Other Eminent Virtuoso's in his Time.

With COPPER PLATES.

Publiſh'd by W. Derham, F.R.S.

LONDON:

Printed by W. and J. Innys, Printers to the
Royal Society, at the Weſt End of *St. Paul's.*
MDCCXXVI.

T O

The Right Honourable

JULIANA,

Countess-Dowager

O F

BURLINGTON,

This Collection of Papers,

As well for her Perfonal Virtues and Merits,
as for her fingular Favours to me, are,
with greateft Refpect and Gratitude, humbly dedicated by

Her Ladyſhip's

Moſt obliged

Humble Servant,

W. Derham.

AMENDMENTS.

PAG. 153. Infert in the Margin at l. 14. *V. Poſt Works.* p. 564. P. 226. l. *ult.* after *looſe,* add *in the ſame manner.* P. 227. l. 4. for *Tab.* III. read I. Ib. l. 27. read *Weight* K K. Ib. l. 29, 30. r. *Hook* E F. G *the Ring to be hung on the Hook* F. P. 228. l. 31. for *believed,* r. *received.* P. 230. *ult.* r. *Height.* P. 231. l. 25, 26. r. *Height.* P. 233. l. *antepen.* r. *Tab.* I. P. 234. l. 17. r. *theſe.* Ib. l. *penult.* after left add, *and by that Means.* P. 237. l. 3. after Stick add, *ſmaller, and tapering upwards towards great* D, *which is an hollow very light Ball of Wood.* Ib. l. 26. read *Tab.* II. P. 238. l. *penult.* r. *Tab.* I. P. 239. l. 12. r. *Tab.* I. Ib. l. 20. r. *proportion.* P. 240. l. 4. for *ſo,* r. *by.* Ib. l. 18. after or add *that.* P. 321. l. 28. r. *Condore.* P. 328. l. 16. r. *abiegno.* P. 336. l. 17. r. *Lignum-Aloes, Civet, Storax & Ladanum.* P. 245. l. 5. read *mea.*

TO THE

READER.

HE principal Author of these Papers being a Person of great Repute, I thought the Publication of them would be very acceptable to the Curious ; and therefore was willing to undertake the Work, although I found it would be very laborious, by reason the Papers were very numerous, and in great Confusion.

After Dr. Hook's Death, both his Papers, and some of his Figures and Modules (but I fear not nearly all) fell into the Hands of my ingenious Friend Richard Waller, Esq; out of which he selected those that he published in 1705 ; and intended others for the Press : But dying before he had accomplished that Design, a Part of the Papers were entrusted

To the READER.

trufted to me, by Mr. Waller's Lady, and Jonathan Blackwell, *Efq; In which I expected great Matters from fuch illuftrious Names, as I found among them: But when I came to perufe, and examine them, I found only here and there fome, that anfwered my Expectation; which the Reader hath in the following Collection. In which he may probably expect fome of the many* Lectures, *which the Doctor read in* Grefham College, *and thofe of Sir* John Cutler's *Inftitution. But the beft of thefe Dr.* Hook *himfelf, or Mr.* Waller *publifhed: So that what I have in my Hands, will be of little Ufe to the learned World, moft of them feem to have been intended by the Doctor, for half an Hour's Amufement to a fmall Auditory, rather than for the Prefs.*

As for Order, or Method, little could be obferved in fuch a confufed Variety of Subjects, as thefe Papers contain. And therefore the beft I could do, was to rank them, as near as I could, according to the Order of the Time in which they were written, or communicated.

And as for other Papers interfperfed with Dr. Hook's, *they are, for the moft part, of fuch confiderable Perfons, that the Reader will expect no Excufe for my inferting of them.*

But if any remarkable Obfcurities or Imperfections fhould be met with, it is what I could not help, by reafon fome of the Papers

were

To the READER.

were torn, *some obliterated, some written in an Hand scarce legible, &c. and I was not minded to give my own Sense, lest it should be thought that I had imposed my own, instead of the several ingenious Authors Senses.*

But after all, many of those Imperfections, and Obscurities, are owing to the Miscarriage of some of the Papers, which either never came to Mr. Waller's Hands; or, if they did, were lost, or mislaid, before they came to mine, the Papers being put into different Hands, after Mr. Waller's Death. And whereas Figures, or Modules, *would have explained divers of the Papers, that are published, and have enabled me to have imparted others, altogether as valuable; but finding few, or none, but what are here published, neither among the Papers themselves, nor in the Repository, nor Papers of the Royal Society, I was forced to be content.*

For a Conclusion of this Preface, I shall answer two Accusations that have been, or may be charged upon me: One is, That I have long detained these Papers from the Publick: The other, that I have engaged myself in Matters lying out of my Way. To both which, one Answer may serve, namely, That I have made the collecting, and publishing these Papers, my Diversion, at Leisure Hours: By which Means, and by reason the Papers, out of which these

were

To the READER.

*were selected, were very numerous, and many
of them came late to my Hands, their Publication hath been the longer delay'd. And
as for the Diversity of this from the Business of my Profession: I confess it is not
direct Divinity, but yet I think it, by no
Means, unfit for a Clergy-man's Diversion.
For as it is necessary for a Clergy-man (as
well as others) sometimes to divert, and un-
bend his Mind, from his more serious Stu-
dies, so what Diversion more innocent, or
proper, than that which promotes Knowledge, and Experience, and is a Discovery
(if never so small) of any of the Works of
the infinite Creator? To the promoting
which End, the Publication of these Papers
was, in some Measure, intended by*

W. DERHAM.

CURIOUS

PHILOSOPHICAL

Obfervations and *Experiments*

OF

Dr. *ROBERT HOOK,*

AND

Other Eminent VIRTUOSO's in his
Time.

Of the Invention of the BAROMETER,
in the Year 1659.

IN one of Dr. *Hook's* Papers (not here publifh-ed, becaufe imperfect) I find this Remark, *viz. The Inftrument, for finding the different Preffure of Air upon the Parts of the Earth fubjacent, was firft obferved by the Honourable* Mr. Boyle, *who,* upon the Suggeftion of Sir Chriftopher Wren, *erecting a Tube of Glafs fo filled with* Mercury, *as is now*

ufually

usually done in the common Barometer, in order to find out, whether the Pressure of the Moon, according to the Cartesian *Hypothesis, did affect the Air; instead of finding the Fluctuation which might cause the Phenomena of the Tides, discovered the Variation of its Pressure to proceed from differing Causes, and at different Times, from what that Hypothesis would have predicted. That Propriety of the Air (for ought appears) was never discovered till that Time, which is not yet thirty Years since,* &c.

To this I *W. D.* shall add another Remark I find in the Minutes of the *Royal Society, February* 20. 167⅞, *viz.* Upon a Discourse of some Experiments to be made with the Barometer on the Monument, *it was queried, how this Experiment of the differing Pressure of the Atmosphere came at first to be thought of? And it was related, That it was first propounded by Sir* Christopher Wren, *in order to examine Monsieur* des Cartes's *Hypothesis, Whether the passing by of the Body of the Moon did press upon the Air, and consequently also upon the Body of the Water. And that the first Trial thereof was made at Mr.* Boyle's *Chamber in* Oxford.

The Time, when these Observations were made, was about the Year 1658, or 59; at which Time Mr. *Boyle* having a Barometer fixed up, for the observing the Moon's Influence upon the Waters, happened to discover the use of it in relation to the Weather, and to assure himself, that it was the Gravitation of the Atmosphere which kept up the Quicksilver to such an Height, as the learned Abroad, particularly *Torricelli*, had suspected before.

But although this Use of the Baroscope is owing to Sir *Christopher Wren*, and Mr. *Boyle*, yet, to do every Man Justice, I shall
give

give the Hiftory of this excellent Inftrument, from the Extracts of a very ingenious Friend.

T H E firft Inventor of it was *Torricelli*, at *Florence*, in 1643. From whence Father *Merfenne* brought it into *France* the Year following, 1644. And Monfieur *Pafcal* being informed of it by Monfieur *Petit*, the Engineer, they both tried it in 1646, at *Rouen*, with the fame Succefs as it had been tried in *Italy*. Some Time after which, an Experiment was made with a Tube of forty fix Feet, filled with Water, and alfo with Wine: Which Experiment Monfieur *Pafcal* gave an Account of in a Piece printed in 1647; in which Year he was informed of *Torricelli*'s Solution of the Phenomenon, by the Weight of the Air; and devifed, for the examining it, the famous Experiment with two Tubes, one within the other, which he mentions in a Letter written in *November* 1647. And laftly, in 1648 the fame Monfieur *Pafcal* made his Experiments on the Tops and Bottoms of Hills, Buildings, &c. which laft Experiments Monfieur *Des Cartes* laid Claim to; affirming, that he defired Monfieur *Pafcal* to make them two Years before, and predicted their Succefs, contrary to Monfieur *Pafcal*'s Sentiments.

Monfieur A z o u t alfo laid the fame Claim, but it is the moft probable that Monfieur *Pafcal* had the beft Title.

T H I s Experiment which *Torricelli* made with Quickfilver, *Galileo* had in effect tried with Water in long Tubes by Pumping; with which he found he could never get the Water to afcend above thirty three Feet: But the Caufe he could never hit of.

A f t e r the *Torricellian Experiment* had been much celebrated in divers Places, at laft *Otto de Guerrick*, Conful of *Magdeburgh*, was informed

formed of it by Father *Valerian* at *Ratisbon,* who claimed it as his own Invention: But this was not till the Year 1654. After which *Guericb's* Experiment (called the *Magdeburgh Experiment*) was much talked of.

From this ſhort Hiſtory of the Barometer, not only the Inventor and Improvers of it appear, but in ſome Meaſure alſo the excellent Uſes of it : Particularly the Gravitation of the incumbent Atmoſphere, (one of the nobleſt philoſophical Diſcoveries) the Changes of the Weather, &c.

W. Derham.

The Lord Kingkardine's *Obſervations of the* Pendulum *Clocks at Sea, in* 1662.

THE Lord *Kingkardine* did reſolve to make ſome Trial what might be done, by carrying a *Pendulum* Clock to Sea; for which End, he contrived to make the Watch Part to be moved by a Spring inſtead of a Weight; and then making the Caſe of the Clock very heavy with Lead, he ſuſpended it, underneath the Deck of the Ship, by a Ball and Socket of Braſs, making the *Pendulum* but ſhort ; namely, to vibrate half Seconds, and that he might be the better inabled to judge of the Effect of it, he cauſed two of the ſame Kind of *Pendulum* Clocks to be made, and ſuſpended them both pretty near the middle of the Veſſel, underneath the Deck ; thus done, having firſt adjuſted them to go equal to one another, and pretty near to the true Time ; he cauſed them firſt to move parallel to one another, that is, in the Plane of the Length of the Ship, and afterwards he turned one to move in a Plane at Right Angles

Angles with the former ; and in both thefe Cafes it was found by Trials made at Sea, at which I (*i.e.* Dr. *Hook*) was prefent, that they would vary from one another, though not very much, fometimes one gaining and fometimes the other, and both of them from the true Time, but yet not fo much but that we judged they might be of very good Ufe at Sea, if fome farther Contrivances about them were thought upon, and put in Practice. This firft Trial was made in the Year 1662 ; whereupon, thefe being found to be able to continue their Motion without ftopping, feveral other Clocks of this Nature were made and fent to Sea, by fuch as fhould make farther Experiment of their Ufe. And we have an Account which was given from Sir R. *Holmes*, who tried them in failing from St. *Thomas* Weft-ward about 800 Leagues, and then tacking about fteer'd about 300 Leagues N. N. E. towards the Coaft of *Africa*, and by obferving thefe Clocks only, he was able to judge much better than the Mafters of the other Veffels that were in Company, who differed from his Account, fome 80, fome 100 Leagues, fome more Leagues; and whereas feveral of them thought themfelves near to *Barbadoes*, he judged by his Clocks that he was not far from *Fuego*, one of the Iflands of *Cape Verde*, and the next Day by Noon reached that Ifland. But yet this was not fo exact as was expected ; however, it performed fomewhat towards this Effect of finding Longitudes fomewhat more than ordinary, and enough at leaft to give inquifitive Men Occafion to fpeculate, and make farther Trial. And though there hath been no very confiderable Improvement of that Inftrument, or Experiment fince that Time by any, and tho' I fear it may at beft be infufficient to perform what is neceffary to this Matter, yet I queftion not but that there may be fome other Way that

may

may perform it to a much greater Degree of Perfection, as I fhall hereafter endeavour to prove.

Dr. H o o k's Experiment of weighing Air. *Shewed to the* Royal Society, *Dec.* 3. 1662.

Two fmall Glafs Balls, blown and fealed with a Lamp, each of them about an Inch and half over, were fufpended at the End of a Beam, and counterpoifed with a fmall leaden Weight ; and then a Grain being taken away from the Counterpoife, fo that the Balls preponderated by a Grain, the Beam was hung into the Globe, and the Mouth of it clos'd, and the Forcer was wrought ; whereupon, as the Air was condenfed in the Globe, the Balls by Degrees grew lighter and lighter, and the oppofite Counterpoife at length did more preponderate the Globes, than they had before the Condenfation ; but upon the letting out of the imprifon'd Air, the Balls again recovered their Prepollency, and remained as they were when firft put in.

The Experiment affords us a manifeft Proof of the Weight and Spring of the Air, and after what Manner they work upon the Bodies inclofed in it. 1ft. That though the Air be a heavy Body, yet it not only preffes downwards, as fome have erroneoufly thought, and fo have imagin'd it fhould break People's Necks, and roul and prefs down the Grafs, and all kinds of weak Plants, as *Deufingius* fuppofes ; or fhould prefs a Difh of Butter, or fome fuch foft Body, quite flat, as Mr. *Hobbs* imagines. But 2dly, it preffes upwards and fideways, as much as downwards ; whence every Body, fufpended in it, does fuffer, from this ambient Fluid, a greater
Preffure

Preffure againft its under Side to thruft it upwards, than againft its upper Side, to force it downwards; and does in all Things of Staticks act according to the fame Laws, and after the fame Manner, that other heavy fluid Bodies work upon the Body they incompafs. And this Experiment, in fhort, is nothing elfe but a Variation of *Archimedes*'s Experiment of examining compounded Metals. For the two Bodies that weigh againft each other, being of a very differing bulk, though pretty near of the fame Gravity when in the Air, when they are incompafs'd with a more denfe and heavy Fluid, that which is more bulky muft neceffarily lofe more of its Weight or Power downwards than the other, fince it is a known Law of the Staticks, that a Body, remov'd out of a lighter into a heavier Medium, lofes fo much of its former Gravitation, as the Weight of a Part of the heavier Fluid, equal in Bulk to the inclofed Body, amounts to.

T H E Ufes that may be made of this Experiment may, be many, and thofe, I think, not the leaft confiderable.

Firft, I T may ferve as an Inftance, to fhew by what Means the Vapours and Exhalations are raifed up into the higher Parts of the Air; for if by any Means the Vapours, or Waters rarify'd, obtain a greater Rarity, and confequently a leffer Gravitation than the ambient Air; the Preffure of that muft neceffarily buoy and carry them up fo far, till the Abatement of Preffure on the Parts of the ambient Air, by reafon of their fublime Stations in the upper Regions, and till the Abatement of Heat, that kept the Vapours rarify'd, has reduc'd both to an Æquilibrium, where they are ftay'd and fufpended; which affords us a fecond Ufe, namely, to explain how the Clouds or Exhalations are fufpended and carried to and fro directly at fuch a Height, and no lower nor higher. For fince

it

it is found by Experiments made by *Torricellius,* that several others, whom I now forbear to name, and the Pressure of the Air at the Top of Mountains is differing from what it is in the Valleys,therefore the Rings of Pressure (if I may so call those Parts of the incumbent pressing Atmosphere) seem not at all to be regulated by the Form of the Earth's Surface; that is, are not at all parallel to the Surface of the Earth, but they seem to be regulated rather by the Distance of the Parts of the Air from the Center of the Earth, or rather are parallel to the Surface (if there be any) of the Air, or to the Superficies of the Sea. And, indeed, I have very often observed, not without Wonder, that in cloudy Weather all the under Surfaces of the Clouds have been exactly terminated with a Spherical Concave Surface, no one being raised above or deprefs'd below such a determinate Surface. And I have after observed the Vapours often rise like Smoak upward, till they come to such a Height, and then to cease ascending, and spread themselves in Breadth almost like Oil upon the Water : The Reason of all which is, probably, nothing else but that at such a Height the Air is reduc'd by the Decrease of Pressure to such a Degree of Rarity, that it is unable to raise the Vapours any higher, and below it is able to raise them. The Reasons how the Vapours come to retain that Degree of Rarity, *&c.* is an Enquiry more proper for another Place.

Thirdly, T h i s may hint us a Solution of a late Observation made by an excellent Person, and a Member of this Society, that in Fogs with an Easterly Wind, the Pressure of the Atmosphere was observed to be very great. The Reason of which Phenomenon might, perhaps, be this, that the Cold and Pressure of the Air being then very great, the Density and Gravity of it might thereby

by become fo confiderable, as to raife up many Bodies, even in the Form of Water, and keep them fufpended fomewhat above the Surface of the Earth, though by reafon of the Want of Heat to rarify thofe fmall Parts into aerial Vapours, it were not able to carry them to any confiderable Height.

A Brief Account of the Experiments tried before the Royal Society, *with* Glafs Balls, *November* 19. 1662. 1. *Of driving out the Air by bare Heat.* 2. *Of driving it out by Vapours of Water and Spirit of Wine.* 3. *Of their breaking of themfelves.* 4. *Of their breaking by a Knock.* 5. *Of the Quantity of Water they admitted.* 6. *Of the Weight of Air they admitted.* 7. *Of the fhrinking and ftretching of them.* 8. *Of their breaking outward.*

A SMALL Pipe of white Glafs, melted over a Lamp, is blown into a pretty large Bubble, the fmall Neck or Pipe of which being, whilft the Ball is yet red-hot, fuddenly and carefully fealed up, I obferved that thofe Bubbles being left to cool, fome of them that were either not very equally or over thin blown, would, in the cooling, break inward, with a very brisk and loud Noife, fome fooner whilft yet hot, others later when even quite cold; but this latter yielded the loudeft Report. Some, that were ftrong and even blown, remained intire when quite cold: The Balls of which I obferved to endure a much greater and more violent Blow, before they would
break,

break, than others much of the ſame Make,
which were left to cool without ſealing up. But,
when by a pretty brisk Blow they were broken,
they yielded, beſides the Noiſe of the broken
Pieces, ſometimes a ſmart, ſometimes a more faint
Noiſe. Some of theſe Bubbles whilſt thus hermetically ſeal'd, being pois'd in a pair of exact
Scales, and then the little ſeal'd End nipp'd off, a
Sibilus or hiſſing Noiſe might very ſenſibly be heard
for a ſmall Space of about a Second ; after which the
ſame Scales and Counterpoiſe being left free, the
Bubbles were always obſerved to preponderate,
ſome a $\frac{1}{4}$ of a Grain, others $\frac{1}{3}$, others more. The
End of ſome other of theſe being broken off un-
der the Water, the Water was obſerved to aſcend with a very great Impetuoſity, and to look
white, until ſuch Time as it had fill'd the Bubble
or Ball, about $\frac{1}{3}$ or $\frac{1}{4}$ of the whole; ſome more,
ſome leſs, according as they were more or leſs hot
when ſeal'd up. Then holding the Bubble over
the Flame of a Candle, till the Water was boil'd
or exhal'd away, I immediately ſeal'd up the
ſmall End again, and obſerved ſome of them to
break with a much louder Crack than thoſe that
had been ſealed up when red-hot. Breaking others
under Water, I found a much greater Quantity
of Water to enter, inſomuch as to fill almoſt the
whole Ball, leaving a very little Bubble of Air at
the Top : Others, that I weighed, I found to increaſe ſomewhat more in Weight, by the Admiſſi-
on of the Air, than they had done before by the
other ſealing. After this, having emptied out the
Water, I put into ſeveral of them a ſmall Quantity of indifferently well rectify'd Spirit of Wine,
and taking the ſmall Stem in my Hand, I held the
Ball over the Flame of the Lamp, till the Spirit
with great Impetuoſity was evaporated, and driven
out through the ſmall Neck, in a Kind of miſty

Steam;

Steam; which ceafing, I immediately feal'd up the Neck, and letting the Bubbles cool, I found them to be much of the fame Kind with thofe that I had feal'd up with Water, both as to the Noife they yielded when broke, and to the admitting of Water, and for the weighing of Air; only in this thefe two laft Ways differ'd from the firft, that whereas the red-hot Glaffes when cold were clear, thefe, though they appeared clear when hot, were, notwithftanding, all tarnifhed over, with a Kind of Dew in the Infides when cold; which Dew would quickly difappear, if they were again heated pretty hot. There were feveral other Circumftances, which, becaufe they will be more notable in other Experiments, I here omit.

T H E Reafons of which Phenomena I humbly conceive to be thefe. *Firft,* That the elaftical Power of the exceedingly heated Parts of the Air, that are within the Glafs when red-hot, being very much intended, a very fmall Parcel is able to prefs and keep out all the reft of the ambient, contending Atmofphere; and whilft it has that Ability, the Paffage being fhut, the ambient Air is hinder'd from rufhing in that Way, though the Air within growing colder, and fo lofing its Elater, could not have been able to have hindered it. * Now the Preffure of the included Air againft the Sides decreafing with its Elater, and that with the Heat, and the Preffure of the ambient, remaining the fame, that curious arched Vault of the Glafs is forcibly prefs'd and crufh'd together, and fo the Particles are put into a clofer Texture. And that they are fo, I found by this Experiment. I fitted a pretty large Bubble with a flender Neck into a Bolt-Head, whofe Neck was drawn very

---

* Query, *Whether the Bubbles fhrink?*

fmall

fmall, and left only big enough to contain the Neck of the Bubble, and whofe Bottom was cut off, that thereby I might include the Ball. Having fo fitted the Ball and Bolt-Head, I fhut up the Bottom again with Cement, and filling up the Space left in the Bolt-Head with Water, till it reach'd into the Small of the Neck, I nipp'd off the feal'd Top of the Bubble, whereupon the Water in the fmall Neck rofe about a Barley Corn's Breadth, which could proceed from nothing elfe than its Return to its former Dimenfions, before it was fealed up; which affords us a noble Inftance of Compreffion, where that fo hard and well compactedBody of Glafs is comprefs'd into leffer Room, and that by no greater a Force than that of the Preffure of the Air; whence we may conclude that the Parts of that Body are not fo clofe joined together, but that there may be Pores or Receffes left between them, into which they may be protruded, and fo be made to lie clofer to each other, which whether Water and other fluid Bodies may not do the like, Trial will inform.

T H E Experiments fuggeft thefe Queries.

W H A T may be the Caufe of Noife or Sound?

B Y what Means Heat rarifies and expands Bodies, and Cold condenfes?

W H E T H E R the Caufes of the almoft fimilar Phenomena of the Glafs Drops, may not be deduced from thefe Principles: Or what may be their Caufes?

T H E Strength of a Knock, or what may be the Force of falling Bodies?

W H A T is the true Weight of Air in Winter?

W H E T H E R Bodies, that will not melt, may be expanded by Heat?

T H E Difference of the external and internal Preffure increafing by the Decreafe of the included Air's Elater, if fome Parts of this Arch (if I may

fo

fo call it) be weaker or irregular, the ambient Preffure breaks it in: Even as in Architecture the fame would happen in thofe larger Arches, if in either of thefe Particulars they deviated from the Rules of that Art. But if fufficiently ftrong and equal, the ambient Preffure makes the chryftalline Vault the firmer, as in Arches of Stone is commonly obferved. The Caufe of the Noife I dare not yet determine, but I think it worth a further Enquiry, whether it proceed not from the *Impetus* wherewith the broken Pieces of Glafs are dafhed againft one another, though the Noife feem of another Kind ; or from the fudden rufhing of all the Parts of the ambient Air towards the Middle of the Ball, whereby all the other Parts of the circumambient being likewife moved towards the fame Middle, the Drum of the Ear may likewife be moved, and fo a Sound heard : Or 3*dly*, Which I think the moft plaufible, from the fudden and violent rufhing towards the Center, and (by there meeting each other, or at leaft the broken Particles of Glafs) there finding as fudden and violent a Recoil or Repulfe, one of which two laft (if not a third, namely, the fudden flying out of the Air) feems to be the Reafon of the Noife of a difcharged Shot of Powder. The Alteration, as to Weight, does clearly enough proceed from the Admiffion (which the Hiffing plainly enough fpeaks) of the heavy Particles of Air. A manifeft Experiment that Air does gravitate in Air. The violent rufhing in of the Water argues the forceable Preffure of the external, as the Multitude of Bubbles do the languid Refiftance of the included Air.

An Account of some Trials for the finding how much, ascending and descending Bodies press upon the Medium through which they pass : Made before the Royal Society, Dec. 24. and Dec. 31. 1662.

A Glass Tube about fourteen Inches long, and an Inch and half over, being open above, but shut beneath, was hung by a Piece of Tape fastened about the End of it, to the End of a Beam ; then being fill'd with Water, and a

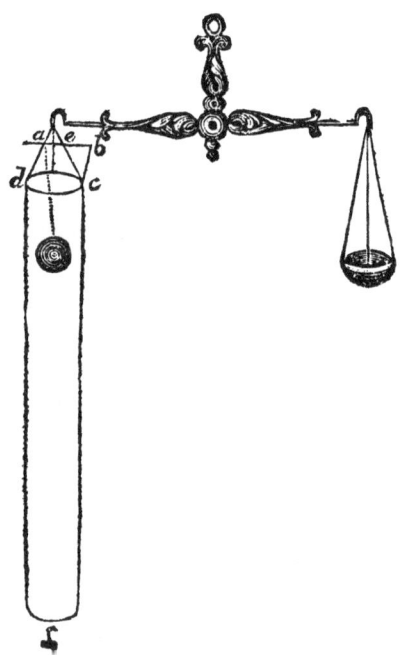

round Glass Ball somewhat more than an Inch in Diameter (which was made heavier than Water, by Quickfilver included in it) being hung by a
String

String of Silk fo far within the Tube, that it was quite covered with Water. The other End of this String was tied to a Wire, that was faftened to the End of the Tube. This Tube, I fay, thus accoutred, being hung at the End of an exact Beam, was counterpois'd with fomewhat more than 36 Ounces Troy. Then the Scales being in a very exact Equilibrium, the Silk String, by which the Ball hung, was fuddenly cut afunder with a fharp Pair of Sciffers. And the Beam, all the while the Ball was defcending through the Water, and after it came to the Bottom, kept its former horizontal Parallelifm. This was repeated a fecond Time with the like Succefs.

At the fame Time in the fame Tube, as it hung in this Pofture, there was let down to the Bottom of it a fmall Piece of Lead, which had a fmall Loop of Wire, through which a Silk String being put, a round Glafs Ball much lighter than Water, and about the former's Bignefs, was, by that String, drawn down, and kept at the Bottom of the Water, and the other End of the String was faftened about the former Wire. This done, the Scales were brought to an Equilibrium, and then, as before, the Thread was cut, and the Ball quickly afcended to the Top ; in which Time the Beam was obferv'd to be very much turned from its Equilibrium, and upon Trial fix Grains, detracted from the Counterpoife, was requifite to bring them to an Equilibrium. This laft Experiment was twice repeated, but in the latter Trial the Parallelifm of the Scales was not at all difturb'd, as in the former Experiment; which gave Occafion for a Conjecture, that the former odd Phenomenon was caufed by fome extraordinary Accident.

In Profecution of this Enquiry, *Dec.* 31. Trial was made by a Variation of the former Experiment;

riment; for the Thread of Silk that the Ball hung by, was not tied to the former Wire, but to a *Suſtentaculum* above the Beam ; then the Scales being brought to an Equilibrium, and the String cut as before, the deſcending Ball made that End of the Beam, to which the Tube hung, to be exceedingly deprefs'd, and being come to the Bottom it kept the Beam in that Poſture.

FURTHER, that it might be known how much heavier that End was than the other, whilſt the Ball lay at the Bottom, the Beam was brought to an Equilibrium ; after which, fix Grains were taken from the Counterpoiſe of Weights. Then the Ball being tied by a String as before, and the Scale wherein the Weights hung being kept up to a convenient Height, that the Beam might hang parallel to the Horizon, and the String cut as before, the deſcending Ball was obſerved manifeſtly to deprefs the Tube End. Trial was made a third Time by counterpoiſing and ordering all Things, as in this ſecond Trial, and detracting only three Grains, notwithſtanding which, the deſcending Ball manifeſtly deprefs'd the Tube End; which laſt Trials were a Confirmation of the firſt Experiment, when the Ball was hung to the Wire.

THESE Experiments ſeem to hint this Axiom, That every Body, whether aſcending or deſcending in a fluid Body, does add ſo much Weight or Preſſure to that fluid Body, as its own Weight amounts to, and not as much as the Weight of ſo much of the Fluid as is equal in Bulk to what the moved Bodies amounts to.

THIS I ſhould have put as an Axiom, did not ſome Difficulties ſuſpend my Aſſent.

Firſt, SINCE the ſwifter a Body is moved, the greater Reſiſtance it finds from the Medium through which it paſſes, and conſequently the ſtronger is its Preſſure againſt that Fluid ; and ſince

deſcend-

defcending Bodies grow fwifter in their Motion, the lower they defcend, it feems rational to judge, that the defcending Ball's Preffure, on the Water, fhould be increafed with its Swiftnefs.

NEXT, fince the Body that hinders its Motion is a Fluid, it feems fomewhat difficult to conceive, how the Preffure of a defcending Body can be communicated to the Bottom, fince the Parts of the Fluid are circulated. And no lefs difficult is it to fay, on what Part of the Bottom the Preffure refts ; whether on the whole, or only that Part immediately fubjacent to the falling Ball ; for which Way foever is taken, there are feveral Difficulties fomewhat hard to be explicated.

Thirdly, IF the Weight of the defcending Body be all the while fuftained by the Fluid, and confequently by the Bottom, how comes the Body, when it touches the Bottom, to prefs with more Force than its own Weight; as is evident, in Bodies defcending through the Air.

Fourthly, SINCE the Preffure of a fluid Body, againft the Bottom, is greater, or lefs, according to the Height of the Surface of the Fluid above it : It feems that an afcending Body, in Water, does manifeftly contradict this Axiom.

COROLLARIES, deducible from thefe Experiments, certainly made, may be fuch as thefe:

Firft, THAT Exhalations and Vapours prefs not lefs upon the Surface of the Terraqueous Globe, when they afcend, than when they are falling ; nay, than when they are fallen : The Certainty of which, I think, were worth examining.

NEXT, That the Preffure of any contained fluid Body, againft the Sides of the Veffel, will be abated by opening an Hole at the Bottom ; though the Height of the Water be continued the fame. That is, that the Preffure of a Perpendicular

VOL. I.   Height

Height of running Water, is not the fame with that of ftanding Water.

Thirdly, I t fhould feem, that the Preffure of a River, againft the Pillars of a Bridge, is lefs whilft the Water is running between them, than when that Paffage is ftopp'd, though the Height in both remaineth the fame.

Dr. H o o k's *Enquiries for* Greenland.
Jan. 14. 166⅓.

W h a t, and how much, was the Heat of the Sun in the midft of Summer, compared with the Heat of it in *England ?*

W h a t is the moft conftant Weather there, whether clear, cloudy, rainy, mifty, foggy, &c ? Or what moft ufual at fuch and fuch Times of the Year ? Next, what Conftancy or Unconftancy there is of the Winds to this or that Quarter of the Horizon, or this or that Part of the Year ? What the Temperature of each particular Wind is ob-ferved to be ; and particularly, whether the North be the coldeft, if not, what Wind is ? What Wind is obferved to bring moft Ice, and what to make a clear Water at Sea ? What Currents there are, how faft, and which Way they fet ? Whether thofe Currents are not ftronger at one Time of the Moon than another, whether always running one Way ? What is obfervable about the Tides, Spring or Neap ? Whether the Sea Ice be falt or frefh ? What Rivers there are in the Summer ? What Fowl are found to live there, and what Beafts ; how they are imagined to fubfift in the Winter ; how they breed and feed their young ? What Ve-getables grow there, and whether they yield any Fruits ? How deep the Cold penetrates into the Earth ? Whether there be any Wells, or deep Pits, or Mines, wherein the Water will remain

unfrozen

unfrozen at the Bottom ? How the Land trends ? And whether the Parts, under or near the Pole, be there thought to be Sea or Land ? Whether the Perſon made any Experiment, about the Load-ſtone or magnetical Needle, or any mathematical Obſervations, about the Height of the Sun and Lu-minaries, or their apparent Diameters, or Refracti-on, or the like ?

W ʜ ᴀ ᴛ Fiſh moſt frequent thoſe Seas, and any thing about their fiſhing, with the uſual Bigneſs of Whales, *&c.* their Strength, the Anatomy of their Entrails ? Whether any People do or have been known to ſtay there all the Winter, and how they do or have ſhifted ? How near any has been known to approach the Pole ? What Notice he has taken of the Moon, *&c.*

Dr. H o o ᴋ*'s Enquiries for* Iceland.
Jan. 21. 166⅗.

H o w deep the Ground is frozen ?
  W ʜ ᴀ ᴛ Wind is coldeſt ?
W ʜ ᴀ ᴛ Rivers and Springs they have ?
  T ʜ ᴇ Anatomy of Whales, or other very large Fiſhes.

A ʙ ᴏ ᴜ ᴛ the Lungs of Whales and Contrivance of Reſpiration in other Fiſhes and Morſes ?

C ᴏ ɴ ᴄ ᴇ ʀ ɴ ɪ ɴ ɢ the Fountain that is hot e-nough to ſcald a Fowl.

W ʜ ᴇ ᴛ ʜ ᴇ ʀ the burning extraordinarily of *Hecla* portend foul Weather ?

R ᴇ ꜰ ʀ ᴀ ᴄ ᴛ ɪ ᴏ ɴ, whether the ſeven Stars are ſeen in the *Pleiades ?* Whether *Mercury* can be oftener ſeen than in *England ?* The differing Heat of Summer and Winter : How near the Moon may be ſeen to the Sun ?

A ɴ exact Obſervation of the Eclipſes that happen.

T ʜ ᴇ

T h e Saltnefs of the Sea-water, by boiling, how much Salt it yields?

T h e Height of the Quickfilver in the Torricel Experiment.

W h a t Wind blows moft and ofteneft?

T h e ufual Temperature of the feveral Winds there.

A b o u t Corruption and Prefervation of Bodies.

W h a t Bodies will keep in the Snow, what not?

T h e burning of the Mountain, other Obfervations with the Needle in feveral Places about *Hecla,* or the other fiery Mountains, and in other Places of that Ifle.

T h e Figure of Snow, whether Hexangular, whether always larger than in thefe Parts?

T h e ufual Bignefs of Hail-Stones and Figure.

W h a t is obfervable about Meteors, as *Ignis Fatuus,* Star-fhooting, Thunder, and Lightning.

W. h a t Kind of Subftances are caft out of the burning Mountain.

A b o u t Haloes and Rainbows, any thing extraordinary.

W h a t kind of Ores, Stones, Clays, Minerals, *&c.* it yields.

W h e t h e r there be any of the *Selenitis,* or *Mufcovy* Glafs to be found there.

T h e Declination, Inclination, and Variation of the Magnet in feveral Parts of the Ifle, with the Diftances and Latitudes of thofe Places, as near as may be.

W h e t h e r the fame Point of a Magnet, that is a Pole of that Stone here in *England,* will be fo there.

W h e t h e r the fame Part of a *Terrella,* that, put upon Quickfilver, will lie toward the Earth here in *England,* will do fo there likewife.

W h e-

W h e t h e r the attractive Virtue of the Magnet increase or diminish there, in respect of what it is found here.

W h i c h Pole is there strongest.

W h e t h e r Iron be more or less apt to rust there than here.

W h a t living Creatures, tame and wild, live and thrive there.

A n y thing of that Kind strange or remarkable among the Beasts, Birds, Insects, or Fishes; as about their Generation, living in the Winter; for what they are or may be made serviceable; either for Burthen, Swiftness, Furrs, Feathers, Meat, &c.

W h a t Kind of Vegetables thrive best in that Island, as Trees, Shrubs, or Plants, and what Kind of Grounds they thrive best in; what Kinds of Vegetables the Sea yields, differing from our *English*. In what their Husbandry differs from ours, and whatsoever of that Kind is remarkable.

W h a t Woods it yields good for Building, Shipping, or other necessary Uses.

W h a t notable Virtues are attributed to this or t'other Plant; whether for Divination, Physick, Dying, Smell or Taste, &c.

T h e Seeds of as many as may be gotten together, with their Names.

H o w several Creatures subsist in the Winter.

W h a t are the predominant Colours of Animals.

W h a t general Change is made on the Shipmen, that does not seem immediately to proceed from Cold, as what Diseases they are most subject to.

T h e Nature, Disposition, Manners, and Customs of the Natives.

T h e i r Apparel for Warmth, Houſing, Victuals, Firing, Bedding, Cookery, and other Ob-
ſervables, either Actions or Utenſils, &c.

A n y notable Effects produced by Cold, &c.

T h e Height of the Iſlands of Ice, their Depth;
whether it be freſh Water; whether it ſeem to be
made up of Snow, and ſeem to lie in Plates one
above another.

W h e t h e r Spirits appear; in what Shapes;
what they ſay or do; any thing of that Kind very remarkable and of good Credit.

H o w much the Celeſtial Bodies are elevated
by Refraction above their true Place.

W h a t Currents there are, the Time of the
Tides in ſeveral Ports; their great riſing and falling in ſeveral Places; any thing notable concerning them.

W h a t Condition the Body is in that is pre-
ſerved by Snow, whether ſhrunk or ſwell'd, or
chang'd in Colour or Taſte, &c.

W h e t h e r Quickſilver will congeal.

A b l a d d e r full of *Engliſh* Air carried thither, and one of that Iſland Air brought back.

Dr.

Dr. HOOK's *Propofals, for finding out the Refiftance of the Air, to Bodies mov'd through it.*

TRYAL fhould be made with Pendulums of all Sorts, whofe Weights fhould be made of feveral Sorts of Materials; as of Metal, Stone, Wood, Feathers, Wool, &c. and thofe fafhioned into feveral Shapes, as round, elliptical, fquare, oblong, flat, to move flat-ways and edge-ways, and the like; then to have one common Standard, or Pendulum, by which the Celerity and Durati-on of all the other are to be meafured.

TRYALS fhould be made with feveral of thefe Pendulums, in the exhaufted Receiver, where there is a much lefs Quantity of Air; and like-wife in the Receiver, where the Air is very much condenfed; and the Differences meafured, as be-fore, and recorded, then compar'd with one an-other, and then with thofe in the free Air.

TRYALS fhould be made with Bodies of feve-ral Subftances, and each of thofe of feveral Shapes, which fhould be let fall from feveral Heights; and the Times of each of their Defcents to be exactly meafured by a Pendulum, and recorded.

TRYAL likewife fhould be made by fhooting, Horizontally, feveral Kinds of Bodies, with a Crofs-Bow, or the like, from the Top of fome high Place, and fo obferving the Time before they touch the Ground. And the

TRYALS fhould be made by fhooting Bodies perpendicularly upwards, and fo obferving both the Time of their Afcent and Defcent.

TRYALS likewife fhould be made by fhooting Bullets, or other Bodies, Horizontally; and fo to obferve with what Force they hit a Body, ac-cording as the Body is nearer, or further, from

the

the Inftrument that fhoots.   And thefe Tryals
to be made with Inftruments of feveral Strengths.

Dr. H o o ᴋ's *Experiment before the* Royal
Society, *February* 11. 166⅓. *about the
Refraction of* Ice *and* Cryftal.

HAVING obferved it to be almoft a general
Rule in Nature, that of pellucid Bodies,
thofe are found to have greateft Refraction towards the Perpendicular, which are moft maffy
and heavy in Bulk, I chofe a very pure and pellucid Fragment of Ice, about an Inch thick,
which had very few, if any, perceptible Blebbs
or Bubbles in it.   Then I took a large cylindrical
Cryftal-Glafs, about fix Inches over ; and filling
it with very fair Water, I put into it this clear
Piece of Ice, which did manifeftly fwim, with feveral of its Parts, above the Water ; and though
I feveral Times deprefs'd it with my Finger, yet
would it incontinently rife, as foon as I had remov'd my Finger.   Then I took it out, and with
a very fharp edg'd Knife, I fhaved one End of it,
(which is very eafy to do) into the Form of a very blunt Wedge, fo that the two Sides of the Edge
compos'd an Angle of about ninety Degrees ; then
fmoothing thofe fhaved Sides, by rubbing them a
little with the Palm of my Hand, I put it into the
Water with the Edge downwards, and holding it
pretty near that Side of the Glafs, which was next
my Eye, I cou'd plainly perceive, by looking
through that Edge, that an Object, placed againft
the oppofite Side, was manifeftly refracted.   For
faftening a fmall Piece of Lead, fo that the lower
End of it reach'd about an Inch under Water, I
could very plainly fee that lower End, a little below

low the Bottom, when, looking through the Ice, the Bottom of it appear'd above the Edge of the Ice; that is, I ſaw the ſame Object in two Places. Now becauſe the Refraction of the Ice made it appear higher than really it was, it ſhews that the Refraction in the Ice was leſs than Water; which will more plainly appear by the Figure: Where H I K L repreſents the cylindrical Glaſs, that

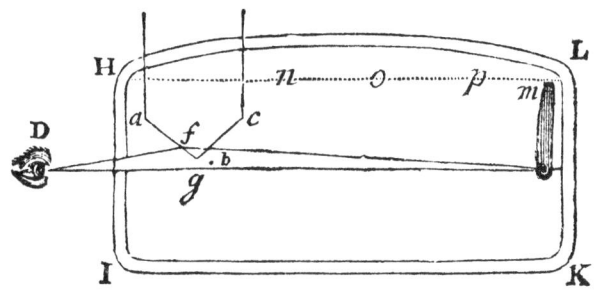

held the Water; *m e*, a Piece of Lead hung a-gainſt the Side of the Glaſs; *a b c*, the blunt Edge of the Piece of Ice; D, the Eye; *n o p*, the Surface of the Water; *f e*, the refracted Line, in which the Point *e* appeared to the Eye; *g e*, the unrefracted. This I ſeveral Times have re-peated, and always found the ſame.

THE Uſe of this Experiment may be, 1ſt, For to make an Exception from that general Rule of M. *Des Cartes,* in the ninth Section of the ſecond Chapter of his *Diopticks*; where he ſays, *Quanto firmiores & ſolidiores exiguæ partes corporis alicu-jus pellucidi ſunt, tanto facilius lumini tranſitum permittunt.* For, it ſeems, by this Experiment, not to be the greater or leſs Fluidity, or Firmneſs of Body, that cauſes a Difference in Refraction, but a more rarify'd or condens'd Texture.

Next,

Next, I т affords us two Arguments againſt their Opinion, who affirm Cryſtal to be generated of Ice. For, *Firſt*, As to its Weight, this is found to ſwim upon Water; whereas the other ſinks. *Next*, The Refraction of Cryſtal is obſerv'd to be greater than that of Glaſs; whereas this of Ice I find to be leſs than Water.

Thirdly, Т н ı s leſs Refraction of Ice, I take to be a good Argument, that the Lightneſs of Ice, which cauſes it to be born up of the Water, is not cauſed only by ſmall Blebbs or Bubbles, but from the uniform Conſtitution, or general Texture, of the whole Maſs.

Dr. H o o к's *Method of making Experiments.*

T н е Reaſon of making Experiments is, for the Diſcovery of the Method of Nature, in its Progreſs and Operations.

W н o s o e v e ʀ therefore doth rightly make Experiments, doth deſign to enquire into ſome of theſe Operations; and, in order thereunto, doth conſider what Circumſtances and Effects, in that Experiment, will be material and inſtructive in that Enquiry, whether for the confirming or deſtroying of any preconceived Notion, or for the Limitation and Bounding thereof, either to this or that Part of the Hypotheſis, by allowing a greater Latitude and Extent to one Part, and by diminiſhing or reſtraining another Part within narrower Bounds than were at firſt imagin'd, or hypothetically ſuppoſed.

Т н е Method therefore of making Experiments by the *Royal Society*, I conceive, ſhould be this.

Firſt,

Firſt, T o propound the Deſign and Aim of the Curator in his preſent Enquiry.

Secondly, T o make the Experiment, or Experiments, leiſurely, and with Care and Exactneſs.

Thirdly, T o be diligent, accurate, and curious, in taking Notice of, and ſhewing to the Aſſembly of Spectators, ſuch Circumſtances and Effects therein occurring, as are material, or at leaſt, as he conceives ſuch, in order to his Theory.

Fourthly, A F T E R finiſhing the Experiment, to diſcourſe, argue, defend, and further explain, ſuch Circumſtances and Effects in the preceding Experiments, as may ſeem dubious or difficult : And to propound what new Difficulties and Queries do occur, that require other Trials and Experiments to be made, in order to their clearing and anſwering : And farther, to raiſe ſuch Axioms and Propoſitions, as are thereby plainly demonſtrated and proved.

Fifthly, T o regiſter the whole Proceſs of the Propoſal, Deſign, Experiment, Succeſs, or Failure; the Objections and Objectors, the Explanation and Explainers, the Propoſals and Propounders of new and farther Trials; the Theories and Axioms, and their Authors; and, in a Word, the Hiſtory of every Thing and Perſon, that is material and circumſtantial in the whole Entertainment of the ſaid Society ; which ſhall be prepared and made ready, fairly written in a bound Book, to be read at the Beginning of the Sitting of the ſaid Society : The next Day of their Meeting, then to be read over, and further diſcourſed, augmented or diminiſhed, as the Matter ſhall require, and then to be ſign'd by a certain Number of the Perſons preſent, who have been preſent, and Witneſſes of all the ſaid Proceedings, who, by Sub-

fcribing their Names, will prove undoubted Tefti-
mony to Pofterity of the whole Hiftory.

Mr. OLDENBURGH's *Letter to Dr.*
HOOK, *Aug.* 23. 1665. *Concerning the*
Plague then, and Grafs in Sheep's and
Oxen's Lungs.

SIR,

I CANNOT but commend you for being fo
careful of yourfelf in this dangerous Time, as
not to venture to come amongft us, efpecially
when you find yourfelf any ways out of Temper.
The Sicknefs grows ftill hotter here, though I
find by all my own, and other Men's Obfervations, that very few of thofe Houfes whofe Inhabitants live orderly and comfortably, and have by
Nature healthy Conftitutions, (you muft take all
thefe together) are infected ; and I can fay, (God
be praifed for it) that as yet not one of my Acquaintance, except an under Poft-Mafter, who
lived clofely and naftily, and had all Sorts of People coming to his Houfe with Letters, is dead:
So that, generally, they are Bodies corrupted,
and Perfons wanting Neceffaries and comfortable
Relief, that fuffer moft by this Contagion.

THAT Obfervation, you mention of Mr. *Boyle's,*
is this, that one of thofe two Phyficians, Dr.
Clerk, and Dr. *Lower,* had affured him, that he
had feveral Times found, in the Lungs of Sheep,
a confiderable Quantity of Grafs, in the very
Branches of the *Afpera Arteria* ; and the other
had related to him, that a few Weeks fince, he,
and a couple of Phyficians more, were invited to
look upon an Ox, that had, for two or three Days,
almoft

almoſt continually held his Neck ſtraight up, and
was dead of a Diſeaſe, the Owner could not conjecture at; whereupon the Parts belonging to the
Neck and Throat being opened, they found, to
their Wonder, the *Aſpera Arteria*, in its very
Trunk, all ſtuffed with Graſs, as if it had been
thruſt there by main Force; which gives a juſt
Cauſe of marvelling and enquiring, both how
ſuch a Quantity of Graſs ſhould get in there, and
how being there, ſuch an Animal could live with
it ſo long.

Extract of a Letter from Ballafore, *Jan. 6.*
166⅞. *From Mr.* Henry Powell, *to
his Father Mr.* William Daniell, *upon*
London-Bridge: *Giving an Account of
an Earthquake,* &c. *after the Appearance of the Comet then.*

THE ſame Star appeared in our Horizon, a-
bout the ſame Time 'twas ſeen with you:
The Effects, in Part, have already been here, by
unſeaſonable Weather, great Mortalities amongſt
the Natives, *Engliſh*, and others. We have had
ſeveral Earthquakes unuſual here, which, with hi-
deous Noiſes, have, in ſeveral Places, ſwallowed up
Houſes and Towns; but about ſeven Days Journey from *Ducca*, where were at that Time three
or four *Dutch*, they, and the Natives, relate this
Story. That in that Place the Earth trembled a-
bout 32 Days and Nights, without Intermiſſion;
at the latter End, in the Market-Place, the Ground
turn'd round as Duſt in a Whirl-wind, and ſo
continued ſeveral Days and Nights, and ſwallowed up ſeveral Men, who were Spectators, who
ſunk and turn'd round with the Earth, as in a
Quagmire;

Quagmire; at laſt the Earth worked up, and caſt up a great Fiſh, bigger than hath been ſeen in this Country, which the People caught ; but the Concluſion of all was, that the Earth ſunk with 300 Houſes, and all the Men, where now appears a large Lake, ſome Fathoms deep : About a Mile from this Town was a great Lake full of Fiſh, which, in theſe 32 Days of the Earthquake, caſt up all her Fiſh on dry Land, where might have been gathered many, which had run out of the Water upon dry Land, and there died ; but when the other great Lake appeared, this former dried up, and is now firm Land.

Extract of another Letter from the ſame Mr. Powell, *to the Perſon abovementioned, from* Caſſumb, *Sept.* 27. 1666.

MINE, laſt Year, adviſed of the unknown Earthquakes which afflicted moſt of theſe Parts, in ſome to the deſtroying of whole Towns, *viz. June* 1ſt, in *Agra*, the King's Seat, at three in the Afternoon, ſuch a Darkneſs poſſeſs'd the Country, that none could ſee his Fellow in the Streets, nor his Hand, though never ſo near his Eyes, which continued half an Hour, and then diſſolved in Rain. It has pleas'd God to ſend this Year ſuch Rains and Overflowings of the Rivers, that in many Places whole Towns, with Cattle and Men, have been carried away, to the Deſtruction of many Thouſands. About the latter End of *Auguſt*, there was ſuch a Storm about *Pattava*, that it roll'd, as it were, that great City, their Houſes, in Heaps, deſtroyed many People, and continued three Days and Nights, in which we have loſt a Salt-petre Boat of Value, and the

Dutch

Dutch another ; alfo both ours and the *Dutch* Houfes, in all thofe Parts, are blown down : We expect the fame, it being ufual with us about the Middle of *October* yearly, but fuch Inundations and Storms were never before heard of.

An Account of a petrified Bone. An oddly-coated Stone Bottle : And a double Goofe-Egg. Produced before the Society, by Dr. Brown of Norwich, Feb. 27- 166⅞.

THIS Bone was found laft Year, 1666, on the Sea-Shore, not far from *Winterton* in *Norfolk.*

IT was found near the Cliff, after two great Floods, fome thoufand Loads of Earth being broken down by the Rage of the Sea, as it often happeneth upon this Coaft, where the Cliffs confift not of Rock, but of Earth.

THAT it came not out of the Sea, may be conjectured, becaufe it was found near the Cliff ; and from the Colour, for, if out of the Sea, it would have been whiter.

UPON the fame Coaft, but as I take it, nearer *Hasborough,* divers great Bones are faid to have been found ; and I have feen a lower Jaw containing Teeth of a prodigious Bignefs, and fomewhat petrified. All, that are found on this Coaft, have been found after the falling of fome Cliffs ; where the outward Cruft is fallen off, it clearly refembleth the Bones of Whales, and great cetaceous Animals, comparing it with the Skull and and Bones of a Whale, which was caft up on the Coaft near *Wells,* and which I have by me.

THE Weight thereof is fifty five Pounds.

THIS

THIS Bottle was filled with a green *Malaga*, above ſeven Years ago, and ſet up in a Nictrio of a Wine-Cellar-Wall in *Norwich*, where it contracted this *Mucor*: It was full at firſt, and is not yet empty.

A GOOSE-EGG, with another in it, or at leaſt over it; the outward Egg containing nothing but the White. The like I have obſerved in Hen's and Turky's Eggs. I would not omit to ſend it, becauſe though it ſometimes happeneth, yet few have the Advantage to ſee it, eſpecially in a Gooſe-Egg.

Mr. Charles Towneley's *Relation with Obſervations of the late Eruption of Water out of* Pendle-Hill. *Communicated by* Richard Towneley, *Eſq*;

AUGUST 18. 1669, betwixt 9 and 10 o' the Clock in the Morning, there iſſued, out of the North-Weſt Side of *Pendle-Hill*, a great Quantity of Water: The Particulars of which Eruption, as I received them from a Gentleman living hard by, are theſe. The Water continued running for about two Hours; it came in that Quantity, and ſo ſuddenly, that it made a Breaſt of a Yard high, not unlike (as the Gentleman expreſs'd it) to the *Eager* at *Roan* in *Normandy*, or *Ouſe* in *Yorkſhire*; it grew unfordable in ſo ſhort a Space, that two going to Church on Horſeback, the one having paſſed the Place where it took its Courſe, the other being a little behind, could not paſs this ſudden Torrent. It endanger'd breaking down of a Mill-Dam, came into ſeveral Houſes in *Worſton*, (a Village at the Foot of the Hill) ſo that ſeveral things ſwam in them. It iſſued
out

out at fome five or fix feveral Places, one of which
was confiderably bigger than the reft, and brought
with it nothing elfe but Stone, Gravel, and Earth.
He moreover told, that the greateft of thefe fix
Places clofed up again, and that the Water was
black, like unto that of Mofs-Pits; and laftly,
that fome fifty or fixty Years ago, there happened
an Eruption much greater than this, fo that it
much endamaged the adjacent Country, and made
two Cloughs or Dingles, which, to this Day, are
called *Oburft* (or, in our *Lancafhire* Dialect, *Eraft*)
Cloughs. Thus far this Gentleman related; what
follows take from my felf: Going, fince this, to
fee what I could of this Accident, I found nothing that did contradict the abovefaid Relation.
What I obferved more concerning this and other
Eruptions, is, that paffing under the North-Eaft
End, commonly call'd the *Butt End* of *Pendle,* I
faw feveral Breaches in the Side thereof, at feveral Diftances from the Top; from thefe, Stones,
mix'd with Earth, had been tumbled down, and
lay in fuch a confufed Order, as if they had been
brought thither by fuch a like Eruption as this
laft; and enquiring of a Country Fellow, who
was our Guide, he confirmed the Conjecture, and
told us, thefe Breakings out of Water were very
frequent, fo that he wonder'd we took fo much
Pains to go and fee this late one. I went to look
amongft the Rubbifh of Stone and Earth, of one
of thefe Breaches, to fee if I could find any thing
like Ore, but could find nothing. Having pafs'd
the End of the Hill, and coming to the other Side,
we, after a fhort Time, difcovered the mentioned
fix Breaches, of which two feemed to be very
near the Top of the Hill, and in the fame horizontal Line; the others at feveral Diftances from
the Top. I went only to the biggeft of thefe
Breaches, in which I obferved thefe Particulars:

The Water had taken away the Soil, (which was but about two Foot deep) and bared the Rock, betwixt ſome twenty and thirty Yards in Breadth, and downwards a conſiderable deal more : It appeared evidently, that the Water came from betwixt the Swarth and the Rock, for, at the Top of the Breach, we ſaw ſeveral Holes, whereat the Water had iſſued forth, others were cloſed up with the Fall of the Earth ; whereſoever the Water had taken away ſome two Foot deep of Earth, the Rock appeared : Amongſt the Rubbiſh I found nothing that could be ſuppoſed to come out of the Bowels of the Hill, but only ſuch Stones as might lie looſe on the Rock, amongſt the Earth that covered it. This is what I obſerved in the Breach, which, for Bigneſs, was moſt remarkable, and preſume, I ſhould have found nothing worth Notice in the leſſer ones. Though the Noiſe of this Eruption was ſo great, that I thought it worth my Pains to enquire further into it ; yet, in all theſe Particulars, I find nothing worthy of Wonder, or what may not be eaſily accounted for. The Colour of the Water, its coming down to the Place where it breaks forth, between the Rock and Earth, with that other Particular of its bringing nothing along but Stones and Earth, are evident Signs that it hath not its Origin from the very Bowels of the Mountain, but that it is only Rain-Water, coloured firſt in the Moſs-Pits, of which the Top of the Hill (being a great and conſiderable Plain) is full, ſhrunk down into ſome Receptacle fit to contain it, until at laſt, by its Weight, or ſome other Cauſe, it finds a Paſſage to the Side of the Hill, and then a Way betwixt the Rock and Swarth, until it break the latter, and violently ruſh out. The great Eruption, mentioned to have happened ſo many Years ago, perhaps, is that taken Notice of by *Cambden* in his

Brittannia,

Brittannia, pag. 613. *Verum hic mons damno quod subjecto agro jam pridem intulit maximam aquarum vim eructans, & certissimo pluviæ indicio, quoties eius vertex nebulâ vestitur, maximè insignis est.*
I know not whether it may not be worth Notice, that going to the Top of the Hill, and obferving a confiderable Part thereof, efpecially towards the Skirts, where Turfs had been gotten, I found that the Rock reach'd within a Yard or two of the higheft Part; confidering this, with what I obferved at the mention'd Breach, and feveral other Places, I think it is very probable, that the whole Mountain, as great as it is, is one continued Rock; and it may be a Queftion, Whether all other Hills be fo or no? But this I leave to further Enquiry.

Extract of a Letter from the Prefident Cornelis Frans, *and the Council in* Ternata, *to the Heer* William Maatfuiker, *and the Council in* Banda, *dated the* 12th *of* Auguft, 1673. *Concerning Earthquakes there.*

WE hereby acquaint you with two Wonders, the like not before heard of. The firft, that on the 20*th* of *May,* being *Saturday* Evening, that great and high Hill *Gammaknotra,* about thirteen Miles from hence, is, for the moft Part, flown up in the Air, which caufed the next Day, being *Whitfunday,* fo great a Darknefs, that we could hardly fee one another; and this was accompanied with a great Earthquake, and all the Land, both here, at *Manado, Chianco, Jafangy,* and *Mindanao,* a hundred Miles from hence, and God knoweth how much further, was covered

with

with Afhes a Foot thick, and fo much was fallen in the Sea, that a fmall fluit Ship, in going and coming from *Manado*, was feveral Times hinder'd in her failing, through the great Quantity of Afhes driving, and fome Houfes and Negeries, at the Foot of the Hill, were quafh'd with the Weight of the Afhes fallen on them.

T H E fecond Wonder is, that on the 12*th* prefent, in the Night, between 11 and 12 o' the Clock, a fudden Earthquake furprized us, with fuch terrible Shakings, as poffibly the like was never known, which encreafed fo violently, that the Hill of *Ternata*, on the South Side, was rent from Top to Bottom ; the King's *Mandarfabas* Stone-Houfes were caft down ; Parts of Hills funk ; all the tiled Coverings, with feveral Walls, caft down ; and the Sea was in that Manner difturbed, that the Ships, here in the Road, expected all to have been caft away ; and Quantity of Fifh was flung on the Shore, with many other ftrange Paffages. And that which is worfe, the faid Earthquake continueth to this prefent Time ; and here is nothing to be feen but bad Spectacles of Ruin. By a further Letter from the faid Prefident of *Ternata*, of the firft of *September*, the before-going Relation is confirmed, and that the Earthquake yet continued, fo that the Night before, the Houfes were thereby terribly fhaked ; all which is more at large exprefs'd in a Relation printed at *Batavia*.

To whiten Bees-Wax, April 3. 1674.

IN *March* or *April* melt yellow Wax without boiling; then having feveral Pewter Difhes ready, dip the Outfide Bottom of each Difh in fair Water; then dip them into the Wax, and take up a very thin Plate of Wax, the thinner the better: Take them off, and expofe them upon the Grafs, to the Sun, Air, and Dews, 'till they be milk white, turning them often. Try fome of them by fprinkling Water on them with a Cloth. *Query,* Whether white Lead may not this Way be made with very thin Plates.

Dr. John Carte'*s Letters to* Dr. Grew, *of the* Belland, *caufed by the Fumes of Lead, and other curious Obfervations.*

I THOUGHT it might be worth while to give you a fhort Account of a Diftemper in *Der-byfhire,* very common among thofe, who are employed in the Smelting-Mills, *i. e.* the Houfes where they melt the Lead down from the Ore; it is by the Country People called the *Belland,* but for what Reafon I cannot learn; it is hard to give a concife Definition of it, becaufe it feldom appears but under the Difguife of another Difeafe.

THIS *Belland* frequently imitates the *Tormina Ventris Scorbutica,* but in a moft exquifite Manner, which is ufually accompanied with extreme Coftivenefs, and a continued Suppreffion of Urine: Sometimes appears like an *Afthma Convulfivum,* fometimes a continued and obftinate *Dyfpnæa,* and often feizes the *Genus Nervofum,* either

in

in a paralytick Refolution of the Parts, or in Spafms.

I T has a different Effect upon Men, according to their Age; if they come not to the Work of the Mills, till they are full grown, or of a middle Age, they fuffer moftly the aforementioned Pains of the Belly, or difficult Breathing. But if taken in while young, and growing, they are fubject to the Palfy; their Limbs (efpecially their Fingers) being often irrecoverably refolved: Or fometimes have their Fingers fo contracted, as to render them (perhaps for ever) incapable of working. Both which I have feen.

I C O U L D not be informed of any Specificks, they had for this Difeafe; but that a Decoction of *Coloquintida,* in Ale, was very common among them. I remember once, an old Man complained to me of the *Belland,* it oppreffed him in the Nature of an *Afthma*; I advifed him to fulphurate Medicines, which did relieve him. The Contraction of the Fingers I have known cured, by often putting the Arms into hot Grains after Brewing.

I H A V E not obferved, whether any of thofe, that are paralytick by the *Belland,* die Hectick, as Dr. *Pope* relates of them, at the *Mercurial* Mines in *Firmly,* but it feems not improbable that they may.

T H I S Diftemper is not only incident to Men, but other Creatures, as Horfes, Cows, Dogs, Cats, Hens, Geefe, *&c.* but, efpecially, Cats are fubject to it: Indeed few Creatures, that are young, will live near thefe Mills without the *Belland.*

D O G s do in their Fits howl and tumble up and down, foaming like *Epilepticks*; this the People impute to the Pain of their Bellies.

I K N O W a fmall Rivulet, on which fome of thefe Mills ftand, wherein Trouts have been caught, which have been fuppofed affected with

het

the *Belland*, by the Irregularity of their Growth, their Heads being great and mishapen, their Backs crooked, their Tails very small, which, I am apt to think, might proceed from their feeding on the *Smitham* or *Duft* that is washed down at a Flood : For not only the Fumes, but also the Washings of Lead Ore, and the *Waste* (as they call it) *i. e.* the Duft that remains, after the Ore is melted, is very noxious to most Sort of Creatures, and for this Reason, they,that live near the Mills, dare not water their Horses at the River, upon a Flood.

THESE poisonous Fumes are not only hurtful to Animals, but also injurious to Vegetables ; for if the Smoak be driven much upon any one Place, it deftroys all the Grass of it.

Now that the *Belland* in Men, or other Creatures, proceeds mostly from the Smoak, will be easily granted ; but what these Fumes is impregnated with, is the Queftion : Some fancy them to be Antimonial, but then, methinks, they should have the same Effect with the Flowers of that Mineral, and I never heard that any of them were inclined to Vomit. I am much more apt to think, that the *Mercury* in the Ore is the Caufe, both becaufe they, that work in the *Mercurial* Mines, are fubject to the like Symptoms, efpecially the Palfy ; and alfo I am told, that this *Belland* often begins with a Swelling of the Glands about the Throat, which, perhaps, if not prevented, might terminate in Salivation. But why *Mercury* should operate fo varioufly upon Bodies, differing in Age, is a Queftion will hardly be folved, till it appear more plainly, whether it be nearer a-kin to Alcalies or Acids : Its Effect is eafily foretold in Bodies that abound with Acids, whether Scorbutick or Venereal ; but in younger Perfons whofe Humours are more infipid, and their Blood freer from both fix'd Salts and Acids, it may, perhaps, fix
itfelt

itſelf upon the Nerves, as the cooleſt Parts, and impede the Motion of the Spirits ; but I had rather hear others Reaſons about the Cauſe of theſe Things, than trouble you with my own.

So m e other Things I have been informed of by the Work-men, as that a little Spar mix'd with the Lead Ore, promotes its Fuſion, I ſuppoſe, as the yellow Marchaſite, that's found with Silver, makes that Metal flow the ſooner : That if there be any Holly-Wood in the Fire, it hinders the fluxing of the Ore, which is certainly cauſed by the glutinous Sap of that Wood.

T h a t the Smoak is obſerved to follow the Water very much : I ſuppoſe the Coldneſs of the Water does condenſe the Fumes, as is ſeen in reviving *Mercury* from *Cinnabar.* A blue Film is obſerved on the Surface of thoſe Waters, where the Smoak falls.

T h a t a Man may by wetting his Finger in his Mouth, or common Water, draw it through melted Lead or Iron, without any Prejudice.

Sir, T h e s e Obſervations will ſeem barren, yet as good as I could make among theſe booriſh People of the *Peak,* few of which can give a rational Account of either what they do, or ſuffer, in ſuch Matters.

I am,

Mancheſter,
Octob. 27.
1678.

Sir,

Tours, &c.

Part

Part of a Second Letter to Dr. Grew, *Dec.* 6. 1678.

S I R,

SINCE I writ to you about the *Belland,* I have been in *Derbyfhire* ; all, that I could learn far-ther of it, was, that they are lefs fubject to that Diftemper in thofe Smelting-Mills, that ftand in an open and moveable Air, or that have large Chimnies, and are not built clofe : I met with a Gentleman who told me, a Servant or two of his had it very feverely in their Bellies, and were cu-red by taking the Salt that comes from the Sul-phur-Well at *Knaresborough* ; this Remedy is, I think, one of the likelieft I have heard of.

Dr. HOOK's *Defcription of his Weather-Wifer* ; *about* Dec. 5. 1678.

THE Weather-Clock confifts of two Parts; *Firft,* that which meafures the Time, which is a ftrong and large Pendulum-Clock, which moves a Week, with once winding up, and is fuf-ficent to turn a Cylinder (upon which the Paper is rolled) twice round in a Day, and alfo to lift a Hammer for ftriking the Punches, once every Quarter of an Hour.

Secondly, OF feveral Inftruments for meafuring the Degrees of Alteration, in the feveral Things, to be obferved. The firft is, the Barometer, which moves the firft Punch, an Inch and Half, ferving to fhew the Difference between the greateft and leaft Preffure of the Air. The fecond is, the Thermometer, which moves the Punch that fhews the

the Differences between the greateſt Heat in Summer, and the leaſt in Winter.   The third is, the Hygroſcope, moving the Punch, which ſhews the Differences between the moiſteſt and drieſt Airs. The fourth is, the Rain-Bucket, ſerving to ſhew the Quantity of Rain that falls ; this hath two Parts or Punches ; the firſt, to ſhew what Part of the Bucket is fill'd, when there falls not enough to make it empty itſelf; the ſecond, to ſhew how many full Buckets have been emptied.   The fifth is, the Wind Vane ; this hath alſo two Parts ; the firſt to ſhew the Strength of the Wind, which is obſerved by the Number of Revolutions in the Vane-Mill, and marked by three Punches ; the firſt marks every 10000 Revolutions, the ſecond every 1000, and the third every 100 : The ſecond, to ſhew the Quarters of the Wind, this hath four Punches ; the firſt with one Point, marking the North Quarters, *viz.* N : N. by E : N. by W : N. N. E : N N W. N E by N. and N W by N. N E. and N W.   The ſecond hath two Points, marking the Eaſt and its Quarters. The third hath three Points, marking the South and its Quarters.   The fourth hath four Points, marking the Weſt and its Quarters.   Some of theſe Punches give one Mark, every 100 Revolutions of the Vane-Mill.

T h e Stations or Places of the firſt four Punches are marked on a Scrowl of Paper, by the Clock-Hammer, falling every Quarter of an Hour. The Punches, belonging to the fifth, are marked on the ſaid Scrowl, by the Revolutions of the Vane, which are accounted by a ſmall Numerator, ſtanding at the Top of the Clock-Caſe, which is moved by the Vane-Mill.

Dr.

Dr. Hook's *Contrivance of a Vessel, to measure the Quantities of Rain falling : Being a Part of his* Weather-Wiser *in the preceding Paper.*

PROBLEM.

To make a Vessel, which, when it hath received a certain Quantity of Water, shall empty itself.

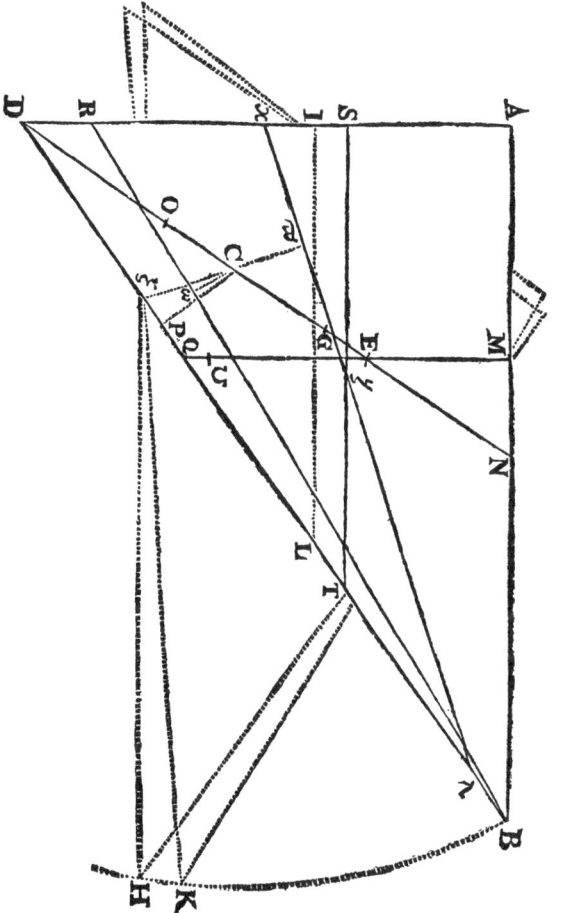

L e t the Veſſel be a Triangular Priſm, as z poiz'd like a Balance upon a Foot, ſo that the leſſer End may only deſcend, and not the greater, by means of the Stop D. And let one of the Sides be ABD. From N, the Half of AB, draw the Line DN; and from M⅓ of AB, draw MQ parallel to AB; therefore E ſhall be the Center of Gravity of the Triangle ABD. And becauſe AB is an open Side of the Veſſel, ſome Point between E and D, as G, ſhall be the Center of Gravity of the whole Veſſel; taking a Point at P near Q, towards D, erect PC, and let C be one of the Centers of Motion, upon which, and the like oppoſite Point in the other Side of the Veſ-el, it ſhall turn as a Balance. *Secondly*, By adding Weight in O oppoſite to G, equiponderate the whole Veſſel upon the Center of Motion C; therefore DCN will be a Balance, whoſe Center is C, and the Weights of equal Moment are G and O. *Thirdly*, Draw the Line ST parallel to AB, ſo that C may be the Center of Gravity of the Triangle DST.

Firſt, I s a y, if the Veſſel be fill'd ſhort of ST, the Side D ſhall preponderate; if higher, the Side B; becauſe C is the Center of the Balance DCN, and the Centers of Gravity of all the like Triangles, leſs then DST (as DIL) are upon the Arm DC, and the Centers of all the greater upon the Arm CN. Hence it follows, that becauſe it is ſtopp'd from deſcending at D, the Veſſel ſhall reſt till the Water riſe above ST, when the Side, towards B, ſhall preponderate.

2dly, I s a y, if the Veſſel be inclined towards B, the Part B ſhall ſtill preponderate; let ABD be inclined, (C the Center as before;) ſo that the Water, that lay before at ST, lies now as κζλ, and let πCς be a perpendicular Line, becauſe the

Triangles

Triangles DST, D $_{κλ}$ are equal, but $ζ_κ$S, the Triangle nearer the Perpendicular, is taken away, and $ζ_{λτ}$, being farther off, is added on the Side towards B ; therefore that Side preponderates, and the more the lower it defcends, becaufe the Center of Gravity, of the Triangle $ζ_λ$T, runs farther and farther from the Perpendicular, till it runs over at B.

3*dly*, I s a y, that when a Part given of the Water is poured out, the Refidue ftill preponderates, while it remains inclin'd. Let the Water be reprefented by the Triangle DRB in the Motion of pouring out, Part being run over ; the Center of Gravity of the Water, is $υ$ in the Line MQ: and C $_ω$ at right Angles to BR, will be the Perpendicular, as CP will be the Perpendicular when B is defcended fo low, that DB becomes horizontal, (that is, when all the Water muft be poured out) therefore CP is between CO and $υ$, but by Conftruction the neareft Point of MQ is without CP towards B, therefore $υ$ preponderates ; therefore the Veffel ftill inclines, till all be poured out. Therefore that, which was required, is perform'd.

S C H O L I U M.

I f it be requir'd that the Veffel, after it is empty, fhould return again to its former Pofition, there muft be added to the Point O yet more Weight at K, enough to reftore the emptied Veffel, in which Cafe a Triangle may be drawn as DBR, whofe Weight upon its Center $υ$ fhall equiponderate to K in O ; it feems therefore, that the Veffel fhould defcend no lower than till BR be horizontal. But becaufe nothing that moves towards an Equilibrium refts there, but is carried further by the imprefs'd Force which it gains in de-

fcending

ſcending to this Equilibrium, as it appears in all Manner of pendulous Motions. And becauſe K may be leſs than any Magnitude aſſigned, therefore, notwithſtanding the Counterpoiſe of K, it will deſcend ſo low, as to pour out all ; that is, having gain'd an impreſs'd Force in its Deſcent from B to K, there is no Reaſon but it ſhould continue it beyond the Equilibrium to H and further.

Beſides this, I find two other Contrivances of Dr. *Hook's,* among the Minutes of the Royal Society of *April* 1670. for meaſuring the Rain that falls, in theſe Words : Mr. *Hook* ſhew'd an Experiment in Mechanicks, which was a Way how to take notice of all the Rain that falleth, and was deſigned as a Part of the Weather-Clock. The Contrivance is the ſuſpending the Bucket that was to receive the Quantity of Rain, that fell at any time (whether more or leſs) ſo that according to the Quantity therein contain'd, the Place thereof ſhould either be higher or lower, but certainly be determin'd. This was perform'd by a Counterpoiſe to the ſaid Bucket. The Counterpoiſe was contriv'd two Ways ; either by a String of leaden Bullets, ſo order'd, that when the Bucket was quite empty, all the Bullets reſted upon a Table ; but when there fell as much Water into the Bucket, as equall'd the Weight of one of the leaden Bullets, then the Bucket deſcended one Space, and one Bullet was lifted up ; when twice as much, two Bullets ; and when three times as much, three Bullets were lifted up ; and ſo forward, till all the Bullets were lifted up, and the Bucket had deſcended to its Place of Emptineſs ; whereupon the Chain of Bullets preſently deſcended, and lifted up the Bucket into its empty Place.

But

But becauſe this Motion proceeded by Jumps, and was not equable, therefore a ſecond Contrivance was alſo ſhewn, which was this,

The Counterpoiſe to the Bucket, when empty, was a Cylinder immerſed into Water, *Mercury,* or any other Fluid. Which Cylindrical Counterpoiſe, according as the Bucket receiv'd more and more Water, was continually lifted higher and higher out of the Water, by Spaces always proportioned to the Quantity of Water that was contained in the Bucket. And when the Bucket was fill'd to its deſigned Fulneſs, it immediately emptied itſelf of the Water, and the Cylinder plung'd itſelf into the Water, and raiſed the Bucket to the Place where it was, again to begin its Deſcent.

This Contrivance, here made uſe of, was declar'd to be very uſeful for making a new and uſeful Beam, for examining the Weight of Bodies, without any Trouble of adjuſting, the Riſing of the Cylinder immediately ſhewing the determinate Weight of any Body, put into the Scale, without any farther Trouble.

Mr.

Mr. TOINARD's *Observation of the Diffe-rence of Longitude between* Paris *and* Breſt, *with Obſervations of* Jupiter's *Sa-tellite Eclipſes, in* 1679.

H.   "

10 *Dec.* 79. A Paris à   12 50 08

A Breſt a   12 22 37

Breſt ſelon le grande Carte de France
de Samſon del' an 1650 eſt plus oc-
cidentale que Paris de degr.   -   08 10 00
Qui valent le temps de   -   -   00 32 40
Mais l'obſervation faite 10 Dec. donne de
difference   -   -   00 27 31
Par conſequence la Carte qui eloigne
Paris de Breſt de   -   -   00 32 40
Dont il faut ôter la veritable difference 00 27 31
Se trompe de   -   -   -   00 04 22
Qui valent plus d'un degre & un cart.

3 *Dec.* 79. A Paris l'immerſion du Grand
19   Satellite à   -   -   10 53 23
A Paris l'immerſion du Pre-
mier a   -   -   09 16 03

Son gros camarade environ demie heure & demie
caſt apres.

Mon

Monſieur T O I N A R D's *Obſervations of the Eclipſes of* Jupiter's *firſt Satellite in* 1680.

Satellitis *Jovis* primi ſeu proximi immerſiones in umbram *Jovis* Pariſiis, 1680. Stilo novo.

	D.	H.	M.		D.	H.	M.
Sept.	12	16	20	*Octob.*	16	7	30
	14	10	50		21	14	55
	21	12	45		23	9	25
	23	7	15		28	16	50
	28	14	40		30	11	20
Octob.	5	16	35	*Nov.*	4	18	45
	7	11	5		6	13	15
	9	5	35		8	7	40
	12	18	30		13	15	10
	14	13	0		15	9	40 vel 35

Dr. H o o k's *Account of the great Hailſtones that fell in* London, *on* May 18. 1680.

A т about 10¼ Hour in the Morning, in *Greſham College*, I obſerved the falling of a great Shower of Hail ; concerning which, I obſerved theſe Particulars.

T h e Day before, it rain'd almoſt, all the Day, a gentle Rain, and, by turns, the fore-part of the Night. At about three or four o' the Clock in the Morning, was very much Thunder and Lightning, with an exceeding violent Shower of Rain ; whether any Hail then fell, I know not, being in

V o l. I. Bed ;

Bed ; but, by fome Circumftances, I believe
there did, for there were found, in the Morning,
feveral great Spots of Wet, which, 'tis probable,
proceeded from Hail-ftones that fell down the
Chimney. It continued to rain, and now and
then to thunder much, till about Nine ; then it
clear'd up, and the Sun fhone very clear, and

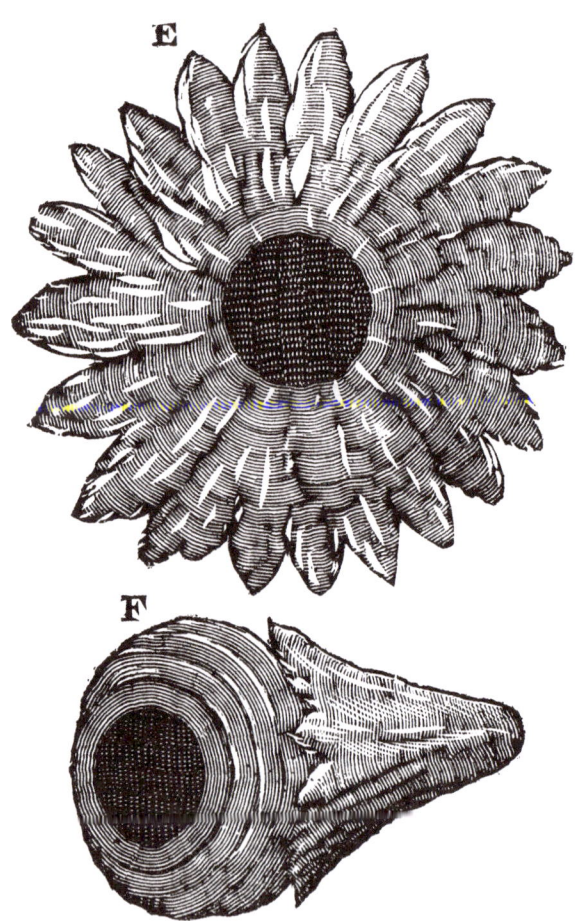

there

there was ſcarce a Cloud to be ſeen ; about ten it
began to thicken, and I heard the Thunder to
the South Eaſt ; at about half an Hour after ten,
it grew very dark, and thunder'd very near ; and
ſoon after there began to fall a good Quantity of
Hail-ſtones, ſome of the Bigneſs of Piſtol Bullets,
others as big as Pullets Egs, and ſome above 2½ Inches, and near three Inches over the broad Way ;
the ſmaller were pretty round, and white, like
Chalk, or Sugar Plums ; the other of other
Shapes : Some of the moſt remarkable were theſe.

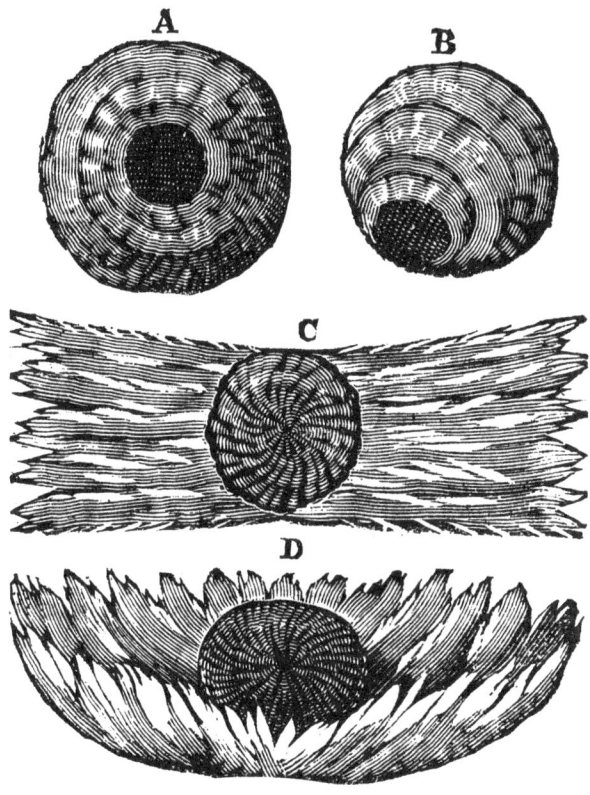

BREAKING many of them, I found them to be made up of Orbs of Ice, one encompaffing another; fome of them tranfparent, and fome white, and opaque; fome of thefe were to the Bignefs of near an Inch in Diameter, and were orbicular every Way. Some of them had the white Spot in the Middle, as A; others towards one Side, as B; and the Variety of white and tranfparent Spots very differing; thofe, which exceeded thefe in Bignefs, were made by an additional Accretion of tranfparent Icicles, radiating every Way from the Surface of the White Ball, like the Shooting of Niter, or toothed Sparre. Thefe in fome ftood, as it were, feparate in diftinct Icicles, which were very clear and tranfparent, and had no Blebs or Whitenefs in them. Others of them were all concreted into a folid Lump, and the Interftices filled up with Ice, which was not fo clear as the *Stiriæ*, but whiter; and thereby one Side, which, I fuppofe, was the undermoft, was flat, almoft like a Turnep; and the Radiations appeared to proceed from the Ball in the Middle, more towards the upper Side, and moft toward the Sides; the Edges and Top were more rough, and the Ends of the *Stiriæ* appeared prominent; which the Figures will better exprefs.

THE Extent of this Shower I cannot yet certainly learn, but have, by the Information of feveral, underftood it was feen above ten Miles off. I was alfo told by feveral Perfons, that, a little before the Hail fell, there were was heard a great Noife out of the Sky, like the Shooting, or Emptying, of a Cart-load of Pebbles, as if they had fallen one upon another in the Air.

FROM the Manner of their Figure, I conceive, their Accretion was made by a Congelation of the Water, as they fell; that the fmall white Globule in the Middle, about the Bignefs of a Pea, was

the

the first Drop that concreted into Hail ; this, in falling through the Clouds beneath, congealed the Water thereof into several Coats or Orbs, till some of them came to the Bigness of Pigeons Eggs, some white, some transparent, according to the several Degrees of Coldness it passed through, whilst they congealed ; that the last Accretion was made by a more violent and sudden Cold, in the lower Part of the Cloud, where they passed through almost a continued Body of Water. Other Varieties of their Forms, which were very many, I conceive, must be made by their meeting with one another in their Passage.

Notwithstanding Mr. Waller *hath published the Substance of this Paper, in Dr.* Hook's *Life,* p. 22. *yet the Original may not be unacceptable to the Reader, by reason of the Figures, which the Doctor hath given of those monstrous Hail-stones; which I, my self, saw falling, in great Numbers, in* Great Lincoln's-Inn-Fields, *and notic'd to have fallen on* May 19. 1680. *one of which a Servant brought me in his Hand, as large as a Turnep, and of the same Shape, which I instantly measured with a String, and found the Compass of the widest Part to be above thirteen Inches; which, I confess, seems somewhat incredible ; but, I think, I did it with great Care, and was not mistaken.*

W. Derham.

The

The Reverend Mr. PASCHALL's *Letter to Dr.* HOOK, *of an Earthquake in* Somerfetfhire, *dated* Jan. 4. 1680. *from* Chedfey *in that County.*

Worthy Sir,

YESTERDAY about feven in the Morning, I, being about to rife, took Notice of what feemed a fmart Guft of Wind, which was followed with a Jog of our Houfe, and that immediately with a very fenfible Shaking of the Houfe, and particularly the Bed in which I lay. I doubted the Fall of fome large Piece of Timber, or Stone-Work, and caufed the Servants to make diligent Search all about for the Caufe of it ; though not without Sufpicion that it might be an Earthquake. Before Night I became fully fatisfied that it was fo, for my Neighbours, many, obferved the like in their Houfes, though no Hurt was done. This Day, I hear, that it was in other Parifhes, one within a Mile of me, lying in the fame Level ; another above four Miles from me, lying on the further Side of an Hill, and which is a firm Rock. This Afternoon comes a Letter from an Acquaintance of mine in *Bridgwater,* (two Miles from me, and on the other Side of their navigable River) which fays thus, " I fuppofe you heard of
" the Earthquake, which happened with us this
" Morning about feven a Clock : It fhook our
" ftrong Stone-Houfe fo, that I began to look
" whether the Walls were fcattered or crafed, with
" a Noife, as if fome very great Thing had fallen
" upon the Ground. One or two in *Baftover* (a
" Part of that Town on our Side of the River)
" were ready to leap out of Bed upon it, &c."
The Air was very calm, as being a frofty Morning, upon the Snow lying, which fell the Day
before.

before. It lafted but a very fhort Time. I do not remember, for thefe eighteen Years of my Abode here, to have known any fuch thing ; but I call to mind the Obfervation of *Acofta*, and others, that they do moft commonly happen in Places near the Sea, and fuch is our Country ; of which I meet many Arguments which perfuade, that it was, in thefe Parts of it, formerly gained from the Sea. If you fee my Lord of *S*—-, I prefume, it would not be unacceptable to his Lordfhip, to have an Account hereof, feeing, 'tis likely, it will be a Matter of publick Difcourfe.

An Extract of Mr. Leuwenhoek's *Letter from* Delf, *Jan.* 6. 1680. *Concerning the Minutenefs of fome* Animalcules *in the* Waters.

———SINCE I perceive you are pleafed with fome of my Speculations, I have, herewithall, tranfmitted alfo a Copy of that hafty Calculation, which, at the Defire of the Honourable *Conftantine Hygens van Zutichem*, I drew up for him in Writing ; which was this which follows.

SIR,

I HAVE been often confidering of the exceeding Smallnefs of thofe Veffels, of which the Sinews and Mufcles of thefe fmall Creatures muft be furnifhed with : That which put me upon this Speculation was, the Query put to me, Whether I could, by my Microfcope, difcern the Particles of which Water doth confift ? To which I frequently gave this for an Anfwer, That there are in the Water living Creatures, many Millions of

Times

Times ſmaller in Bulk than a ſmall viſible Sand. Further, That each of theſe Creatures, though I have not, as yet, been able to diſcover their Paws, Finns, or the like Inſtruments, by which they move, muſt, neverthelefs, be furniſhed with ſome Kind or other of Organs, fit to produce that Motion. And that theſe Organs muſt be made up of Veins, Arteries, or Veſſels, to convey Nouriſhment to them, and in Sinews or Strings, to ſtir and move by, &c. If ſo, then by theſe Veſſels the Water muſt find its Paſſage, and conſequently the Particles of Water muſt be conſiderably ſmaller than theſe Veſſels, otherwiſe it could not freely paſs them ; now the whole Animal itſelf, being but ſcarcely viſible, we muſt conclude, theſe their Veſſels muſt be wholly inviſible, and how much more inviſible muſt be the Parts of Water that move in them ; inſomuch that I am very confident, that no Man will ever be able to attain, by the Help of Microſcopes, to diſcover and diſtinguiſh the Particles of which Water doth conſiſt.

Now that there are ſuch Creatures, that are ſo many Times ſmaller in Bulk, than a ſmall viſible Sand, I prove by theſe following Obſervations and Calculations. I do generally ſuppoſe (becauſe, as far as my Sight was able to help me, in taking the proportionate Bigneſs, I ſo judged it) that about 3 or 400 of the ſmalleſt of theſe Creatures, laid one by another in a Line, may make the Length of the Diameter of a middle-ſiz'd Grain of Sand. I here ſhall make uſe only of the leſſer Number, *viz.* 300; which multiplying cubically, I find the Product to be 27000000 ; whence it follows that there will go the Quantity of 27 Millions of theſe Animals, to make the Bulk of one ſmall Grain, &c. If we then ſuppoſe that eighty of theſe Sands, laid one by another, will make

but

but one Inch in Length, then there will lie in the Space of a Cubical Inch no lefs than 512000 of thefe Sands, each of which being fuppofed to be as big as 27000000 of thefe Creatures, the Inch Cubical will contain no lefs than 13824000000000, almoft fourteen Millions of Millions.

I HAVE confidered alfo of the fmall Veffels, that ferve to compofe the Parts of our Bodies, and conceive them to be Pipes a thoufand Times fmaller than an Hair of a Man's Head ; and by a Brafs-Rule, curioufly divided into Inches, and each Inch into thirty Parts, endeavouring to find, how many of thefe Hairs Breadths would make an Inch, I found that twenty Hairs would lie one by another in the thirtieth Part of an Inch, and therefore 600 in an Inch ; and meafuring my Body, I found that one Part, with another, equall'd a Cylinder of eight Inches Diameter ; fo that thefe Proportions confider'd, I find, that one of thefe Veffels muft be 360000000 fmaller than a Pipe of an Inch Diameter, and, confequently one Part of the Body being equal to a Cylinder of eight Inches Diameter, which is 64 Times as big as one of an Inch, the Cylinder of the Body is bigger than the Cylinder, of one of thefe Veffels, no lefs than 23040000000. Now if the Veffels of the Bodies of thefe fmall Creatures, in Pepper Water, fhould hold the fame Proportion to their Bodies, how can we conceive the Parts of the Water fhould be difcovered, that fhould move in thofe Veffels.

Reifilius

Reifelius *his Letter to* Dr. Grew, *concerning a Man's periodical Lofs of his Speech, from* Stutgard, March 6. 1680.

A T enim, ne fine fymbolo coram altari veftro fanctiffimo appaream, appono hic Cafum quendam mere naturalem quidem, ut mihi videtur, propter multos fimiles affectus periodicos Cephaleos pictum, Convulfiones, Colicas, ut de Febribus nihil dicam, fed rarum tamen, ob tam conftantem tamq; multis annis durantem periodum, uti obfervatus fuit a Collega meo examinante in prefentia Principis noftri aliorumque magnatum *Novembri* menfe praeteriti anni. Cujus caufam, cum neque mihi neque aliis detur affequi, ab Societatis Regiae Judicio difcere gaudeo & rogare audeo. Quomodo a fermentatione quadam ut in Febribus aliifq; morbis deducenda fit haec affectio, haereo. An a coeli meridiano vigore trahendum malum feu bonum, dubito, cum olim aliis horis & inordinate notata fit haec affectio. Symptomata tamen, quae quondam antegreffa, morbofum quid innuunt. Hic talis eft. *Georgius Algaier*, *Georgii Algaieri* Cauponis *Jefingae* propè *Kircfemium* in Ducatu *Wirtenbergico*, filius temperamenti Cholerici, annorum 25, jam ante annos quindecim fefto S. *Stephani* ftatim poft coenam, adeo male toto in corpore fe habuit, ut nullibi fe continere potuerit. Anxietas cordis erat tanta, ut, nifi per Vomitum ingentem levatus fuiffet, fuffocari fibi videretur. Hora poft vomitum unica praeterlapfa melius agebat, at per totum trium menfium decurfum valde triftis & melancholicus, interdum etiam quafi terrore percuffus evadebat. Elapfo hoc tempore, primo faltim per unum fere momentum vocem & loquelam, quam prius accurate ratè callebat, amifit, ut ne verbulum quidem, neque ullam vocem emittere poffet.   Quotiefcunq; verò loquela amittebatur, toties tum, (quod tamen ultra dimidium annum non duravit) è ventriculo, aliquid furfum, fauces verfus reperi fentiebat. Atq; uti primum vocis & linguæ fuppreffio faltem momentanea, fic eadem poft indies indiefq; crefcere incipiebat, ita ut a momento ad femihorium totum, duas, tres, & ultimò ad 23 Horas, inordinate tamen, duraverit.   Tandem typum adeo conftantem habebat locutionis reftauratio, ut jam per 14 annos, non nifi fingulis diebus ab hora 12 meridiana, per horæ integræ fpatium, ad primum fcil. Pomeridianæ punctum ufque loqui poffit.   Nec falli poteft hominis Horologicus fenfus horarum tranfpofitione, cum vel nullis campanis fonantibus terminum horæ duodecimæ ufque ad primam femper & quam accuratiffimè obfervet.   Notandum etiam quod dum loqui poteft patiens, aliquantifper balbutiat, quin & tum extra tum intra locutionis tempus linguam ipfam non fatis volubiliter queat movere. Præter amiffionem vocis & loquelæ nulla de actione queritur, fenfus tum interni tum externi funt integri.   Accuratiffimè femper audit, unde vel geftibus vel literis (fcribere enim fcit fatis intelligenter, ut ipfum hoc imitatione expreffum propria manufcriptum teftatur nomen *Georg Ulgryer Jefingus*) ad interrogata cuivis quantum poffibile eft refpondet.   Vixit aliàs hactenus omnimodò fanus, nec ullum morbum, excepta Febri, qua ante tres menfes vexabatur, quotidiana; cujus paroxyfmum inordinatè jam mane, vefperi, jam etiam noctu fuftinuit, paffus eft; neque typus amiffæ loquelæ ex Febri minimùm mutatus. Vivit etiamnum poft febrem fanus omninò & incolumis. Hucufq; Cafus Muti periodicè loquentis.

Mr.

Mr. P I G O T *in his Letter to* Dr. H O O K, *from* Oxford, *Nov.* 26. 1681. *saith,*

M R. Caſwel, *in his Travels with Mr.* Adams, *obſerved* Lidford-Bridge, *in or about* Dartmore *in* Devonſhire, *whoſe Plane is level with the Ground, yet* 59 *Feet above the Water, that runs ſwiftly under it.*

At Droitwych *in* Worceſterſhire, *he viſited the Salt-Springs, which he found, upon Taſte, to be far ſalter than the Sea. They have three in the Town, cloſe by a freſh River Side, and could have more, but that the Merchants will not permit any more, to keep up the Trade. He tells me, the poor dirty Women, that work at the Salt-Houſes, are never troubled with Lice, Fleas, or Flies.*

Mr.

Mr. Leewenhoeck's *Letter in* Nov. 1681, *of the Structure of Hair ; of the Excrements,* &c.

S I R,

I HAVE fhewed that the *Cortex* of the Hair of an Elard Hart, &c. was compos'd of Globules. I found the fame of my own Hair. I have fince found it like the Bark of a Tree of Globules, but irregular from the fqueezing of the Hair. The Subftance of the Hair is made of Threads ; fome judge the Hair hollow, others to have Marrow ; but viewing a Hog's Hair, I found the Hollownefs of thofe Hairs from Cleft. Hair grows by Protrufion, not as Plants, being thruft continually forward, from within the Skin outward ; what was within moift, expos'd to the Air, dries and fhrinks, and the outwardSkin hardening,the inwardThreads, upon fhrinking, cleave into one or more Clefts, which feems like Marrow. In a Piece of Hog's Hair the Threads appear plain, even in a common Microfcope, but bigger in Proportion to the Circumference for Eafe ; the Threads were but few, from the Roughnefs of the Razor. By thefe may appear their Miftake, who affert Hairs round; 'tis rather true, they have all differing Figures. A Friend vifiting me after a Fit of Sicknefs, whereby he had loft all his Hair, complain'd of a great Itching all over his Skin, yet his Stomach was good, which the Doctor attributed to a Sharpnefs of Blood ; which I rather afcribed to the filling of the Body, and from the new growing of the Hair in the Pores, whence it had fallen by his Difeafe, the Pores of which being clofed, the new Hairs, thrufting againft the *Cuticula,* caufed the Itching. I myfelf have been fo troubled in the Spring,

which,

which, I conceive, to proceed from the same Cause, being my self hairy, and shedding them yearly, as, I conceive, most Men do ; this I observed in two Parts of my Body, also in three Places of my Hand, where I have shorn off the Hair, and found, that some Hairs grew, others not ; some fell out, and I could pull them out without Pain; also, that these, which fall out, have thin sharp Roots ; those which stay, thick ; also such, as have no Hairs on their Body, have Pores, and an issuing Matter, not so fit for Hair. This appears like black Specks, and are supposed Worms ; and some Doctors of *Aken*, did prescribe this Man to stand with his Back to a Fire made of Oak, and anoint his Body with Honey, that by Means of the Sweetness and Warmth the Worms may come out, and so be cut off with a Razor, as the Gentleman Patient himself told me ; hereupon I try'd to press, both out of my own, and out of another Man's Nose, these supposed Worms, which seem'd, from their Shape, much to favour the Opinion, seeming to have a Head which proceeded from that Part of the Hair, which was next the Air, it being browner than that within the Skin, but no two like one another. I observ'd all its Parts, but found nothing like an Animal ; but in several I found small Pieces of Hair, some 25, others 100 Times thinner than a common Hair. Hence I concluded, the supposed Animals are only the Places of those Hairs fill'd with the usual Food of Hairs ; my Opinion is confirm'd by new-born Children, over-grown with Hairs, which, I suppose, from too much Nourishment, grow hairy, but when they want that Supply, the Hairs fall out, and grow not again.

In a Looseness I view'd my own Excrements, and took notice of what I eat and drank; it consisted of clear, yellow, roundish, irregular Particles, also of vast
Quantities

Quantities of Globules, like thofe of Blood, fix together equal to $\frac{1}{3}$ of the whole; others but $\frac{1}{18}$ of a Blood Globule: Thefe I found in a tranfparent Liquor, in which were many Animals, as big as a Globule of Blood, their Bodies oblong and flat, with many Feet underneath, with which they moved quick; like a Pifs-a-bed againft a Wall, tho' they moved their Paws quick, yet they went but flow. Once I found but one in the Bignefs of a Sand, at other Times, 4, 5, 6, 7, or 8. I have feen other fhap'd Animals, (but of the fame Bignefs) like River Eels; thefe were very numerous, and fo fmall, that 5 or 600, extended in Length, would not reach the Length of a River Eel; thefe wriggled like a Snake, very quick, like a Pike fhooting through the Water. At another Time, I faw Multitudes, 200 Times lefs than a Blood Globule, the *Axes* being about one to fix, and I am confident, I have feen above 1000 living Animals, in the Bignefs of a Sand, fwiftly moving, and of three or four feveral Sorts. Some have thought, thefe Animals might pafs into the Blood; but, I conceive, the Paffages of the Blood are fo fmall, that though the Animals were 1000 Times lefs, they could not pafs. My ordinary Excrements, mixed with a clear Liquor, had no Animals; but when thinner than ordinary, it had. I found alfo Parts of the Food I had eaten, undigefted, as the Pipes of Afparagus, the fofter Parts being digefted.

T H I s Summer, in our Meadows, I have obferved the Dung of Cows, Horfes, &c. frefh, but found no Animals. It confifted of Multitudes of Globules, fome $\frac{1}{6}$, others $\frac{1}{18}$ of a Blood Globule, in a clear Liquor. In *May* laft, riding my Mare hard, I obferved the laft thick Part of her Urine, and found, the thick Afh Colour of it was caufed by a great Variety of differing Globules, fome as

big

big as thofe of Blood, and thefe compofed of fix.
The firft of thefe were like a clofe-grown Bunch
of Grapes, and though not perfect round, yet I
call them Globules.

Dr. Hook's *Letter to* *Dr.* Trapham,
of Enquiries for Jamaica, *Feb.* 18. 168⅔.

SIR,

I T will be a great Obligation to the *Royal Society,* if Dr. *Trapham,* or any other ingenious
and knowing Perfon in *Jamaica,* will pleafe to
communicate any curious Obfervations they fhall
make, concerning any Part of Nature; as concerning the Temperature and Qualities of the
Air, the Seafons, Winds, Storms, Hurricanes,
Rains, Hails, Dews, Mifts, Fogs, &c. the Heats,
Colds, &c. of the Seafons; the Qualities of
Springs, Rivers, Lakes, &c. the Defcription of
any of the Animals, Birds, Beafts, Fifhes, Ser-
pents, Infects, or of any of their Qualities or
Ufes, for Food, Phyfick, Pleafure, &c. The
Defcription of their Vegetables; as of their Herbs
and Shrubs, whether of the Land or Sea; of the
Trees; their Ufe in Food, Phyfick, Building,
Dying, Perfuming, Firing, Joinery, Turning,
Bows, &c. The Defcription of any of their
peculiar Stones, Minerals, Ores, Metals, Clays,
Earths, Sands, &c. of what Nature, what Ufe
made of them, &c. Alfo to inform them concerning any accurate Obfervations, that have been
made of any Eclipfes of the Moon, and particularly that of the 11*th* of this Inftant *February*;
of the Variation of the magnetick Needle, from
the Meridian, or North Point; of the Times of
the Tides, both Spring and Neap, and of the
Height

Height it rifes; of the Currents, what, when, which Way; of the Depths and Soundings of the Seas thereabouts, and whatever of this Kind shall be communicated ; or if any curious Jewels, Shells, Seeds, &c. shall be fent, the Society will not only pay the Charge of Freight, but any other way gratefully acknowledge the Favour that the Communicator shall defire, either by recording it in their Regifters, or publishing it in their Hiftories.

Mr. Lewenhoek*'s Letter to Mr.* Oldenburg, *receiv'd from* Dr. Crowe, *Aug.* 14. 1682. *Of the Fibres of the* Mufcles, Dura Mater, Brain, *and* Moxa.

Excellentiffime ac Eruditiffime Vir.

GRATISSIMAS, præteritæ menfis decima prima ad me datas, literas accepi, in quibus humanitatem Nobilitatis Veftræ, dominorumque philofophorum vifam, grato animo agnofco.

IN literis 22da Februarii fcriptis, nobilitas veftra inquit, amicorum quofdam optare, ut fumma cum exactitudine obfervarem fibras mufculorum carneas, ut & corticem, medullamque cerebri.

IN literis meis, *Anno* 1674. prima Junii datis, dixi : Fibras mufculorum carneas ex valdè parvis confiftere globulis : Sed quo nobilitati veftræ reliquifq; amicis magis fatisfaciam, omnes præcedentes meas obfervationes rejeci, firmiterque propofui, de novo, clarè ac perfpicuè eas, oculis meis mihi perfpiciendas, fumere.

I n t e r alias, *carnem vaccinam* accepi, quam,
acutissimo cultello, in frusta concisam, per microscopium â membranulis suis separavi, quo per-
racto, tum primum mihi nudè ac dilucidè apparu-
it, tenuissima illa membranula, cui fibræ carneæ
quasi involutæ, aut intextæ jacent, cujus etiam
Anno 1674. prima *Junii* in literis meis memini,
dicens: Membranulas illas ex tot striis ac fibris
consistere, quasi nudo oculo omentum alicujus
bestiæ aspiceremus. Easdem membranulas jam
propius observans, totas illas solummodo consistere ex fibris transversim inter se mixtis, comperi, quarum quædam, in oculo meo, decies, vicies,
& tenuissimæ quinquagies tenuiores pilo. Cogi-
tabam, num quædam crassioris generis, quæ in
ramos se dispergebant, non essent vasa lympha-
tica.

S u b l a t i s, à prædictis fibris carneis, prænominatis membranulis, eas nudè ac perspicuè vidi,
quæ in hac carne erant ad crassitiem communis capilli. Ubi spissæ ac densæ, rubicundæ erant; ubi
tenues ac dispersæ jacebant, magis apparebant
pellucidæ.

V a r i a observandi methodo usus sum, particulas carnearum harum fibrarum videndi, perpe-
tuoq; inveni, eas ex talibus compositas partibus,
quibus aliam quam globulorum figuram appropri-
are nequeo. Imo & minima fibrarum carnearum
frustula, grano arenæ aliquoties minora, coram vi-
su meo in plurimas divisi partes. Præterea etiam
observavi, carne adhuc recenti & humidâ, quod,
tum compressis vel fricatis carnis globulis, illi globuli resolvantur & conjungantur, quasi oleaginosam, vel aliquo modo concretam, videremus, materiam aquosam.

H i globuli, ex quibus fibras carneas consistere
dixi, adeo exigui sunt, ut (juxta oculum, meumq;
visum, judicium ferens) dicam 1000000 non con-
fecturos

fecturos unici arenæ grani, aliquo modo grandioris, quantitatem.

E т quamvis in mentem veniat, me antea Nobilitati Veſtræ ſcripſiſſe, particulas, ex quibus caro, adeps, oſſa, capilli, *&c.* conſiſtunt (quæ a me globuli vocantur) non eſſe veros & propriè ſic dictos globulos, ſed figurâ globulis proximos, eadem tamen hic repetam: Ex. gr. Imaginetur quis ſibi, ſe magnam veſicarum ovinarum, vel aliarum, aquâ repletarum habere quantitatem : Hæ veſicæ quamdiu, ab omni parte, ab aere circumdantur, rotundæ erunt: Sed imaginemur nobis, eas promiſcuè & indiſcriminatim in vas aliquod injici: Quo facto, veſicæ, globoſam, quam in aëre habuerant, rotunditatem, non ſervabunt, ſed a ſe invicem compreſſæ, nullum vacuum (ſic loquendo) locu relinquent: Et ſic quævis veſica aliam, ob flexibilem ſuam mollitiem, accipiet figuram. Sed quæ in vaſe ſupremæ jacebunt, in quantum ab aëre amplectuntur, globoſam retinebunt rotunditatem : Idem de globulis carnis, propter eorum mollitiem, ſit judicium.

Piam Matrem obſervavi, comperique membranam hanc, variis ſanguinis vaſibus intertextam, præter ea quæ nudo oculo, cerebro injacere, cernere poſſumus, (præcipue ſeparatione Piæ Matris & cerebri facta,) & inter ea venæ admirandæ & incredibilis tenuitatis : Et quantum dijudicare poſſum, membrana illa ex admodum exilibus conſiſtit fibris. Ulterius vidi, prædictas multiplices venas, per membranam hanc diſperſas, ramos ſuos per cerebri ſubſtantiam quoque diſpergere: Eo modo, ac ſi nobis imaginaremur diverſas & ſuperficiei terræ palmitibus ſuis injacentes vites (quas venis Piæ Matris comparo) eaſque ubique ex palmitibus ſuis radices, in plurimos diſperſas ramos, alte in terra egiſſe. Terram hic mihi imaginor ſub-

ſtantiam

ftantiam cerebri, & radices, venas per cerebri fubftantiam difperfas.

Accedens jam ad partes ipfius cerebri, adhuc affirmarem, id, præcipue ubi paululum com-
preffum ac compactum, non nifi ex globulis,
& non ex aliis confiftere partibus : Sed ubi rarum ac tenue, cultro concifum aut feparatum,
fefe oftendebat dilucidiffima materia, quafi oleum fuiffet, quam videns imaginabar mihi cultro
id caufatum, globulofq; cerebri difruptos aut
fractos. Verum enimvero perfeverans in obfer-
vando, non tantum beftiarum, fed & pifcium, &
præcipue quidem Afelli majoris cerebrum, clare
perfpexi materiam illam oleaginofam, non fuiffe
cultro ex difruptione globulorum caufatam, fed
reverà effe materiam feparatam, cui prædicti cerebri globuli quafi injacebant. Ulterius vidi, fed
clariffimè in cerebro Afelli majoris, prædictam oleaginofam fubftantiam, reapfe etiam ex globulis,
fed multo minoribus, quam ipfius cerebri, confiftere.

Primo nominati globuli cerebri, meo judicio,
circumcirca, globulis fanguini ruborem afferenti-
bus (ex quibus fanguinem confiftere dixi) magnitudine æquales funt. Hi majores globuli, ex maxi-
ma parte cerebrum conftituentes, refpectu globulorum fanguinis, valde irregulares vel inequales exiftunt. Hujus rationem exiftimo vel globulorum fir-
mam inter fe, aut cum vafibus conjunctionem, vel
eorum mollitiem, adeo ut fe feparari non finant,
quin (fic loquendo) a fe invicem difcerpantur, ubi
e contra globuli fanguinis in fluidiori materia mo-
ventur, & propterea etiam, globofam fuam rotunditatem, quando in latiori fpatio exiftunt, re-
tinent.

In animum fubit, me antea temporis obfervaffe
cerebrum *Anatis*, & tum judicaffe, cerebrum ex
parte confiftere ex filis, aut admodum exilibus
vafibus

vafibus. His filis vel vafibus poftea mihi fæpius oc-
currentibus, tum temporis & idem judicabam, ea
tantum produci per firmiffimam globulorum (ex
quibus cerebrum folummodo confiftere putabam)
inter fe unionem, & qui minima extenfione fic in
fila mutarentur. Sed obfervationes meas per integrum menfem continuans, clare admodum vidi,
multiplices valde, & fupra modum exiguas venas
(de quibus antea certus effe non poteram) eas in
beftiarum cerebro exiftere, & revera venas effe,
licet cognitu admodum difficiles. Verum obfer-
vante, exactiufque infpiciente me Afelli majoris cerebrum, multiplicia illa minima vafa, aut venulas,
quæ fupra modum pellucidæ, clare mihi oftendi :
& multas, licet in ramos difperfas, & quindecies vel
vigefies filo bombycis exiliores, tamen cognofcere
potui : Horum dictorum vaforum vel venarum
maximam multitudinem, in quantitate cerebri ad
magnitudinem arenæ, vidi : Præterea & vafa fanguine repleta, vel quærubicunda apparebant, ut etiam vafa ad craffitiem unici fili bombycis, & infuper
pellucida vidi.

Hasce meas obfervationes circa beftiarum cerebrum perfequens, vafa ante nominata, admodum
perfpicue quoque oftendere potui, eaque fumma
cum admiratione vidi, partim ob ingentem multitudinem, partim ob fupra modum fummam eorum
exilitatem. Si enim juxta oculum meum judicium
feram, dicere teneor, quod, fi globulus, fanguini
ruborem afferens, in octo effet divifus partes, & u-
naquæque octava pars effet firma & folida,
ne una quidem harum partium hæc vafa tranfire
poffet. Et quamvis diverfis vicibus prædicta cerebri vafcula mihi perfpicue ob oculos pofueram, in
obfervationibus meis circa illa tamen continuavi,
& quo penitius ac fæpius obfervarem, eo exactius
admodum multiplicia illa vafcula, cum ipforum
ramis (qui adeo infirmi ac debiles, ut minima con-
trectatione difrumperentur) dignofcere potui.

INTER

INTER dictos globulos, ex quibus cerebrum ex parte confiftit, globulos fanguinis jacentes vidi, qui, ob perfectam rotunditatem, clare a globulis cerebri diftingui ac dignofci poterant : hos fanguinis globulos opinabar e fanguinis vafibus per cerebrum difperfis, & cultro concifis, effluxiffe.

INTER corticem & medullam cerebri, aut parvam, aut nullam fere, differentiam, obfervare queo : præfertim cum paululum rariorem, & tenuiorem, eam mihi videndam fumo : tantum dicam, venas, aut vafa corticem cerebri permeantia, aliquo modo fubfufci vel fubnigri effe coloris, ubi e contra vafa medullæ cerebri erant dilucida ac pellucidiora.

IN cerebro, fed plerumque in cortice, tam exiles ac rubicundas, ex majoribus procedentes, venulas vidi, ut capere nequeam, quomodo globuli fanguinis eas permeare poffint : & ultra, quo pacto globuli paulo rariores, & feparatim obfervati, ferme nullius faltem admodum modici effent coloris, ubi e contra fanguis in hifce vafibus ruberet. Imo & per ipfas venas, in fubftantiam cerebri proximam, color ille rubicundus penetrarat, eamque infecerat. Sed animo revolvens, me in obfervandis Pediculis fæpe vidiffe, quando Pediculum efurire feceram, ipfique prope fame confecto, jam fanguinem fugendum darem, ipfum non potuiffe confumere fanguinem, aut etiam ejicere ; quo evenit, ut globuli fanguinis rubicundi liquefierent, & in materiam fluidiorem refolverentur, & fic per totum Pediculi corpus, imo per ipfas ungulas & cornua difpergerentur, omnibufque partibus ruborem afferrent. Caufam non confumpti fanguinis opinabar, inteftini aut parvarum in Pediculo venarum exficcationem, defectu alimenti caufatam : quo debitus ac ordinarius fanguinis motus fuit impeditus, nec jufto modo per totum corpus vehi potuit. Sed memini, hanc fanguinis mutationem, in fanguine, in vitro per aliquod tempus, fervato, aliquando etiam a me obfervatam.

Et

Et idem in parvis cerebri venis accidere poſſe opinor (quamvis adeo exiguæ ſint, ut globulus, rotunditatem ſervans, penetrare nequeat) ut reſolutis globulis, & venæ rubræ appareant, & cerebrum adjacens rubore tingatur.

MEDULLAM ſpinalem Vituli, Ovis, Gallinæ, ac Aſelli majoris etiam obſervavi, quam ex iiſdem cum cerebro partibus conſiſtere comperi, cum hac ſolummodo differentia, quod præter globulos, quos cum cerebro ſpina medulla communes habet, in hac ingens globulorum oleaginoſorum & pellucidorum numerus, ac diverſæ magnitudinis jaceret. Quidam enim quinquagies majores reliquis, ac præterea admodum molles, ac fluidi. Cæterum medullæ ſpinales multis ac ſupra modum tenuibus inſtructi erant venis aut vaſibus. Præterea hic per medullam ſpinalem diſperſæ erant fibræ coloris ſubfuſci, & ad craſſitiem capilli, quædam vero tenuiores : quibus viſis imaginabar mihi in initio, num quævis fibra forſan non eſſet vena : ſed ſumma cum exactitudine penitius inſpiciens atque obſervans, comperi, quamvis fibram non eſſe vas, ſed ſingulas earum conſiſtere ex aliis valde exiguis fibris aut vaſibus ſibi invicem adjacentibus, inter quas fibras pellucidiſſima videre erant vaſa ad craſſitiem fili bombycis. Hic tum opinabar, an hæc vaſa non eſſent ea, quæ ſpiritibus animalibus per medullam ſpinalem vehendis inſerviunt.

HÆC ſunt, clariſſime ac nobiliſſime vir, quæ poſt ultimos, indefeſſos, & exactiſſimos labores, hucdum in cerebro, &c. detegere valui.

UTI dixi antea, quo pacto multæ venæ ſibi invicem adjacent conjunctæ quaſi una tantum eſſent vena, ſic illud mihi non tantum occurrit in medulla ſpinali, & interdim quoque in cortice cerebri : Verum etiam in fructibus, & ſeminibus, præſertim in Caſtancarum venis. Ut & in cortice & putamine

tamine Amygdali: in secunda nigri piperis membrana ; In putamine Avellanæ nucis duro, & membrana quæ intus in concavo ei adhæret, & in molli cortice cui nucleus injacet involutus : ubi quidem 15 aut 20 tenuissima vasa sibi invicem adjacentia vidi. Etiam in membrana nucleum Juglandis immediate amplectens. Omnia hæc vasa ex continuata tortuositate composita sunt, eo modo ac si nobis imaginaremur tenuissimumaliquod filum æreum aut ferreum crassiori pressim circumvolutum (in formam qua fustis vel baculus fissus iterum fune colligatur) postea extracto crassiori filo, tenuissimum illud quod ei circumvolutum fuerat, omnes gyros ac circumvolutiones retinebat. Eodem modo (ut dixi) tenuissima in prænominatis seminibus & fructibus vasa contorta vidi. Præterea in Malo & Piro tenuissima sibi invicem adjacentia vasa observavi.

A N N U S jam præteriit, cum in ædibus suis, nobilis dominus Constantinus Hugenius a Zulichem, mihi monstraret *Moxam,* addens, quo pacto inustione istius herbæ podagra sanaretur · Aliquantulum hujus sic dictæ herbæ Moxæ mecum domum retuli, carpoque manus impositum juxta præscriptum urendi modum, combussi (ex curiositate nimia, nam podagra non divexor) quo extraordinariam hujus combustionis effectum detegerem, observavi autem cuti, in loco ustionis, injacere materiam flavam ac oleaginosam, quam principio judicabam pec combustionem cutis causatam. Verum hanc cutis inustionem intermittere coactus fui, non ob dolorem, sed sanationis difficultatem : si enim tam facile sanare possem, ac vulnus ex incisione cultri, (quod colligatum ac consutum sanatum æstimo) sæpius hanc inustionem iterarem. Per microscopium Moxam examinavi, firmiterque sentio Moxam non esse herbam ex optimæ terræ pharmacis artificiose paratam, ut autumat dominus Busschoff in tractatu de Moxa p. 52 sed solummodo vaporem

porem aliquem ejectitium alicujus fructus, ficuti in malis Perficis, Cydoniis &c. lanofam videmus fubftantiam cortici adhærentem. Cogitaram etiam me de fructibus quibufdam collecturum herbas moxæ quodam modo fimiles, fed hucufque efficere non potui.

Moxa, quoad figuram, goffypio refpondet: ficuti enim inter pilos, capillofve, & lanam, nulla, nifi quoad craffitiem & longitudinem, differentia, utpote ex globulis confiftentes,& ad rotunditatem inclinantes: æque parva inter moxam & goffypium differentia, & illa & hoc enim duobus planis gaudent lateribus. Eandem figuram, lanofum illud quod interne rubri corticis caftaneæ convexo adglutinatum, oftendit: in hoc tantum differens, quod moxa multo fubtilior fit goffypio, hoc caftaneæ lanofitate. Moxam, cum inuftio manus non placeret, juxta & goffypium, forfice parumper diffectum, quo facilius ignem perciperet, chartæ anguftæ impofui, & hæc moxæ & goffypii combuftio fibi invicem exacte refpondebant, adeo ut mecum ftatuam, fi inuftio quendam, circa fanationem podagræ, producat effectum, illud non evenire per aliquam moxæ propriam qualitatem, fed tantum per inuftionem ipfam, & fi goffypio inuftionem faceremus, nos tantum effecturos quantum moxa.

Ulterius moxæ, goffypii & lanofitatis caftaneæ æqualem fumfi quantitatem, quam juxta fe invicem pofita combuffi, comperique quodvis horum trium poft fe reliquiffe materiam aliquam oleaginofam, fed moxa plurimam ; caufam imaginabar, quod, quamvis quantitas moxæ quoad oculum non major, revera plus materiæ effet in moxa, utpote quæ fubtilior molliorque, goffypio arctius conjunctas haberet partes, & propterea majorem olei quantitatem poft combuftionem reliquerat. Adeo ut credam dominum Biffchoff a Chinenfibus moxæ qualitates, præparationemque extollentibus, effe feductum ac deceptum.   Etiam

E T I A M animo recolens commune chirurgorum dictum, goffypium (ut Holl. dicitur) effe ignitum, hoc eft, inflammationem caufare, & noxam afferre vulneribus, quando iis colligandis applicatur. Malignitatem, goffypio adfcriptam, in hoc confiftere judico, *viz.* quod, ut antea dictum, duo plana, & per confequens quævis particula, duo acuta habeat latera Hæc acuta latera tenuiora, fubtiliora & duriora globulis fibrarum carnis, propterea (cum goffypium vulneribus applicatur) non tantum caro adhuc fana, fed materia incarnationi novæ inferviens, & molliores carne fana globulos habens, vulneratur ac læditur imo conciditur & refolvitur. Sed contrarium cum linteo evenit, utpote cujus partes rotundæ & arcte fibi invicem junctæ, majus corpus efficiunt, ideoque globulos carnis & materiæ incarnationi infervientis tam facile non, aut in totum non lædunt.

H æ c funt, nobilis vir, quæ excellentiæ veftræ dominifque philofophis hac vice per literas nunciare volui : Submiffe & fubnixe rogans, nobilitas veftra velit dominis philofophis multam meo dicere nomine falutem, dataque occafione, refcripto, has bene perlatas, & quo pacto hæ meæ obfervationes aut conveniant cum antecedentibus, aut in quantum (fi) ab illis difcrepent, fignificare. Nunquam occafioni deero, qua demonftrare potero

Excellentiffime Clariffime Vir,

Quod Sim Nobilitatis Veftræ

Addictiffimus Cultor,

Subfignaverat

A N T H O N I U S L E W E N H O E C K.

Dr. John Carte'*s Letter to Dr.* Hook, *of Worms like* Millepedes, *in the Stomach,* &c.

S I R,

I SEND you the following Cafe, which, in fome of its Circumftances, is not very common : A Girl about eight Years old, who has never been very healthful, but of late hath looked more pale than ordinary, and troubled with Pain at her Stomach, yefterday, upon taking a purging Powder, vomited a Sort of Infects, to the Number of about a Hundred, very much refembling little *Millepedes* ; I faw fome of them, and three, that were living, I put in a Box, and a little Duft to them, but they followed the Fate of the reft, and died prefently ; I have fent you fix of them. The Child had taken Worm-Seed over Night, but had a very troublefome Night, could fcarce be held a Bed, complaining both of the Pain and Sorenefs of her Belly, fancying the Worms had eaten it thin in one Place, and would eat a Hole in it. The Length of one of the biggeft, (though there was but little Difference) was ⅕ of an Inch: I view'd them through a fmall Microfcope, which did not reprefent them fo clearly, as to diftinguifh them from the common Wood-Lice, only their Bellies were more tranfparent, and their Heads of a more confufed Figure, which laft I thought afterwards might be caufed by the rowing up of the *Antennæ* or Horns, which I obferved fome of the common *Millepedes* to do, when they die That among them which was black, was accidentally fo, by dropping a little Ink upon it.

T H E Child, after her vomiting, had a Stool, in which were feveral very fmall white Worms, about an Inch long, which are not uncommon, but

fhews

shews that the *primæ viæ* abounded with such putrid Humours, as are usually productive of a verminous Brood : She is now very hearty, and eats her Meat well, and free from all the former Symptoms.

I H A V E heard some Stories of the like Nature, but am not forwards to relate them, because they totally depend on the Credit of others : One Man I know, who, many Years ago, was reduced to a thin consumptive Habit, and, upon taking *Mercurius Dulcis*, voided by Stool an incredible Number, or rather Quantity, of small Animals, which (according to the Description I had of them) were less than these, and of a rounder Figure.

IF these were bred in a *Folliculus* of their own, that Part must apostemate, and so a purulent Matter be evacuated with them ; but I rather think, they must be generated in the common Passage, and I remember I have often seen Abundance of Animals bred in humane Excrements, but was not so curious to observe their Figure.

I T is hard to imagine, how Worms should live in the Stomach, amidst that acid Humour, which, whether it be the Cause or Effect of Digestion, has the Force of a *Menstruum* ; but it must be supposed, that in such Bodies, the Ferment is alter'd, if not destroy'd : You observe lately, that Birds are very industrious to kill Infects before they eat them ; I am apt to think, if they pass'd immediately into the Gizard, there was no Need of killing them first ; but the *Ingluvies* supplying the Want of Teeth, and only macerating what other Creatures chew, has no Acidity that would offend them.

S I R,

S i r, I write this Account haſtily, becauſe I would have you ſee them as ſoon as might be.

Mancheſter, *I am,*
Auguſt 25,   *S I R,*
1682.

      Your humble Servant,

         J. C a r t e.

N. B. *The Child had not taken any* Millepedes, *nor uſes to eat Earth or Dirt, which I have known ſome diſtemper'd Children do.*

A Letter from Mr. J. Y o n g e, *to* Dr. H o o k, *of divers curious Matters ob-ſerv'd by him.*

A W o m a n, about 36 Years old, had from her Childhood been ſickly, more eſpecially tormented in her Belly with a Pain, accompanied at firſt, every three Months, and afterward every three Weeks, with a round Swelling like her Fiſt, in her left *Hypocondria,* ſenſibly moving to and fro', and plainly to be felt : Horrid Pain would then deprive her of Senſes, twelve or twenty-four Hours ; and then ſhe would recover again, be without Pain, and the Tumour vaniſh, without being followed by any Evacuation, of either Wind, Water, Excrement, &c.

 T h o s e Paroxyſms, for many Years, kept a due Courſe of three Weeks ; ſhe was generally coſtive, found that Milk irritated her Pain, that Fleſh and all ſalt Meats diſagreed with her.

       N o t-

NOTWITHSTANDING this, she married about twelve Years since, and had one Child. During her Breeding, her Pains observ'd the Course, and abated nothing of their Vehemence, which equall'd, if not exceeded that of Child-birth.

UNDER this Plague she liv'd, till about *February*, 1680, the Pain seem'd fix'd on the left Side, on the Region of the Spleen, and seem'd as if proceeded from the Lodging of some heavy Thing, and begot such Pain, as she could not lie down in her Bed. Thus she continued in a miserable Condition, using Purges, Clysters, &c. which were advised by charitable People, she being very poor. The 15th of *November*, 1681, she became quit of all the Pain in her Side, and then felt somewhat to burthen, and, as it were, stop the *Intestinum Rectum*, causing frequent Motions to Stool, but no Evacuation, but a little Slime like a *Tenesmus*. The Suppression of her Evacuations that Way, for six Days, so press'd on some of the urinary Channels, that her Urine also stopp'd. In this doleful Condition, she sent for me, when, giving me the abovesaid History, I guess'd somewhat extraordinary must be in the *Rectum*. Accordingly, examining by a Probe, I felt a hard Substance like a Stone, which, with a strong Pair of *Forceps*, I extracted, and then cleansed out the Bowels with a Clyster; she remained void of any Pain, and is so to this Day.

THE Thing extracted, was of a round Figure, somewhat oblong, with some Depressions, such as a Man's Fingers make on Pitch, Plaister, or Wax. In Weight, was one Ounce and a Quarter; was five Inches round, swam on Water, though seem'd a Stone. Its Outside was black as Jet, smooth as Varnish, but no thicker than a Man's Skin; next to it, it was stony, or gritty, like Brick, the Thickness of half a Crown. After some Months, I

cut

cut it in two with a Hatchet, and found that next to the gritty Shell, it was full of a woolly, hard Subſtance, like rotten Rags, or Sponge, or chew'd brown Paper, within which, lay a Lump of the Bigneſs and Form of a ſmall Prune. Cutting that in two alſo, I found it a Prune, or Plum indeed, the Pulp of which was dry, and hard as Paſteboard, as was the Kernel in the Shell, that lay in the Middle of it.

Whence it's manifeſt, that all theſe Accidents, that had ſo long moleſted this poor Woman, proceeded from this Plum, or Prune, ſwallowed above thirty Years before ; which, probably, ſtuck in ſome folding of a Gut, or a Cavern, or Cell of the *Colon*, increaſing its Dimenſions by the Adheſion of new Matter, till (no one knoweth how) it tumbled down to the *Rectum*, and I drew it forth. But how the Surface became petrify'd, and ſo uneven, and varniſh'd over with a black ſmooth Matter, is to me a Wonder.

Before I broke it, I thought it might be a Gall-Stone, (tho' ſhe never had the Jaundice) having lately ſeen a Gentlewoman, almoſt dead in that Diſeaſe relieved by the Evacuation of one, almoſt as big as a Pullet's Egg, and another from a Man, as big as a Nutmeg. Both followed (tho' coſtive before) with a Lask, diſcharging prodigious Quantities of Choler. The Authors are innumerable, that mention this latter Sort, though I meet none ſo great, *Vide La. Riverius Obſ..ab Henrico Ruffeo com. obſ.* 4. *Tho. Bartholin Acta Med. A.* 71, 72. *obſ.* 100. *J. Fernelius lib.* 6. *de part. Morb. & Sympt. J. Skenckius, Obſ Med. Sennertus, &c.*But few ſpeak of any, that appear generated in the Guts, *Vide Miſcel. Curioſa vol* 6. *obſ.* 20.

There

THERE lately died, in *Cornwall,* a Woman of about 154 Years of Age; I have employ'd a Friend to give me a particular Account of her Manner of living, *&c.* which I will not fail to tranſmit to your Hands.

HERE was lately, alſo, an Ewe kill'd, that had a full grown Lamb lapp'd up in the *Omentum,* among the Guts, without the Womb; queſtionleſs it was a Conception *in tuba Fallopiana,* which, when growing big, broke forth into the Bowels. But that the *Pedunculus* ſhould hold, and where the *Placenta* was faſtened, is ſtrange: In the *fundus uteri,* it cou'd not be, and if any where elſe, how was the nutritious Juices, *&c.* conveyed to it. It was ſeparated from the *Uterus,* and the Bowels thrown away before I knew it, ſo that I could not make that Examination: This Accident is not ſo new, but that Inſtances of the like are given by Monſ. *Bayle,* Mr. *Blegny, de Graeff, Elſchotius, Riolanus, Rheynbuſe,* &c.

A CHILD was lately heard, by ſeveral People, to cry in its Mother's Womb, ſome Days before the Birth; do not Children then breathe by the Lungs, before they are born?

I find ſuch another Relation (if not the ſame) of a Lamb in the Omentum, *told by Mr.* Younge, *in the* Phil. Tranſ. *Numb.* 323.

WILLIAM DERHAM.

Obſer-

Obſervata quædam Anatomica in Veſpertilione *diſſecto* 22 *die Sept.* 1682. *Per* T.
Molyneux, *M. D.* Dublinij.

EXTERNAM hujus animalis figuram verbis
deſcribere, ſupervacaneum fore exiſtimavi,
utpote cum in hiſce noſtris regionibus adeo fre-
quens occurrat Veſpertilio, ut cuique volenti, eum
vivum intueri, facillime obtigit ; vel ſaltem omnibus conceditur, ut illius vivam aſpiciant delinea-
tionem, cum apud tot varios autores de animalibus ſcribentes, hoc accurate depictum inve-
nire liceat. Iis igitur omnibus omiſſis, quæ alii de
Quadrupede hoc volanti jamdudum tradiderunt,
ſolummodo hic notabimus quædam hactenus neg-
lecta & inobſervata quæ in illius diſſectione nobis
videre contigit.

Et primo *Penis* in conſpectum venit, inſignis
quidem magnitudinis, habito reſpectu ad exiguum
animalis corpuſculum ; in eo Oſſiculum hujus
figuræ (1.) æmulum delituit, oſſiculum in Mu-
rino pene contentum longitudine duplo ſuperans.

Teſticulos habuit ſatis amplos, extra abdominis
cavitatem prominentes.

Veſiculæ Seminales, ex utroque latere *Veſicæ Uri-*
nariæ ſitæ, ſemine mirum in modum turgidæ con-
ſpiciuntur, Phaſeoli magnitudinem æquantes.

LONGITUDO omnium *Inteſtinorum,* ſcilicet
a Pyloro uſque ad anum vixdum 6. pollices æqua-
bat ; at in Mure diſſecto (cujus ſimilitudinem ex
omni animalium genere maxime præ ſe fert.) Inteſtinorum circuitus 21 pollices ſuperavit, nullâ
habita ratione illius appendicis inteſtinum *Cæcum*
dictæ, quo omnino caret Veſpertilio ; cujus Inteſtinorum brevitatem, notabilem levitatis gratiâ,
a Naturâ conſtitutam eſſe opinor, quæ ab hoc
Quadrupede, per aerem volitare deſtinato, quic-
quid eſſet oneri provide deſumpſit.

Ven-

Ventriculus, *Lien* & *Renes* iiſdem partibus in Mure omnino perſimiles ſunt; at Hepar & Pulmones in duos duntaxat lobos dividuntur.

P E N I T U S mortuo animale *Cor* motum ſuum, *viz.* Syſtolen & Diaſtolen, amplius horæ ſpatio peragebat.

Oculi inſigni convexitate donantur; eos autem in hunc finem ita fabricatos ſuſpicor, ſcilicet ut in tenebris videant; quippe per ſolam noctem & opaca crepuſcula prædam ſuam (Muſcas ſcil.) animal hoc infectatur, quas inter volandum Hirundinis ad inſtar captat.

Blaſius in ſua Anatome diverſorum animalium, ubi de Veſpertilione loquitur, hanc controverſiam inter quoſdam Medicos natam, meminit; ſcil. an *Caudam* habeat; ſed de re ipſis ſenſibus adeo evidenti ortam eſſe contentionem magnopere admiror, quippe æque bene diſputaſſent, an Mus Caudam habeat, cum in eo non magis manifeſtam *oſſis Coccygis productionem* (quæ in omni animale *Cauda* nominatur) quam in Veſpertilione aſpicere liceat.

A N I M A L eſt *Viviparum*, & nihil commune ullæ Volucrum ſpeciei poſſidet, exceptis *alis* & robore *Muſculorum pectoralium* alas moventium; quippe nec Bipes eſt, nec Pennatum, nec Roſtratum, *&c.* ſicut omnia volantium genera: quamobrem a *Clariſſimo Willughbeo*, in ſuo pereleganti libro de Avibus, inter Aves nequaquam numeratur, licet alii ſcriptores Veſpertilionem inter eos collocare haud dubitaverint.

D U M in vivis eſſet animal, in Pixide lignea incarceratum per ſpatium quatuordecem dierum aſſervabam, quo tempore Muſcos ex omni genere & Arancas avide comedebat: * *Corporis* autem *Situs*

---

* *I have ſeen him in this Poſture aſleep, above forty ſeveral Times.*

tus (quem semper eligebat quoties somnum cape-
ret) singulare quid & insolitum videtur ; quippe
spreto molli gramine in fundo suæ Caveæ sub-
strato, pixidis lateri adhæreret, & posterioribus
suis partibus directe elevatis, anterioribus autem
& Capite perpendiculariter deorsum positis, sus-
pensus semper quiesceret: in hac autem insolita
positura corpus suum sustentaret *posteriorum pedum*
beneficio, quorum uterque quinque digitis, acu-
tissimis unguibus armatis, instructus est, & ab his
ligneæ pixidis lateri infixis, pondus totius Corporis
tuto dependebat ; *anteriores* autem *pedes,* unico
tantum digito instructi, ad illius sustentationem
in hoc situ nequaquam contulerunt.

S i quis hujus animalis *Ostiologiam* cupiat, con-
sulat Cap. 26. Partis Secundæ Anatomiæ Blasianæ
variorum Animalium, & Tabulam 41. ubi Ves-
pertilionis *Sceleton* & *Effigiem* videre licet.

The

The Rule of Falfe Pofition, in Dec. 1682.

MUⅬTIPLY the Pofition by the alternate Errors, and if the Errors be of the fame Kind, divide the Difference of the Products, by the Difference of the Errors ; but if they be of divers Kinds, the Sum, by the Sum : And the Quotient, fhall give the Number fought.

FOR Demonftration, what Number is that, which being multiply'd by B 3 will produce the Plane B A 30.

Pofitions.	Pofitions.
Let it be $A - C = 6$	Let it be $A - D = 8$
into $B = BA - BC = 18$	into $B = BA - BD = 24$

$$A = 10.$$
$$B = 3.$$
$$C = 4.$$
$$D = 2.$$
$$BA = 30.$$

The Errors there ore are.

First Error. BA plane $- BA + BC = 12.$
Second Error. BA plane $- BA + BD = 6.$

The Lefs fubftracted out of the Greater, there remains,

$$\overset{12}{B}C - \overset{6}{B}D = 6 \text{ the Difference of the Errors.}$$

Which being multiply'd into their altern Errors, the Products will be as follows.

BC Defect	BD Defect
$A - D$	$A - C$
$\overline{BCA - BCD} = 96$	$\overline{BDA - BCD} = 36$

And Subftraction again being made $\Big\} A = \dfrac{BCA - BDA}{BC - BD} = \dfrac{60}{6} = 10$

Again. Data $\begin{cases} B \\ BA \end{cases}$ quæritur A.

Sit $A - C$	Sit $A - D$
in $B = BA - BC$.	in $B = BA - BD$
BA pl. $BA + BC$ minus.	BA pl. $- BA + BD$
Ergo Errores $+ BC$	$+ BD$

$$\frac{A - D}{BCA - BCD} \quad \text{min.} \quad \frac{A - C}{BDA + BDC}$$

$$A = \frac{BCA - BDA}{BC - BD}$$

Pof.

$$\text{Pof. } A + B \qquad \text{Pof.} \qquad A + C$$
$$\qquad D \qquad \text{Errors} \qquad E$$
$$B \cdot C \; :: \; D \cdot E \cdot$$
$$C D = B C \cdot$$
$$A D + C D = A E + B E$$
$$A D - A E = A D - A E$$
$$\frac{A D - A E}{D - E} = A$$

As the Difference of the Errors to the firft Er-ror, fo is the Difference of the Pofitions to a Number, which, contrary to the Sign of the firft Error, being added or fubftracted to or from the firft Pofition, gives the true Pofition.

W H E N the Errors have different Signs, their Sum is their Difference.

T H E Reafon of the Proportion betwixt the two Errors of Pofition, is, becaufe the Numbers added or fubftracted, and apply'd to the one Term of Proportion, are proportionate to the Numbers added or fubftracted, and apply'd to the other Term, becaufe two Numbers, apply'd or divided by the fame Number, continue the fame Proportion. Likewife, if you add or fubftract like proportional Parts, the Sums or Differences will be in the fame Proportion.

A s the Error of the firft Pofition to the Error of the fecond Pofition, fo is the Error of the firft Operation, to the Error of the fecond Operation. But the Rectangle of the Means, is equal to the Rectangle of the Extreams. Subftract the one, from a Number containing the other, and you leave the true ; only in greater Products contain'd fo many Times, as the Difference of the leffer Error of Operation is to the greater Error of O-peration, becaufe the leffer Error could not take
away

away fo many Truths as the greater Error had made in the greater Product.

$$\begin{array}{ll} \text{From} \quad B\,A & \text{From} \quad B\,A \\ \text{Take} \quad B\,A - B\,C & \text{Take} \quad B\,A - B\,D \\ \hline \text{Remains } B\,A - B\,A + B\,C & \text{Remains } B\,A - B\,A + B\,D \end{array}$$

The Difference $B\,C - D\,B$

$$\begin{array}{ll} A - D & \overline{A - C} \\ B\,C & B\,D \\ \hline B\,C\,A - B\,C\,D & \overline{B\,D\,A - B\,C\,D} \end{array}$$

Subſtract $B\,D\,A - B\,C\,D$

Remains the Diff. $B\,C\,A - B\,D\,A$

Divide it by $\dfrac{\underline{\qquad\qquad}}{B\,C - B\,D} =$ Is the true Pofition (fought.

Dr.

Dr. Hook of Earths, Salts, &c.

March, 14, 168$\frac{1}{3}$.

THE Nature of Clays, Stones, Limes, &c. being difcourfed, I mention'd the Sorts of Stone which were here call'd Freeftone, viz. fuch as could be faw'd with a tooth'd Saw, fuch as *Cone, Rigate, Burford, Ketten,* &c. That Stones were of two Natures, one bituminous, or fulphureous, the other faline and watery ; the fulphureous would calcine into Lime, the faline make Glafs, vitrify or diffolve, and moulder with the Rain, Air, and Froft. That both thefe Sorts are often found in the fame *Portland*-Stone one Part whereof will moulder, the other harden with the Air. That Loam is a Mixture of various Sorts of Clays and Sands, and may be feparated by wafhing. That fuch a Material is ufually chofen for Brick-Earth, as being moft eafily foftened and tamper'd for moulding, and moft eafily and fpeedily dry'd for burning, and moft eafily burnt ; to make it yet more eafy for burning, 'tis ufually dry, and expofed to the Winter Rains and Frofts, for mellowing againft the Spring. That the fineft Clay would make the beft Bricks, were it not for the more than ordinary Labour and Charge in wafhing, working, moulding, drying, baking, as is evident in Pottery, and Tiles, and efpecially in the *Roman* Bricks, which are fome of them of fo fine an Earth, fo well moulded, and fo thoroughly burnt, as to laft even to this Day, as intire and perfect as when firft made, in all probalility. That hungry Clay was hardeft and beft to endure the Fire without melting, but faline, and fine Clays,were moft apt to vitrify : And thence the throwing in of three or four Shovels of Salt into

a

a Pot Furnace when hot, made all the Pots in the Furnace to be glaz'd. That *China* was such an Earth, as was very difficult to be vitrify'd.

C O N C E R N I N G Salts, and other volatile and fix'd Bodies, I mention'd, that there were two Sorts, one that was homogenous to the Air, and would be dissolved into it. This was call'd Volatile; the other heterogeneous, and would not at all be so dissolved and mixed with it; and these were call'd Fixed. Of the Volatile, there are various Sorts, which will be dissolv'd into the Air, by differing Degrees of Heat. Spirit of Wine, or such other fermented Spirits, Camphire, the odorous Gums of Flowers, and Herbs, will be dissolv'd into the Air with a small Degree of Heat; other Bodies more difficultly, and require a stronger and stronger Heat, as they are more and more fixed; so some Salts and Gums, &c. will not rise at all: And these are call'd fixed Bodies, or Alcaly Salts. Of these which are dissolv'd into the Air, some are tasted as it were, by the Nose, others not in the same Manner as in Tinctures made in Waters; some, whereof the Tongue does taste, others not.

C O N C E R N I N G the *Oxford* Trial by blue Starch, which they affirm'd would turn red, with Acids, I said 'twas impossible, Smalt being Glass, but it must be *Litmus*, or *Indico*: But most likely *Litmus*; being a clear, blue Tincture; but *Indico*, a thick Precipitation.

T H E Experiment was very considerable, though plain, giving a further Explanation of Gravity, by making a large Glass vibrate, with a Viol Bow: By which Vibration, a certain Undulation is plainly seen to dart out from all such Places where the Glass vibrates. And it was very plainly visible, that the Water, and Bodies in it, did move towards every such vibrating Part, and from every other Part that was at rest.

Dr.

Dr. Hook's *Experiments of the floating of Lead*, &c.   July 4, 1683.

WEDNESDAY, *June* 27, 1683, I shew'd two Experiments to the Society, which succeeded; of which I gave an Account, *Wednesday*, *July* 4, 1683, as follows.

Of the floating of unmelted Metal, upon the same melted, with the Cause.

I. THERE was melted, in a Crucible, about a Pound and half of Sheet Lead, and whilst it remain'd melted, several small Pieces of the same Lead were gently one by one, by the Help of a *Forceps*, laid upon the clear and bright Surface thereof (the Scum and Litharge being first removed) and it was found that they all swam upon it, and did not sink to the Bottom; but if they were all cover'd or plung'd under the Surface, they would not rise again, but sink to the Bottom, and soon be melted.

THE Occasion of the Experiment, was a Suggestion, that Lead, when it concreted, did (as Water when it congeals to Ice) settle itself into a more rarify'd Texture, than when fluid; by which Means, it became lighter than the melted Lead, and so swam at the Top of it. But though the Effect were answerable to the Assertion, yet the Cause, assign'd, was false; for it was very evident, that the Reason of its swimming, was much the same with that of the swimming of a Needle, or of Water-Spiders, and many other Insects upon the Surface of the Water; namely, a Coherence of the Air to the Surface of the swimming Body; which Coherence of the Air does depress and remove a greater Part of the Fluid, Lead, or Water, than

the

the meer Bulk of the Body itfelf would do ; which, in both thefe Cafes, is very evident ; and was, in thefe Trials, very remarkable; for the Surface of the Lead did plainly bend and fink below its Level, with a Roundnefs where the Piece of Lead lay ; which bending of the Surface, was made the greater by a thin Plate, or Skin of Litharge which the Air does prefently make upon melted Lead, fo foon as ever a former is remov'd or fcummed off.

Of the Condenfation of Air by Water.

II. T H E R E was ftuck into the Side of a Piece of wooden Pipe, for conveying Water, a fmall cylindrical Pipe of Glafs, about a Foot long, and fomewhat better than half an Inch in Diameter ; one End of which Pipe was hermetically feal'd, but the other End was open, and communicated with the Cavity of the wooden Pipe, by means of a fmall Hole bor'd in the Side of that wooden Pipe into which the open End of the Glafs Pipe was thruft hard, having a little Linnen Rag wrapped about it, as is ufual for Taps put into the End of a Barrel, or other Veffel. Then *(there being about a Foot of Air left in the Glafs Pipe)* Water was forc'd into the wooden Pipe by a fmall Force-Pump ; and it was plainly to be feen, that as the Water was more and more ftrongly forc'd into the wooden Pipe, the Air left in the Glafs Pipe, by the Water that enter'd into it by the aforefaid Hole, was condenfed into a leffer and leffer Room ; fo that hereby, the true Degree of the Preffure of the Water could be eafily found and meafured; which was conceiv'd to be an Experiment, or Inftrument of great Ufe for Water-Works, becaufe by means hereof, the Force of Water, in any Pipe, might prefently be known ; namely, both from
what

what Height it defcended, and to what Height it would there again rife. The Rule of doing which, was the next Day, to be brought in.

Dr. Hook's *two Experiments, shewing the Preffure of Water in Pipes, and how to meafure it. Alfo the Expanfion of melted Metals, made before the* Royal Society, July 4, 1683.

JULY the 4th, 1683. I read the Accounts of the two Experiments made *June* 27 ; and likewife further explain'd the Ufes of them, by Difcourfes in other Particulars, namely, that the fecond Experiment was of great Ufe for the trying the Strength of Pipes, for Conveyance of Water. By which Means, I have examined feveral Sorts of earthen and other Pipes and Cements, and have found that earthen Pipes, made of a Material only, as hard as Houfe-Tiles, would endure the Preffure of 100 Foot of Water ; that the Ufe of the other Experiment, was chiefly luciferous, namely, to fhew the Nature of Fluids and Congruity, of which I fhould fhortly have Occafion to difcourfe more at large.

THEN I produced and read the Rule, according to which the Preffure of the Water, in any Pipe, might, by means of a Trial with the former Inftrument, be calculated and reduced to certain Meafure in Feet and Inches. The Means of performing, I fhew'd, were principally two, firft Arithmetically, and fecondly, Mechanically.

THE Arithmetical Rule was this ; that the Length of the Cylinder of the Air in the Pipe, before it was prefs'd upon by the Water in the Pipe, fhould be compared to the Length of the Cylinder

of

of the same Air, when compress'd by the Water of the Pipe, and the Difference noted ; namely, the Length of the Cylinder of Water thrust into the Pipe, by the Pressure. Then to resolve this Proportion. As the Length of the Cylinder of Water thus compress'd, is to the Length of the Cylinder of Water so thrust in; so the Height of the Standard of Water, at the Time of Trial, to the Height of the Cylinder of Water pressing in the Pipe, which is equal to the Height to which the Water of that Pipe, so press'd, will ascend above the Surface of the Water in the small Pipe.

THE Height of the Standard of Water, at the Time of Trial, is easily known by the Height of the Mercurial Standard at that Time; which, being now grown very common and useful, is almost every where to be met with, and may otherwise be easily supply'd ; for as the Weight of Water, to the Weight of Quicksilver, so the Mercurial Standard, to the Height of the Water Standard.

THE Weight of Water, to that of *Mercury*, is by many Trials found to be near as 1000 to 13593, or as 1 to 15, according to his Account following Numb. - - ~ -

THE

T ʜ ᴇ Geometrical, or Mechanical Way, was this. Upon a Table, or Plane

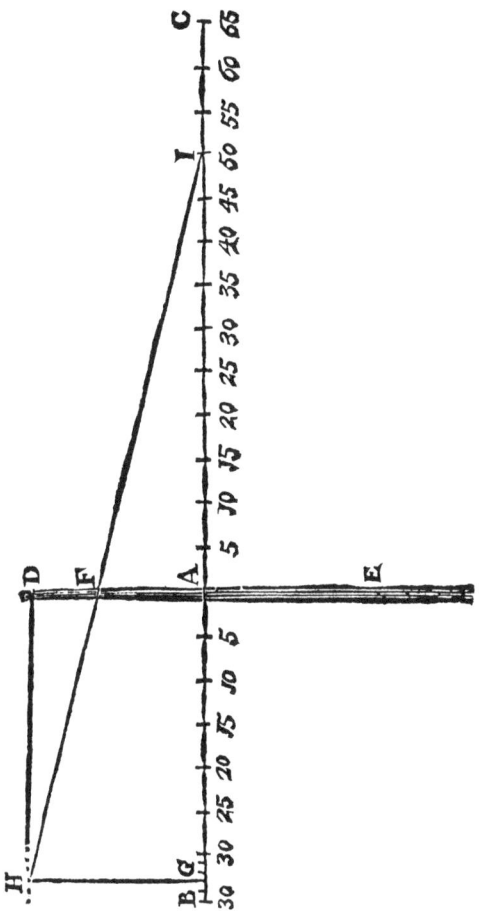

draw a Line, as **B A C** ; then crofs it at Right Angles, with another Right Line, as **D A E**, then divide **A B**, into thirty-fix Parts, and continue the fame Divifion, from **A** towards **C**, fo far as you have Occafion of Foot Heights of Preffure ; as fuppofe to 100 ; then

then ſubdivide one of theſe Parts, lying next to A into twelve equal Parts. Then knowing the preſent Water Standard, count, from A towards B, ſo many Parts and Duodecimals, as it is then Feet and Inches: Croſs the Line, at that Point at Right Angles, with another Line, as G H, and from G, ſet off the Length of the Cylinder of Air in your Glaſs, before Compreſſion ; then ſet off the Length of the additional Cylinder of Water, from A towards D, as ſuppoſe to E, and laying a Rule over the Points H and F, ſee where it croſſeth the Line A C, as at I, then count the Parts and Duodecimals from A, and that ſhall give the Preſſure or additional Height of the Water, above the Level of the Water in your Water Poiſer in Feet and Inches : The Reaſon of all which depends upon the reciprocal Proportion of the Strengths of Air to the Extenſions thereof.

T ʜ ᴇ ſecond Experiment, was made, to ſhew a Way, how to find the true and comparative Expanſion of any Metal, when melted, and ſo to compare it both with the Expanſion of the ſame Metal, when ſolid, and likewiſe with the Expanſion of any other, either fluid or ſolid Body. An accurate Account of which is neceſſary, to compleat a Hiſtory of Expanſion or Gravitation. The Method of trying it was, by having a Veſſel full of melted Lead, and alſo a ſolid Body of Iron to be ſunk into it ; this ſolid Piece of Iron was about $1\frac{1}{2}$ Inch Cubical, and into it, was faſtened a very ſmall Wire of Iron, big enough to thruſt it under the Surface of the melted Lead, and make it ſink therein, (for, of itſelf, it ſwam upon the Lead, as Wood upon Water). This Wire was faſtened perpendicularly, under a Scale, and ſo much Weight put into the Scale as ſerved to make it ſink under the Surface of the Lead ; then taking it out of the Lead, and ſeeing by the additional Weights, put

in

into the other Scale, to counterpoiſe it, firſt in the Air, then in Water, or any other Liquor, the comparative Weight of each of them was eaſily diſcoverable.

The Reaſon of the making of which Experiment was, to hint the Neceſſity there is, in all Experiments fit to be made Uſe of for any Philoſophical Theories, of reducing them to a Certainty of Quantity; without which, no certain and unqueſtionable Concluſion can be made. Now tho' a certain Standard of Weight, Meaſure, Expanſion, Power, Motion, &c. be not made Uſe of; yet if ſome one determinate Meaſure for each of them be pitched upon, 'twill be enough to make the comparative Trials uſeful; though it were to be wiſh'd, that ſome univerſal, natural Standard of Meaſure for all Things were found out, thoſe that have hitherto been thought of, having been doubted of, as to their Univerſality and Certainty, at all Places and in all Times.

Not knowing when the following Experiments were made, I inſert them after the foregoing, by reaſon of ſome Congruity between them.

W. Derham

An Account of some Trials for the finding out the Pressure of the Parts of Water one upon another ; and the elastical Power of the Air.

FOR the making these Experiments, there was prepar'd a long Tube of Glass, seal'd at one End, and being erected perpendicularly, with the seal'd End downwards, it was fill'd with Water, and so fastened against the Side of a Wall ; then there was taken another small Tube of Glass, very even drawn, and small enough to be let down within the former Tube ; this Tube was 12 Inches long, and was seal'd at one End, and divided into Inches, Halfs, and Quarters ; then, to the open End of this Tube, was hung a small, long Plummet of Lead, which would easily slip down to the Bottom of the longer Tube, and draw down the small Pipe with it ; both which were gently so let down by a small Thread, as the Experiment requir'd, which afforded these Observations. The Pipe, when the lower and open End first touch'd the Water, being full with Air, not heated by the touching the Pipe with a warm Hand, or otherwise, was observ'd by Degrees, as it descended, to be in part fill'd with Water, and so much the more by how much the deeper it descended. And observing the Degrees of Condensation of Air in the Pipe produced at several Depths, we found them to be these. At *Gresham* College, the 24 half Inches of Air lost one half Inch of its Extension

at

at   2 Halfs at   3 Halfs at   4 Halfs
at   which is therefore a 5th Part of a Cylinder of Water able to counter-balance the Preffure of the Air.   The whole therefore may hypothetically be judg'd to be - - - - - -

I DID, fince that, erect a Tube fome 13 Foot
high ; and fitting all Things as in the former Experiment, I collected this Table A, whofe firft
Row of Numbers fhews the equal Spaces into which
the Air was extended ; and the laft
fhews the Height of the Water above
the under Surface of the Air.   Since
that, in the fame Tube ftanding in the
fame Place, I reiterated the Experiment, and collected this following
Table B.

A

48	00
47	08½
46	17
45	27
44	36
43	45½
42	58
41	68½
40	80
39	91½
38	105½
37	117
36	130½

ALL which three
Tables, being fo diffe-
rent one from another,
may feem to overthrow
each other, and the Cer-
tainty of this Kind of
Experiment in general.
But as I cannot vindi-
cate the Trials from
fome Errors (it being
almoft impoffible to

B

24	00
23	13
22	31
21	52
20	76
19	101½
18	127½
17½	142

make thefe Kind of Trials fo accurate, that
there fhall be no Miftake committed) fo neither
do I believe, that thefe feeming Contrarieties do
wholly proceed from the Unaccuratenefs in the
Procefs.   For fince the Air is fometimes under a
greater, and fometimes a lefs Preffure, the Degrees
of Force, requifite to promote the Condenfation
further, muft neceffarily be differing.

AND hence by the firft Table, I judge the
Height of a Cylinder of Water, able to balance
the Preffure of the Air, when that Experiment was
made,

made to be   by the Second Experiment
I judge the counter-balancing Pillar, then to be
between 390 and 400 Inches, or near about 33
Foot ; by the third, I guess it to be about 382
Inches, or near about 32 Foot. This Experiment therefore, if accurately made, at several Sea-
sons and Times of the Year, may afford us a very
easy Way of knowing the Pressure of the Air at
that Time, and this more accurately and nicely,
than can be perform'd with *Mercury* the ordinary
Way.   For   whereas   the   Shortening and Increase of the Mercurial Cylinder, is at most not
above 2 or 3 Inches, in this Experiment, the aqueous Cylinder will change fourteen times as much.

N e x t, this Experiment may help us to guess
at the Pressure of the Sea Water against the Air,
let down to the Bottom of it in a diving Engine,
by knowing the Proportion between the Gravity
of salt and fresh Water. But it were very desirable that such, as have the Opportunity of making
Trials at Sea, would be diligent in it. For though
there seems to be no Doubt, but that Water pro-
portionably presseth according to its perpendicular
Height; yet it is not easy to predict, how much
it may vary from that Hypothesis ; which Devi-
ation may be caused, either from the extreme
Cold at the Bottom of the Sea, which may weaken
the Spring of the Air, or from the differing Gravity of the upper and lower Parts of salt Water ;
or from somewhat else, whereof we may be yet ignorant.   Now for the more accurate making of
these Trials, I think it were very requisite to have
some such Engine as this.

T a k e a good strong Glass Bottle, that will
hold about a Gallon ; and let there be fitted to it
a handsome Screw Cover of Brass, and shap'd like
those Covers that are usually put upon Chirurgeons
Bottles, that are made of Pewter.   Let the Cover
be

be very well cemented on, and the Screw be made to go very cloſe through the Top of this Cover ; let there be made ſeveral very ſmall Holes with a Needle Drill, then hang a good Weight under the Bottle, and let it down with this Cover up-moſt, for by this Means, by drawing it up from ſeveral Depths, and weighing the Quantities of Water it brings up, it will be eaſy to know the Weight of the incumbent Column of Water.

T H E R E might be many other Ways, but this I take to be the moſt cheap, eaſy, and certain of any ; nor is there any Danger of breaking the Bottle, either inward or outward ; for as the Bottle de-ſcends, the Water ruſhes in, and as it is drawn up, the Air goes out.

The following Experiments are here inſerted, by reaſon of their Congruity with the foregoing.

W I L L I A M   D E R H A M.

More

More Experiments of Preſſure.

Fig. I.

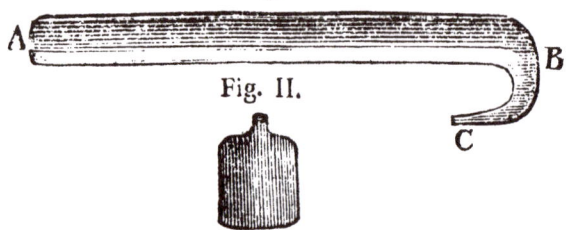

Fig. II.

THERE was taken a Glaſs Tube **A B C,** (*Fig.* I.) about 23 Inches long, and near $\frac{3}{4}$ of an Inch over; this was cloſe ſeal'd up at one End **A,** and the other End **B** was drawn into a very ſmall Pipe **C,** and bended according to the Shape in the Figure. This Pipe was found to weigh $1\frac{13}{16}\mathfrak{z} + 4$ gr. or 874 Grains, being fill'd with ſalt Water, and the Outſide wiped dry (which was conſtantly done in all the ſubſequent Trials) it weighed $4\frac{7}{16}\mathfrak{z} + 10$ gr. or 2140 Grains, whence if we deduct the Weight of the Pipe 874, we have 1266 Grains for the Weight of the Water that fill'd the Pipe. This Glaſs Tube being faſten'd to a Line, to the End of which was hang'd a Plummet of Lead, to make it ſink; 'twas fitted ſo as to be let down perpendicularly into the Water with the ſeal'd End **A** foremoſt, by which Means the ſmall Hole of the Pipe **C** was open downwards (that Hole being made purpoſely ſmall, that the Air could not get out at it whilſt the Water got in, nor the Water get in whilſt the Air was paſſing out.) Then the Glaſs was, for a ſhort Time, ſo held in the Water, that all of it, except the ſmall bended Pipe, was cover'd and incloſ'd with the Water (which was

ob-

obferv'd in every Trial, to the End that the Air,
within the Pipe, might be well cooled) and being let down to the Bottom, and there fuffer'd to
ftay for a fhort Space. Afterwards being drawn up,
loofened from the Line, dried, and exactly
weighed ; its Weight was found $3\frac{13}{16}$ ʒ + 3 Grains
or 1833 Grains ; whence, deducting the Weight
of the Tube 874, we have 959 Grains for the
Weight of the Water it brought up. The Place
was in the Channel to the North of *Quinborough*,
the Depth of the Water 16 Fathom and a Foot,
or 97 Foot, where we made the fubfequent Trials which are rang'd in this Table.

Top full	2140 — 874 = 1266	
At 97 Foot deep — 2	1833 — 874 = 959 .	307
At 97 Ft. deep — 2	1832 — 874 = 958 .	308
At 8 Ft. 3 In. — 2 Ft.	1060 — 874 = 186 .	1080
At 16 Ft. 6 In. — 2 Ft.	1257 — 874 = 383 .	883
At 33 Ft. — 2 Ft.	1500 — 874 = 626 .	640
At 66 Ft.	1737 — 874 = 863 .	403
At 66	1734 — 874 = 860 .	406
At 33 From the Mouth	1530 — 874 = 656 .	610
At 16½ of the Tube	1296 — 874 = 422 .	844
At 8¼	1131 — 874 = 257 .	1009

A BUNDLE of Corks being knit up in a Hand-
kerchief, and faften'd to the Line at 33 Foot from
the fmall End of the Glafs, the Tube was again
let down to the fame Depth, and the Corks, floating upon the Water, fufpended it at that Depth ;
for, a good while afterwards. Then drawing up the
Cylinder, by meafuring, the Cylinder was found to
have taken in juft as much Water, as it had in the
laft Trial, but the Weight of the Glafs was not
examin'd. Other Trials were made the next Day
with the fame Glafs Cylinder, *viz.*

At

At 8¼ Foot from the Top 1172 — 874 = 298.
Just at high Water, the Water being at a stand.

    At 8¼ Foot 1131 — 874 = 257
    At 16½ Foot 1300 — 874 = 426
    At 33 Foot 1510 — 874 = 636
    At 49½ Foot 1635 — 874 = 761
    At 66 Foot 1712 — 874 = 838

T H E Trials did agree, by Measure, with some I had made in the Morning.

A N O T H E R Trial was made of the last Experiment, because it was done when the Water had some Current, and the String seem'd to stream a good Way from the Perpendicular ; to prevent which Inconvenience, the Boat was suffer'd to drive with the Current, by which Means, the Line seem'd to go down perpendicularly into the Water. So the Cane being pull'd up, after it had staid some time at the Depth of 66 Foot, it weigh'd 1719 — 874 = 845. At 82½ Foot, and left to drive perpendicularly 1883 — 874 = 1009.

Wednesday, March *the 11th, in the Afternoon, near the same Place, where the former Trials were made, there were made these following Experiments of Compression.*

U N T O the Neck, or Mouth, of a common Quart Glass Bottle, was fitted a Valve, that opened inwards, and shut outwards ; this Bottle was so let down into the Water, that the Mouth went foremost, by which Means, the Water had, as the Bottle was sinking, a free Passage into the
                                    Body

Body of it, to compreſs the Air; but by the ſhutting
of the Valve, when the Bottle was again drawn up,
it was hinder'd from getting out. This Bottle,
when empty, weigh'd 37 ⅞ Ounces, and 24 Grains,
or 18204 Grains; fill'd with ſalt Water, it weigh'd
78¼ Ounces and 3 Grains, or 37563 Grains; whence,
taking the Weight of the Bottle 18204, we have
19359 Grains, for the Weight of the Water, that
fill'd the Bottle. This Bottle being let down 13½
Fathoms by the Ship's Plumb Line, or 81 Foot,
the Valve was ſo hard ſhut, when it was taken up
again, that it was difficult to be thruſt open.
Though when the ſmall End, or Mouth, of the
Bottle, was ſet upward, the Valve being made of
Braſs, without Leather, was found to leak a little,
by the hiſſing Noiſe the Air made at it. And when
by a Knock, the Valve was beaten down, the Air
made a Noiſe in ruſhing out like that of a Bottle of
Ale when it flies; the Bottle, and the Water it
brought up, weigh'd 65½¼ Ounces, or 31656
Grains; whence, deducting the Weight of the
Bottle 18204, we have 13452 Grains for the
Weight of the Water. This Bottle was again let
down to the Depth of 14 Fathom, or 84 Foot;
and, being drawn up, was found to weigh, whilſt
the compreſs'd Air remain'd in it, 65½ Ounces, and
19 Grains, or 31279 Grains; when the Air was let
out, it loſt 21 Grains of its former Weight, counterpoiſing only 31258 Grains, which was ſuppoſ'd to
proceed partly from the freeing of the compreſs'd
Air, and partly from the Loſs of a little Water,
that the violent Eruption of the Air had blown a-
way; from which laſt Sum, by deducting the
Weight of the Bottle 18204, we have 13054 for
the Weight of the Water.

March

March the 13th, another Experiment was made with another Bottle of the same Fashion, which empty, weigh'd 37$\frac{11}{13}$ Ounces and 12 Grains, or 18162 Grains; fil'd with salt Water to the Valve, it weigh'd 77$\frac{11}{13}$ Ounces and 3 Grains, or 37353 Grains; whence, deducting the Weight of the Bottle 18162, we have 19191 the Weight of the Water that fill'd it   this Bottle being let down 8 Fathom, or 48 Foot, the Bottle, compress'd Air, and Water together, weigh'd 60$\frac{11}{13}$ Ounces and 12 Grains, or 29142 Grains; the Air being let out softly, which requir'd a long time, and the Bottle, and Water afterwards weigh'd, was found 24 Grains lighter, *viz.* 29118 Grains; whence, deducting the Bottle 18162, we have 10956 Grains for the Water. The Experiments are ranged together in this Table.

T H E Bottle, with a bended Copper Pipe at the Top, being let down 8$\frac{1}{4}$ Foot deep, brought up in it 4$\frac{11}{13}$ Ounces, and 24 Grains of Water, the Bottle being weigh'd before-hand with a dead Weight, or counterpois'd; the same Bottle, kept longer at the same Depth, brought up 8$\frac{9}{13}$ Ounces and 25 Grains of Water; the same Bottle, kept yet longer a great deal, brought up 9$\frac{7}{13}$ Ounces and 6 Grains; the Water that fill'd the Bottle, weigh'd 41$\frac{16}{13}$ Ounces and 24 Grains; which different Proportions of Water, taken in, we judg'd to proceed, either from the leaking of the Vessel at the Screw, by which Means, the Water had a Passage into the Bottle below the Mouth of the bended Pipe, which would therefore serve for a Vent-hole for the Air to get out at; or else that the Motion of the Top of the Water being a little uneven, the Pressure upon the Bottle must consequently alter, there being sometimes a greater, sometimes a shorter Pillar of the Water above it; secondly, the Bottle itself was, by the cockling of the Boat,

some-

sometimes lifted higher, then deprefs'd lower, which did also alter the Height of the prefsing Pillar; whence, as the Prefsure was a little increas'd, the Water got in; and, as it decreas'd, the Air got out; and, being held a long while in that Posture, many of those Changes did very much augment the Quantity of Water within the Glass.

Experiments of the Weight of Water.

A WHITE Glass Viol, made in the Manner describ'd in Figure II. with a small short Neck, was, by Trial, found to weigh, when empty, 1425 Grains; when fill'd exactly full with salt Water, it weigh'd 5247 Grains; whence, deducting the Bottle 1425, we have 3822 Grains, the Weight of the salt Water. The same fill'd with fresh Water taken out of the *Thames* at *Greenwich*, about low Water, weigh'd 5164; whence, deducting 1425, we have 3739, the Weight of that fresh Water. And weighing afterwards the Water, wherewith the Strong Ale at *Margat* is brew'd, we found it exactly the same with the Water taken up at *Greenwich*; whence we conclude, the Proportion of these fresh Waters, to this salt, to be as 3739 to 3822; that is, near as 45 to 46.

Trials of the Heat and Cold of the Water.

A SEAL'D Thermometer was let down to the Bottom of the Water, at 16 Fathom and a Foot, with the great Ball upwards, and the Stem downward, to the End that, if the Cold were extreme, it might have so far condensed the Spirit of Wine, as to have admitted the Air to have got in out of the Neck. And so by pulling it to the Top, we might have known the Cold at Bottom; but though the Thermometer was suffer'd to remain

a

a long Time at that Depth, and were ſuddenly pull'd up, we could not find that it had any whit more condens'd the Spirit of Wine, than it was by keeping the ſame Thermometer a pretty while juſt under the Water, at the Top, when we judg'd the Temperature of this Water, both at the Top, in in the Middle, (for, by other Trials, we found the ſame at other Depths) and at the Bottom, to be all the ſame.

N. B. *The Inſtrument deſcrib'd in the* Nuntius ad Abyſſum, *much better for the Purpoſe than this.*

R. W.

Obſervations of Sound.

BEING at a Place of the *Thames,* about four Miles above *Graveſend,* there happen'd to be ſhot off ſeveral ſmall Pieces of Ordnance, by a Ship that was about half a Mile farther up the River ; the Multitudes of the Echoes of each of which Shots, made a Noiſe among the ſeveral Hills, Woods, and Banks, on both Sides of us, juſt like Thunder. And could they have been number'd, they would, queſtionleſs, have exceeded an Hundred. And having ſince had the Opportunity to obſerve the Noiſe of Thunder, it ſeem'd to me to be deducible partly from Echoes ; which would yet ſeem more probable, if we could, by any Experiment, find that the Clouds would rebound or echo a Sound. A Gun being afterwards ſhot off by the Veſſel we were in, when we were near the Mouth of the *Thames,* and ſeveral Ships being on this and that Side of us, we could very ſenſibly hear ſeveral Echoes rebounded from them.

Dr.

Dr. H o o k*'s Contrivance of a very com-
modious Windmill* ; *communicated to the*
Royal Society, July, 11, 1683.

JUL y the 11th, I read the preceding Difcourfe
and Accounts of the two Experiments fhew'd
on *July* the 4th; and further explain'd each of
them by verbal Difcourfes. Then I fhew'd thefe
two Experiments following, which I explain'd by
Difcourfes, fomewhat in the following Manner.

THE Firft, was the Module of a Windmill, in
which were thofe Particulars following confiderable, not to be found in any other yet made ufe of.

1. THAT it had no Need of any Houfe, but
what might be placed, either immediately upon
the Ground, or under the Ground, according to
the feveral Ufes to which it might be apply'd.
Whence follow'd,

2. THAT the Houfe need not be any Impediment to the Force of the Wind, which it ufually
is in all other Windmills.

3. THAT it doth of itfelf turn to all Winds,
and fo needs not the Attendance, Watching, and
Labour of Men to fet it, which is neceffary in o-
ther Mills.

4. THAT the Vanes are contriv'd of the moft
perfect Form, to receive the whole Power of the
Wind, for the Cylinder thereof it is expofed to :
Which is effected by the particular Slope of the
Vanes thereof, whereby the Force of the Wind
becomes equal upon every Part of the Vane, from
the Center to the Tip, or Extremity thereof. An
equal Progreffion of Wind caufing every Point of
the whole Vane to make an equal Arch of Rota-
tion, or an equal Angle at the Axis.

5. FOR

5. F o r that it needeth not fo big an Axis, nor fo ftrong Vanes as other Mills, the greateft Strength of this being in the Way of pulling, the other in the Way of thrufting; and this being capable of being ftrengthen'd by Ropes, like the Tackling of a Ship.

6. F o r the eafy Way of producing a circular Motion below, without the Help of Trundles or Cog-wheels, which are both a great Impediment to its Motion, and do wear, and often need Repair.

7. Fo r the eafy Way of communicating a reciprocating perpendicular Motion, which is ufually perform'd by the Help of Wheels.

8. Fo r the Cheapnefs of it, there being fo many Particulars not neceffary to this, omitted, which are ufually done in other Kinds, and not without Neceffity.

A l l which Particulars confider'd, it makes it to be the moft plain, fimple, cheap, and eafy to be made and ufed, that has been yet made ; and yet the moft powerful in its Effects, and the moft univerfally applicable to all Purpofes ; (as grinding, bruifing, beating, fawing, pumping, placing, twifting, drawing, turning, lifting, &c.) that can be made of equal Bignefs.

I have thought worth while, to infert this Account of the Windmill (although fcarcely intelligible without Figures, or a Module, which I never could meet with) becaufe fomebody, or other, may be fo fortunate to find the Module, or, by the Hints here given, contrive a Windmill like this.

W. D e r h a m.

Dr.

Dr. Hooᴋ's *Contrivance to ſtop great Weights falling*, July 11, 1683.

THE ſecond Experiment was a very plain and eaſy Way, how to ſtay a Weight from falling, ·when the Rope, or Chain, by which it is drawn up or let down, ſhall chance to break. This was effected by a ſmall Arm extended out from the Top of the Weight to the Side, with a Hand, or Pipe, at the End thereof, which graſped, or in-cloſed, another Rope or Chain, extended from the Top to the Bottom ; which Hand, or Pipe, was ſo wide, as to ſlip freely upon the ſaid Rope, ſo long as the Weight was ſuſpended by its own Rope ; but ſo ſoon as that any way fail'd, the Hand graſped the Side Rope faſt, and hinder'd the Weight from deſcending to the Bottom. This was one of the plaineſt, eaſieſt, and moſt ſimple Ways of effecting this End, though the ſame may be effected divers other Ways, as certainly, which I have alſo con-triv'd. The explicating it, by a Scheme, makes it the more intelligible. I repreſents the Weight, *a b* the Arm, moving with a Joint at *c*, upon the o-ther Part of it *k*, faſt into the Weight, *ef* repreſents the Rope,

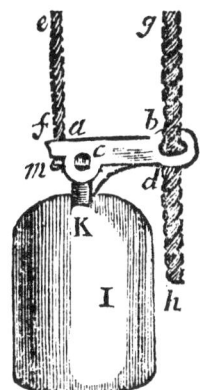

by which the Weight is either drawn up or let down, faſten'd to the Elbow *m* ; by which Means the Wriſt, and Hand of the Arm, is kept at Right Angles with the Part faſt in the Weight, and ſo the Hand ſlips freely upon the greater Rope *g h*, extended from the Top to the Bottom, to which the Weight can deſcend ; *d* repreſents a Spring, by which,

which, fo foon as the Rope of the Weight, which holds by the Elbow *c*, fails, the Arm is extended ftreight; by which the Hand *b*, prefently holds faft the Rope, or Chain G *b*, by being made oblique to the Perpendicular, and, fo creeking the Rope, and fo hinders it from falling; as, by the Experiment fhewn, plainly appear'd.

THE Ufe of which Contrivance, though poffibly it might, to fome, feem very trivial and infignificant, as feeming to be calculated for keeping a Clock, or Chime Weight, from falling, is not altogether fo flight and foolifh; for even for that Ufe it may fometime or other poffibly fave 100 Pound Expence, and the Lives of fome Men. But if apply'd, in general, for the hindering Weights to fall, it may deferve a fomewhat better Value, and be found very confiderable, fince it may be very inftrumental to fave many Mens Lives, and much Charge, and great Inconveniences, which do very often now, for the Want thereof, happen. For whereas, in many Mines, the Men themfelves are often drawn up and let down in Buckets; and generally the Ores, Stones, Waters, and divers other Things, belonging to thofe Works neceffary for procuring Ores, or other Minerals, are fo conveyed; and upon the failing of the Rope, Chain, or other Part of the Engine, do often fall from Top to Bottom, and fo are not only dafhed in Pieces themfelves, but deftroy, and do oftentimes irreparable Injury to Men, or what elfe they meet with in their Fall. By this Means, all fuch Bodies are fecured from the Fall, and kept hanging at the Place where they were when the Rope brake, or other Part of the Engine fail'd, and thereby the Bodies themfelves are preferv'd intire, and no other Harm done by their Fall. The fame Thing is applicable alfo to Men, afcending, or defcending, by Ropes or Rope-Ladders, and to Stones, Timber, or Materials for a high Building.

Dr.

Dr. H o o k's *Way to take the Impreſſions of Medals*, &c. *imparted to the* Royal Society. Octob, 31, 1683.

HAVING been ſhewn, by Mr. *Frazier*, the Impreſſions of ſeveral of the King of *France*'s Medals, in a certain thin tranſparent Subſtance, much like *Muſcovy* Glaſs, but much more tough ; on which, on the one Side, appear'd the perfect Impreſſion of the Medal, in *Entaglio*, or ſunk in; and, on the oppoſite Side, the very Figure of the ſaid Medal in *Baſſo Relievo*, or ſwelling out. And, conſidering what Way this might be done, having formerly taken off the Figure of certain Carvings, by Glue, ſo as to be able to caſt them in Plaiſter of *Paris*, or burnt Alabaſter ; upon making Trial with a Glue made of *Icthuocolla*, diſſolv'd over a gentle Heat, in courſe Spirit of Wine, by laying it upon a fair ſtamp'd Crown Piece, and ſuffering it to lie a conſiderable Time, till it was thorough dry, cold, and hard ; I found that it afforded me the ſame Kind of Subſtance, both for Toughneſs, Tranſparency, and Fitneſs, to receive and retain the Impreſſion of the Coin upon which it was laid, as the Subſtance ſhew'd me, containing the Impreſſion of the *French* Medal. This I ſhew'd the Society, and explain'd to them the Way of doing it. And alſo related, that the ſame Impreſſions might be ſo taken with common Joyners Glue ; but the Plate would not be ſo tough, nor ſo tranſparent.

T H E Preſident mention'd, that there had been a certain *Frenchman* here in *England*, ſome time ſince, who had certain tranſparent Plates like *Muſcovy* Glaſs ; with which, he could eaſily copy out any Picture or Print, by laying it upon the
ſame,

ſame, and writing upon it with Ink, as on Paper ; the ſame being very tranſparent ; and ſo cauſing the Print, on which it was laid to appear very plain through it : And inquiring, whether I could do the ſame, upon my affirming that I could, he deſir'd that I would ſhew the Experiment of it at the next Meeting.

N. B. *Dr.* Liſter *mention'd the Way of contract-ing Seals with Mouth-Glue.*

Dr. H o o k *imparted to the* Royal Society *this Preparation, to copy any* Picture, &c. Novemb. 7, 1683.

I PRODUCED a Plate, made according to the preceding Deſire ; which had the ſame Pro-perties with that which was made by the *French* Gentleman. This was very thin, and as tranſparent as *Muſcovy* Glaſs, or *Selenitis*. It was alſo tough, and would bear Ink as well as any Paper, and ſo was fit to make uſe of, for any Experiments for drawing, or copying Pictures or Maps. The Manner of making it, I explain'd to the Society, to be thus. Firſt, I prepar'd a very thick Ciſe of *Icthuoceolla*, well diſſolv'd in Spirit of Wine, and then clear'd from all its Rags and Foulneſs, by ſtraining it through a clean Cloth ; then taking a Looking-Glaſs Plate, well ſmooth'd and poliſh'd, I rubbed the ſame all over with a fine Rag, moiſtened a little with pure Sallad Oil ; but ſo as only to hinder the Subſtance that was to be pour'd on it from ſticking to it, but not to make it foul or uneven. Having ſo prepar'd theſe Things, I heated the Siſe, and, when again pretty cold, I pour'd it upon the oiled Side of the Glaſs Plate,
and

and fo taking the Plate, and inclining it this Way
or that Way, till the whole Plate was cover'd by
the Sife, I laid the Plate horizontal, and fuffer'd
it to lie fo till it was thoroughly dry.

Dr. Hoo k's *feveral Difcourfes of Improve-*
ments of Scales, Beams, and other Inftru-
ments, for weighing Bodies more nicely ;
and firft, one to find any defired Part of a
Weight, or Body to be weigh'd. Dec. 5, 1683.

I Produced *an Inftrument for the fpeedy and*
exact finding any defir'd Part of any Weight
given, *whether Commenfurate, or Incommenfurate.*
The Inftrument, (being only a Module, and to
ferve only for Explication and Experiment, and
not for conftant and continual Ufe) was a flender
Fifhing-Cane, ftreightened very well, of about four
Foot in Length, and tapering from one End to the
other ; this Material I made ufe of upon a double Account ; *Firft*, for its Stiffnefs ; and, *Second-*
ly, for its Lightnefs, that I might, as near as poffible, make it to be without Weight, and bending, and fo approach to, or reprefent, a *mathe-*
matical Line.   Now the Part, I propos'd to find,
being a *Decimal, Centefimal, Millefimal*, or the
Powers of the *Decimal Fractions*, I divided the Cane
into eleven equal Parts ; at one of which, from the
greater End, I, with a Needle, drew through
it a fmall Silk Thread, by which I fufpended it ;
and by adding Lead to the fhorter End, I pois'd
it, until it came to an *Equilibrium*, and fo it hung
horizontally.   Then I made two *Scales*, with two
Rings, whofe inner Edges were thin and fharp,
by which they might hang upon the Ends of the
horizontal, or *equilibrated* Cane.   The *Scale* and
Ring,

Ring, for the *greater* and *shorter End*, was made ten times as heavy as the other *Scale* and *Ring* for the smaller and longer End. These being thus prepar'd, I hung on the *Scale* upon the *greater* or *shorter* End, at any Distance from the Thread: Then, hanging on the little Scale, upon the lesser End, moving it nearer and farther from the suspending String, till the Beam hung *in Equilibrio*; the which became an Instrument for finding the Decimal, Centesimal, or Millesimal Parts, or Fractions of any Weight given. Suppose a Pound be to be so divided ; Put the Pound into the great Scale, and then counterpoise it with Weight, as of Sand, Water, Minium, &c. in the lesser Scale ; this shall be a tenth Part of a Pound : Remove the Pound, and put the Decimal Counterpoise in the greater Scale, then counterpoise this in the lesser, and this shall give a Centesme of a Pound : Remove the Decimal, and put the Centesme in the Greater, and the Counterpoise to it in the Lefs, shall give the millesimal Part of a Pound, and so onward for the ten thousanth, hundred thousandth, or thousand thousandth Part of a Pound ; which, this Way, may be most exactly found and determin'd : And the like for any other assignable Part whatsoever of commensurate, or incommensurate Proportion, to the whole Quantity, of what Weight soever ; the Beams being accordingly proportion'd in Strength and Dimensions, whether it be for great and massy Bodies, or exceeding minute and curious ; and, by this Means, with some small Addition, the smallest Bodies may be as certainly weigh'd, as the most tractable, even to the thousand thousandth Part of a Grain, far beyond the Reach of the Hand, or the naked Eye. And, as the *Microscope* doth help the *Eye* to make *invisible Bodies*, and *Parts visible*, so may this help the Hand to make the *intractable Bodies tractable* and *ponderable*

rable, and *comparable,* by other Trutinations than
thofe of Sight; which is of confiderable Advantage
in the Inquiry after the feveral Natures of the
Intims of Things, as I may hereafter fhew, more
particularly. In the mean time, I conceive,
there was no great Reafon for any, either to affirm the Experiment falfe or erroneous, or to
flight it for its Plainnefs and Obvioufnefs ; fince a-
ny, that underftands *mechanick Principles,* will
fave me the Labour of making a Demonftration.
And how obvious foever it be now known, yet I
do not find it hath been taken Notice of by any
Writer of *Mechanicks* ; nor did I ever know any
that had ufed it, or taken Notice of it, for this
Purpofe ; and though it may be faid to be a *Stilyard,* yet 'tis as differing from the common Ufe
of the *Stilyard,* as that is from a *common
Beam.* I mention'd alfo, how neceffary an *Inftrument* this was in almoft all *Philofophical Examinations,* efpecially in all Trials that concern the
Limits and Bounds of Powers, in the *Intims of
Bodies.* This *Proportional Balance,* will be of
general Ufe, and to fuch, particularly where
Weights are troublefome to carry and remove ;
and, I fuppofe, the only Reafon, why it has not
been ufed, is, becaufe it has not been thought
of ; though it were altogether as obvious, as to
fet an Egg on End.

*This Inftrument being eafily underflood without a
Figure, I have therefore omitted the giving any.*

A Second Inftrument for weighing; or, a Sort of Effay-Scale.

December 12, 1683, I produced another Experiment, which was alfo an Inftrument for weighing, which might alfo be of very general Ufe ; and that was not only for examining the Weight of any Sort of Gold or Silver Coin, or any other Veffels or Pieces of thofe Metals : But alfo for examining and effaying the Nature of the Metal itfelf, of which thofe Pieces, or Veffels, fhould be made, both as to the Species of the Metal, and alfo as to Finenefs, Purity, or the contrary Qualifications of them. Now though this be to be done by means of ordinary Gold Scales and Weights ; yet, I dare affirm this Way to be altogether as fure as the other, and abundantly more eafy, both for Carriage and Ufe. And there might as well have been Objections made againft the Art of Printing, becaufe a Writer was able, before that Art was found, to have wrote Letters, and Words, as fair as they could, by that Art, be printed. The Invention of the Inftrument was grounded upon the Theory of the Nature of Springs, which I have formerly fhew'd, and explain'd in this Place ; and the Way of examining the Goodnefs or Badnefs, of this Kind of Metal, and of difcovering the Species of the Metal itfelf, was grounded upon the Experiment of *Archimedes*, improv'd and explain'd by *Getaldus* ; which two Theories, being rightly underftood, will take off all Objections againft the Truth and Reality thereof, with all impartial Perfons.

T H E Inftrument was made of a Coyle of Brafs Wire, one End of which, was held in the Hand ; and, to the other End, was faften'd a fmall Net of Hair, in which Net, the Piece of Metal to be

exa-

examined was put ; and then the whole was lifted
up by the Hand, and, by means of a fmall Top of
a Feather, faſten'd to the lower Part of the Wire,
the Length of the whole Spring augmented by
the Weight of the Piece try'd, was obſerv'd, and
by the Diviſion on the ſaid Feather, the Number
of Grains were to be taken Notice of ; this gave
the Quantity or Weight of the Piece itſelf in
Grains. Then, for the ſecond Qualification of the
ſaid Metals, it was to be found by holding the
Piece (now weigh'd, and in the Scales made of a
Net of very fine Hair) into fair and clear Water,
and obſerving by the relaxing of the Spring, how
much the Piece grew lighter ; for thereby the ſpe-
cifick Gravity of the Metal itſelf, compar'd to
that of Water, was exhibited ; and this without
making Uſe of differing, or indeed any Weight at
all.

A Third Instrument for the same Purpose.

JAN. 9, $168\frac{3}{4}$.

Scales copied from the *Royal Society*, Regist.
Numb. VI. p. 136.

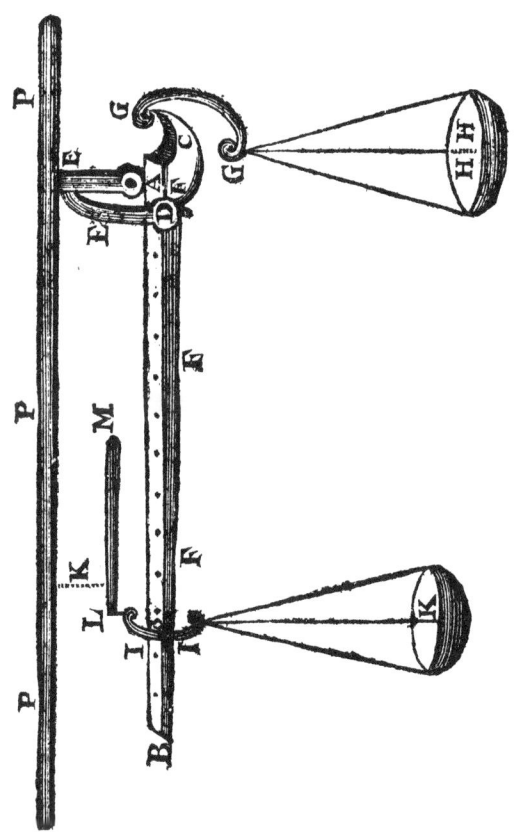

I SHEW'D a Module of a *Beam*, *whereby rea-
dily to find any aliquot, or aliquant Part of any
Weight given.* The Beam was made in the same
Manner

Manner as the firft that was fhewn; namely, that with a Cane; but whereas that was only then divided and defign'd for Decimation, or Decuplation, the longer End of this was divided into 12 equal Parts, and the Face of the Beam was made fo wide, as to be capable of admitting Subdivifion by Diagonals. The fhorter End was one twelfth Part of the longer; at which Diftance, the great Scale was properly fixed, wherein the Weight, to be fubdivided, was to be put: This Scale, when empty, counterpoifed the longer End, without any Scale fufpended on it: And that the removing of a Scale might make no Alteration of the former *Equilibrium,* the Weight of the fame was wholly taken off by a proper Counterpoife, fo that the Scale had no Weight at all upon the Beam. The Way of finding any defirable Part of a Weight given, was thus; If the Part were not fmaller than a twelfth Part, then the fame might be eafily found by one Operation, by placing the Scale at fuch a Diftance from the Axis of the Beam, on the longer End, that the fame fhall be in fuch Proportion to the fhorter End, as the whole Weight is to the Part defign'd; for Inftance, having a Lump of *Ambergreafe,* of an unknown Weight, but 'tis to be divided into three Shares, which are to be in Proportion, one to another, as 345, 234, and 123, to find each of thefe, I thus proceed; adding all the Proportions together, I find they make 702; then, by a Sector, by the Line of Lines, I open the Compaffes to the Length of the fhorter Shank of the Beam; and, by that, open the Sector to 345; then, on the fame Sector fo opened, I open the Compaffes to 702, and fet off that Diftance on the longer Shank of the Beam, and there place the leffer Scale; then putting in the Lump into the greater Scale, I counterpoife it in the lefs, and that gives me me the firſt Share, which is as 345 to 234, and 123 ; this Weight I lay by.

Then upon the ſame opening of the Sector, I take off 234, and ſetting it on the longer Shank, I place the leſſer Scale, aud proceed as before ; and this gives me the Weight of the ſecond Part, namely, 234. Then the Difference between the Sum of theſe two, and the whole, in a common Balance, gives me the third, *viz.* 123.

I f the Part, to be found, be leſs than a twelfth Part, and not leſs than a one hundred forty fourth Part, by ſome previous Diviſion of it, by once weighing, I reduce it to ſuch a Part, as, by the ſecond weighing, I find the Part, to be found, will not be leſs than a twelfth ; and then I proceed as before. This may be perform'd, either by finding two Dividers of the Part, both which ſhall fall within the Compaſs of 12 ; or, if it be a prime Number, then by extracting the Root of it ; which may be done arithmetically in Decimals, to what Accurateneſs ſhall be deſir'd, or by a Line of Superficies on a Sector, or by a Table of Logarithms.

I f the Part to be found, be leſs than a one hundred forty fourth Part, and not leſs than a ſeventeen hundred twenty eighth Part, then it muſt be perform'd by three Dividers, if ſuch can be found, that will fall to be each not leſs than a 12th, or elſe, by the Extraction of the Cubick Root. If the Part be leſs than a 1728th, and not leſs than a 20736th Part ; then, by finding four Dividers, each, within the Compaſs of a twelfth, or by extracting the quadrato, quadratick Root, the Part may be obtain'd by four Operations.

The

The fourth Instrument for weighing.

J AN. 16, 168¼.

I SHEW'D a new *Instrument I had invented, by which, immediately, and without any Trouble, the comparative Weights of any two Bodies given, might be found; if, at least, the Beam were of Bigness enough to bear them.* The Beam was made in the Form of a Crofs, equilibrated upon a fharp Edge in the Center; the Scales were hung upon two Ends (not oppofite, but) next together, which were alfo equilibrated; the fmalleft Weight, in either of the Scales, would make the Arm, by which it hung, to ftand perpendicular, and, confequently, the Arm that bore the other Scale, to lie horizontal. The Bodies to be weigh'd, were each of them put into the Scales, one in the one, and the other in the other; and fo fuffer'd to take their Pofture (which they would prefently do) by putting the Beam in fuch a Pofture, that the Diftances of their Points of bearing, from the Perpendicular under the Center, would be in reciprocal Proportion to their Weight. Dividing then the Arm, on which the greater Weight hung, into ten equal Parts, and each of thofe into ten, and, if the Beam will bear it, each of thofe again into ten, all of which, will make one thoufand equal Parts, I place three Pins upon each of the other Arms, which crofs the aforefaid Arm at Right Angles; the firft two, at the Extremities, the next two, at the Diftance of one tenth from the Center, and the third Pair, at the Diftance of one hundredth; then I provide two Bullets, equiponderant to each other when fitted, the one with a fmall Clew, fomewhat more than the Length of the longeft Diagonal of the two fufpend-

ing

ing Arms, with a Ring at the End to hang upon one of the Pins, the other, with a Ring only. Then, according to the Difference of the Bodies counterpoifing each other, I hang on the Plummet and Line upon that Pin of the Arm over the heavier Body, and is neareft to the Extremity; from which the Plumb Line may fall upon the Divifions of the Arm, and counterpoife it alfo with the Ring and Bullet hung upon the correfponding Pin on the oppofite Arm, then fhall the Plumb Line fhew, upon the Divifions, the proportionate Weight of thofe two Bodies. I need not fhew the great Ufe and Benefit there may be made of this Beam in all Philofophical Inquiries, fince they are obvious enough.

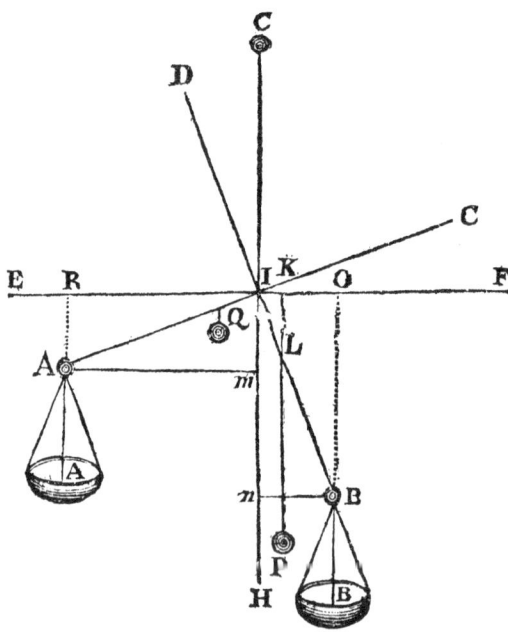

Let **A C**, **B D**, reprefent the Crofs Beam, moving
on

on I, the Scales hanging at A and B. The Weights being put, the heavier in B, the lighter in A, the Crofs pofiteth itfelf as in the Scheme in refpect of the horizontal Line E F, and the Perpendicular G H; and their comparativeWeight is found by their feveral Diftances from the Perpendicular I H, that is, as B N to A M, fo the Weight at A, to the Weight at B. Thus far is clear from the Principle of Staticks. Let K P reprefent the Plumb Line, fufpended at K; I fay then, that I K, is to I L, as B N is to A M, or, as the leffer to the bigger Weight; for A M, is equal to I N, and the Angle N I L, is equal to I L K, therefore K I L, is fimilar to B N I, therefore as K I to I L, fo B N to N I = to A M, fo the Weight at A to the Weight at B . Q . E . D .

The Defcription of a Pair of Japan Scales, and a Japan Stilyard.

J A N. 23, 168¼.

J P R O D U C E D, and fhew'd three feveral Kinds of Beams, for weighing the Gravity of Bodies; the firft, was a Pair of *Japan* Scales and Weights, made and adjufted in that Country, and that with very great Care and Curiofity. The Beam was made of a round Rod of Brafs, tapering a little from the Middle towards the Ends; which were flatted perpendicularly, and had each of them a fmall Hole drill'd through it, tapering both Ways to the Middle, leaving an Edge round the Middle of the Hole, through each of thefe Holes, was put a Brafs Ring of Wire, by which the Difhes were fufpended by four Strings. The Cock, or Tongue of the Beam, was neatly foder'd

into

into the Middle of the Beam, about two Inches broad below, and ⅓ of an Inch at the Top; and about the Middle, between the Beam and the Top, was put the Pin, upon which the Beam play'd; the Hand eof the Beam was alfo made of a Kind of Ring of Brafs, and the lower Part thereof, was flit fo as to receive the Cock, that it might juft freely move between its Sides and no more; and the Pin refted upon two Holes made in the Sides of the faid Handle; the Top of this Handle had a fmall Tongue of Brafs, of the fame Breadth with the Top of the Cock of the Beam, and pointing fo directly at it, when in *Equilibrio*, and fo near approaching it, as juft not to touch it. This Beam was fufpended by a convenient Frame of Wood, as to hold it fteady whilft it was made ufe of; and to find exactly the *Equilibrium*, by giving a little Knock with a fmall wooden Mall, upon the Handle, there was caufed fuch a fhaking, as made every Thing fettle into its due Place; and, by the Ends of the two oppofing Cocks, or Tongues, the Agreement, or Difference, was difcoverable. The Weights were all curioufly, and very exactly, made of Brafs; which, that they might not be adulterated, were, all over the Surface of it, ftamped with the Emperor's Seal, and the Quality of each engraven upon it in the *Chinefe*, or *Japanifh* Characters. Thefe are, by a fevere Penalty, prohibited to be exported into any other Place, and are of great Value in the Country itfelf. The Weights are *Cunderines, Maces,* and *Tales*; ten *Cunderines*, making a *Mace*; ten *Mace,* a *Tale*; and 10 *Tale,* one Pound *Troy.*

The Second, was a *Japan* Stilyard, made upon the fame Principle as our common Stilyards, but with greater Curiofity, and for fmaller Weights, than we generally ufe them, ferving to weigh any Weight from a *Cunderine,* to two Pound *Troy,* or twenty

twenty *Tale.* The Beam was made of a tapering
Rod of Ivory ; the Scale, or Diſh, at the greater
End, was hung by a ſtrong Thread of Silk, which
paſs'd through a Hole in the bigger End of it ; inſtead of Handles alſo, there were three ſtrong
Threads of Silk, at ſeveral Diſtances from the former, which paſs'd through three ſeveral Holes in
the Beam ; and to each of thoſe three handling
Threads, was adjuſted a Line of Diviſions upon
the Sides of the tapering longer Arm ; the Weight
was of Braſs, and ſuſpended on the longer Arm,
by a ſmall Bow of Silk, which might be eaſily
ſlipped to and fro, as Occaſion required.   The
whole Inſtrument was very compleat, and nice e-
nough for the Purpoſes it was deſign'd for, to wit,
for weighing Silver and Gold, &c. in the Way of
Trade.

Tʜᴇ third, was a Stilyard of my own Invention, by which the Weight of any Body, that
could be weigh'd in it, might be found without
the Trouble of removing the Weight, as in the
common Stilyard ; and, by Means of a Plumb
Line, after the Manner of the Balance I ſhew'd
January the 16th, the particular Weight of whatever was put in the Scale, was preſently manifeſted ;
and it had this great Conveniency in it, that the
Diviſions, by which the Weights were determin'd,
were altogether as great at laſt, as at firſt, being
all equal.   The Conveniencies and Uſes, are obvious enough in the weighing, either of great or
ſmall Bodies, all being to be perform'd with great
Speed, and as great Exactneſs, and with much leſs
Trouble.

[*In*

[*In the Minutes of the* Royal Society, *of* Octob. 25, 1677, *I find an Experiment made that Day, by* Dr. Hook, *which, for Congruity, I shall insert here.*]

I T was a very easy Way to examine the comparative Weight of Liquors, and that to so great a Niceness, as very sensibly and manifestly to exhibit such Weight of two Liquors, though they differ'd from one another, but a 100000th Part of their Weight.

T h i s was performed by the Help of a large Glass, of a Pear-like Form, equalling in Bulk about three Pound of Water; which, by Shot included in it, was made almost equiponderant to Water; but yet somewhat heavier, that it might just sink to the Bottom ; but by the finest Hair, tied to the Stalk, could be suspended in the Water. This Hair was tied to the Scale of a Beam ; and this Poise, by a Counterpoise in the other Scale, was made to swim in the Water, so as neither to touch the Bottom, nor the Top. And when so poised, it was found, that a 5th Part of a Grain added to, or taken from the Scale, would make the Glass-Pear rise to the Top, or sink to the Bottom. Whence it was evident, that the whole Glass, weighing about four Pounds (which amounts to 22040 Grains, or 220400 tenth Parts of Grains) and that one single tenth Part of a Grain would turn it. And the Glass, when suspended, being always equal to an equal Bulk of Water, if that Weight be alter'd a 220400th Part, the Poise must be alter'd, and consequently, by Help of the Scales, be made sensible. ·

T h i s

Th is Experiment, and the Nicety thereof being underſtood by the Company, it was deſir'd, that Trials might be made the next Day upon ſeveral Sorts of Water, as Pump-Water, new River Water, *Thames* Water, and Rain-Water, that ſo they might be experimentally ſatisfied of the Exactneſs of this new Inſtrument : Which is new upon this Account, that it hath not been taken Notice of by any of thoſe who have written on this Subject ; as *Ghetaldus, Stivinus, Paſchal,* &c. they having only taken the comparative Weight of ſome ſmall Counterpoiſe within, and out of the ſame Liquor, which they have always perform'd with the ſame Scales, which are no Ways fit for exhibiting the Niceneſs and Curioſity of this Experiment.

O n *November* the firſt following, the Experiment was accordingly made, and it was found, that two Grains of Salt, being put into two Gallons of Water, caus'd the Counterpoiſe to be conſiderably lighter : Which was found to be ſo, upon repeated Trials.

Dr.

Dr. Hook's *Experiment before the* Royal Society, Feb. 6, 168$\frac{1}{4}$, *concerning Magnetism in Drills,* &c.

I THEN produced the *Apparatus* for the Experiment appointed me laſt Meeting, in order to make out my Aſſertion, that the magnetical Virtue in Steel might be excited, and conſiderably increaſed by a Body not generally accounted magnetical ; and therefore, that the affirming a Body to be magnetical, becauſe it excited that Virtue would not always hold good. The Experiment I made, to examine this Opinion, was this. I took a Drill made of Steel ; and, left it ſhould have had any determinate Virtue in it, as to Polarity, I heated it red hot in the Fire, and ſo ſuffer'd it to cool, quenching only the very drilling Point of it in cold Water: When it was perfectly cool, I apply'd a Needle to it, and found, that which End foever I turn'd downward, it would attract the South End of the Needle, and the upper End would attract the North ; and this, as often as I repeated the turning of the Drill, and apply d the Needle to the Ends of it. So that it plainly appear'd to have no determinate Polarity at all, as a Drill, or the like Piece of Steel, touch'd by the Loadſtone. Then I cauſed a Piece of Braſs to be put upon a Table, and holding the Drill very near with the ſame Inclination, and in the ſame Line, that a Dipping Needle left free, when well poiſed, would ſituate itſelf ; I cauſed the Drill to be mov'd with a Drill Bow, ſo as to drill a pretty deep Hole in the ſaid Piece of Braſs, and thereby to warm or heat the Top thereof. Then, examining it again with the Needle, as I had done before, I found that the Drill by this Boring, or Agitation,

gitation, had acquired a Polarity or directive Virtue, as well as an attractive for the Point of the Drill, which, in drilling, refpected the North, whether it were held downwards or upwards, always attracted the South End of the Needle; and the contrary End in like Manner, in either Pofture, attracted the North, in the fame Manner as if the Point thereof had been really touched with the Needle. In the like Manner, I found by trying with a Steel Chizzel by ftriking of its End, when placed in the proper Pofition of the Dipping Needle, that much the fame Effects would be produced.

HEREUPON it was objected, that Brafs itfelf was a magnetical Body, and therefore that this was not a fufficient Eviction; whereunto I replied, that I conceiv'd any other hard Body, placed inftead of the Brafs, would produce much the fame Effect.

I DID therefore propound to have the fame tried with hard Wood, Ivory, Bone, Glafs, or Stone, which have not hitherto been accounted magnetical Bodies, to fee whether they would not be a Means of exciting this magnetical Virtue; for if fo, then either all Bodies, that are hard, muft be faid to be faid to be magnetical, or elfe it will not neceffarily follow, that every Body that excites this Virtue, is therefore to be efteem'd magnetical. And this the rather, becaufe as I have, in Part, fhewn in this Place, and as I fhall hereafter make out more at large, there may be produced in other Bodies, as well as Steel, Iron, or the like, a Quality much refembling that of the magnetical; wherein, notwithftanding, neither the Magnet, Steel, Iron, or the magnetical Virtue, or Power of the Earth, is any Way concerned.

Dr.

Dr. Hook's *Experiment, about the Strength of Ice.*

NEXT, I gave an Account of an Experiment, which I had caufed to be tried in the Prefence of Mr. *Meredith*, and Dr. *Aglionby*, of a Piece of Ice, plain'd true Square, of about fifteen Inches in Length, four Inches broad, and $3\frac{1}{2}$ Inches thick ; this was pretty folid, having no more Blebs in it than common Ice ufually hath. This Piece of Ice, fo fquar'd, was plac'd upon the Engine made on Purpofe for examining the Strength of Bodies, as to bearing. The Places, whereon the two Ends refted, were juft twelve Inches afunder, and the Bar, whereon the Weights refted, was juft placed in the Middle of the Piece of Ice, between the two bearing Cheeks, fo that the Line of Preffure, the Bar being round, was at fix Inches Diftance from each of the bearing Cheeks ; the broader Part of the Ice, was placed horizontal, and the narrower, was placed perpendicular. All Things being thus fitted, we applied the Weight to the two Leavers of the Engine, and began at fifty Pounds ; then mov'd them to 100, 150, 200, 250, and 300, fuffering the Weights to prefs the Ice for fome Time, at every of thefe Pofitions, the Ice ftill bearing them, without breaking, or in the leaft crufhing, either by the bearing Cheeks, on which it refted, or under the round Iron Bar that refted on it ; then removing the Weights to 350, and fuffering them to reft upon it, in a very fhort Time, the Ice broke fhort in two, juft under the Iron Bar, though it did not appear at all to be crufhed, at any of the three bearing Places.

THIS

T H I S Experiment was tried, in order to find, firſt, the Hardneſs of this Body, which is produced by Cold, out of the fluid Body of Water, without the Mixture of any ſenſible ſolid Body, or, is rather the primitive Body, out of which, the fluid Body of Water is made, by a very ſmall Degree of Heat, the Difference between the greateſt Degree of Heat, it will ſuſtain without being thawed, and the leaſt Degree it will ſuſtain without being frozen, being ſo very near the ſame, that one's Senſe will not diſcover it, and even a Thermometer, but very little. So that if Heat and Cold, only, be the Cauſes of theſe Mutations, it is the greateſt Inſtance in Nature of ſo conſiderable a Change of Texture, upon ſo inconſiderable an Alteration of the Cauſes.

Secondly, In order to find the Tenacity or Strength of this Body for bearing, and thence, to give ſome Reaſon, how it comes to bear ſo great Weights, moved, or reſting upon it, without being broken, when it covers the Top of a River or Pond, as has been now ſufficienetly experimented upon the *Thames.* And though the Manner of bearing, when the Ice floats upon the Water, be very differing from the Way of bearing in this Experiment, and ſo the Calculation holds not the ſame in the one and the other; yet this Way of Trial is a neceſſary Ingredient of ſuch a Calculation; ſince, without knowing the Stiffneſs of Ice, as to bending or breaking, and the Hardneſs of Ice, as to cruſhing, ſuch a Calculation cannot be perform'd. The Caſe alſo varies very much from the Manner of the Boundings, and the Bigneſs of the Piece of Ice, whoſe Strength is to be calculated. For in a Pond, where the Edges of the Ice are firſt frozen to the Ground, and ſo the Water underneath being pent in from being able to get out, the Reſiſtence of the Water hinders the breaking
of

of it, even till the resting Weight begins to crush it. And 'tis much the same, where the Surface of the Ice is very large, though it no where toucheth or resteth upon a solid Body at its Brims, there being so great a Length of Water to be moved, before the Water underneath can give Way to the breaking of the Ice. We must also consider the Weight, as bearing in the Center of a round Flake, which is very differing from that of an oblong Shape. To this Calculation we must likewise take in the rising of those ambient Parts of the Ice, which at a Distance encompass the bearing Center, since the Ice can hardly descend in the Center, without at the same Time raising some circumferential Parts, which are more difficult to be broken upwards, than the Center to be broken downwards.

An Experiment of *Dr.* HOOK's, *concerning the swelling of Water by Freezing.*

THE third Experiment I tried was upon Occasion of a Report of Dr. *Crone*, of an Experiment try'd by himself, of applying the freezing Mixture to a Glass of Water, and observing the Water to rise in the Neck of the Glass, before any Part of the Water was frozen. Whence he conceived that the Water itself did actually expand by its Application, before it came to freezing. The Reason of which Phænomena I conceived to proceed only from the shrinking of the containing Vessel, and not from the expanding of the Water, before freezing: To elucidate which, I tried the Experiments I had formerly shewn, to prove the swelling of Glass by Heat, and the shrinking of it by Cold; as also divers other Phænomena, which are manifestly to be ascribed to the shrinking

ing and swelling of the containing Glass Vessel, and not at all to the swelling and shrinking of the Liquor contain'd ; as the dipping such a Glass of Water, in hot Water, will presently make the Water descend in the Neck ; and the dipping the same in Water colder, then the Water in the Glass, or then the Glass it self, will make the same Water rise for some Time in the Neck of the Vessel. However, tho' some Trials were made, whose Effects seem'd, to me, plainly to concur with this Explication, yet the Doctor, and some others, seem'd yet to doubt, whether the Water it self did not actually swell by the Application of the freezing Mixture, before it actually began to freeze ; which if these Trials do not satisfy, there may be several other Ways made use of to find the swelling of the Glass by Heat, and the shrinking of it by Cold. But I conceive no Experiment can be made that will prove Water, without freezing, to be dilated or expanded by Cold, or contracted or condensed by Heat.

Dr.

Dr. H o o κ's *Experiments,* Feb. 13. 1683-4.
fhewing the fpecifick Gravity of Ice, &c.

I Took then a Piece of Metal big enough to fink
the Piece of Ice, I defigned to examine, to
the Bottom of the Water, that fo the compound
Body of Ice and Iron might have a fenfible Gravity in the Water. Then letting it down into the
Water, which I had fet conveniently in a Glafs,
that I might fee this Compound freely to fwim to
and fro clear below the Surface ; the Scales being
conveniently fuftained by a Frame, I counterpoifed
it exactly to an Equilibrium, and found it to a-
mount to 1933$\frac{4}{7}$ of 3000 Parts of a Pound Troy,
which were the Weights to which I reduced this, and
all the other Counterpoifes. Then I fuddenly lifted
up the Ice and Iron into the Scale, and fo counterpoifed it in the Air, and found the fame to be
2567^6 of the fame Parts ; then I took off the Ice,
dry'd the Scale, and let the Iron Weight hang by
the fame String in the Water ; and counterpoifing
it, I found it to amount to 1984$\frac{1}{7}$ of the fame
Parts ; then lifting the Iron out of the Water, and
putting it into the Scale, I found it to be counterpoifed by 2209$\frac{7}{7}$ of the fame Parts. Thence the
Weight of the Water, equal in Bulk to the Ice
and Iron, was 634$\frac{2}{7}$ of the like Parts, and the
Weight of the Water, equal to the Ball, was 224$\frac{6}{7}$;
thence the Weight of the Water, equal to the Ice
was 409$\frac{4}{7}$, and the Weight of the Ice in the Air
was 358$\frac{1}{7}$, and confequently the Weight of the Ice
in Water was 50$\frac{7}{7}$; that is, the Weight of the Ice,
to that of the Water, was very near, as 7 to 8 ;
that is, the Ice was lighter than the Water, by an
eighth Part of the Weight of the Water ; or the
Water heavier than the Ice, by a feventh Part of
the

the Weight of the Ice. So that the Expanfion of the Ice, to the Expanfion of the Water, was as the Weight of the Water, to the Weight of the Ice; that is, as 8 to 7 : So that the Water, by its freezing, becomes expanded one feventh Part of its Bulk, and confequently that 7th Part muft float above the Surface of the Water, and of the Bulk of Ice muft remain immerfed in the Water ᵷ Part of the Bulk of the Ice floating above it.

T H E Ice I made ufe of, in this Experiment, was not very full of Blebs, or Bubbles; nor was it perfectly free of them, but of a middling Nature, which may pretty well hold, as a Standard, or common Meafure of a great Congeries of feveral Sorts of Ice, fome of which may be much more porous, and fome much lefs, as I have had Occafion feveral times to obferve, in this great Froft. The Time, in which I try'd this, was pretty warm, and fo it thawed; and the Water having ftood all the Day, expofed to the Air, was confequently much of the fame Temper; and thence I counterpoifed the Ice and Iron firft in the Water, and then prefently lifted it out of the Water into the Scale, fo that all that levitated in the Water was immediately put in the Scale: The Water was ordinary Pump, or Well-Water, and is accounted a pretty good frefh Water; which Circumftances I mention, as having Significancy, as will by and by appear.

F O R from this Experiment it plainly appears, that the common Opinion that the Ice, upon a fudden Thaw, finks to the Bottom, is falfe, tho' never fo confidently afferted by the Water-men: For in this Experiment, where the Water was pretty warm, in refpect of Ice, and thawed the Ice very faft; yet an eighth Part of the Ice floated above the Water, and Water by Heat, without boiling, will not expand near that Proportion :

nay,

Nay, I have found, that throwing in a Piece of Ice into Water boiling, it still floated, and funk not, much less can it sink in a tepid Water upon a Thaw.

N E X T, from hence we may collect, that in the Northern Seas, at least one Eighth Part of the Bulk of any Body of Ice floats above the Water: I say, at least an Eighth; for possibly it may be one Seventh; for first (as is affirmed by many Voyagers to the Northern Seas) the Ice is found to be pretty fresh, and to have little or no Taste of Brackishness; and so, one Part taken with another, not heavier than this Ice I made use of. Next, the Water, notwithstanding, in which it floats, is salt, and and consequently about a 40th Part heavier than common fresh Water. Thirdly, This salt Water, tho' it do not freeze, is yet pretty near the same Degree of Coldness with the Ice that floats in it, and consequently yet more heavy than the same Water when more tepid. For as I shall hereafter prove, Bodies that freeze not, are yet not less cold than other Bodies that do freeze. Fourthly, That the Sea-Water, near the Bottom, is yet much more cold, and much more salt, than in the same Place it is near the Top, and consequently must much contribute to the floating of a greater Part of the Ice. That the Water is colder at the Bottom, than above, was positively affirmed by Mr. *Roachford*, who try'd it in the *Sound*; and that salt Water is salter at the Bottom, than at the Top, any one may find.

A L L which Particulars consider'd, it will not seem altogether so incredible, or indeed strange, that there should be floating Islands of Ice in the frigid Zones, of so great a Height above the Surface of the Sea: For, supposing it to be globular, above a 4th Part of its Diameter must float above the Water, to make a 7th Part of its Bulk to float,

and

and confequently the Depth of the Ice under Water need not be fo very great, to make fo great a Height above the Water; but if the upper Parts of it above the Water are yet much higher, and more fpongy than folid Ice, as confifting, in great Part, of Accumulations of Snow, then may that Height, above the Water be raifed much higher, and be made poffibly to equalize, if not exceed, even the Depth of the Ice below the Surface of the Water, efpecially if the Bottom of the faid Ifland be flat, as moft probably it is, and as broad, if not broader, than the Compafs of it at the Surface of the Water; as alfo if Parts above the Water be tapering, like a Pyramid, to the Top. Again, If the lower Parts of the Sea, in thofe Parts, are colder than at the Top, as probably it may be in the Spring, the frefher Parts of the Water may be congealed, even at the Bottom, and fo augment the Bulk of it by new Accretions underneath, and fo continue to buoy it up more and more, and fo raife the upper Parts more and more into the Air. And confonant to this we find, that the greateft Iflands of Ice are found in the Spring, after the Winter is paft, and the Air begins to have a Tepidnefs in it; and not fo much, if at all, in the former Part of the Winter, when it freezes more violently at the Top of the Water.

As to the Reafon why Water, when of fuch a Degree of Temperature, becomes fo folid a Body; and why, when of another Temperature, it becomes fo fluid, I fhall not now fpend your Time in explaining, defigning to do it in my General Theory of natural Operations. This only I fhall mention here, by the by, that the Body of Ice, tho' very hard, is very little fonorous, in refpect of Glafs, which to the Sight it fo much refembles: That the Blebs in it are not Vacuities, but a Kind

of

of Air, which has its expanſive Power, or Elaſticity, as well as common Air: That this Air does not, upon the Thaw, retreat into the Water, as it ſeems to come out of it upon the freezing, as by Experiment I have found.

Farther Experiments, made Feb. 20. 1683-4. *by* Dr. Hook, *before the* Royal Society, *concerning the* Phænomena *of* Ice.

THE proceeding Diſcourſe was read, and ſome Matters therein more particularly explained by Deſcription partly, and partly alſo by Experiments.

THE Experiments were firſt to ſhew, that the Blebs in Ice (ſuppoſed by ſome to be Vacuities, like the Blebs in Glaſs Drops) are filled with Air, which has the ſame Properties with common Air. I took then a Piece of Ice, and putting it into Water, which was tepid, as having ſtood in a warm Room, by which the outward Parts of the Ice quickly thawed, and ſo there remained nothing at all of Air ſticking to the Outſide of it; then whelming a Cup-Glaſs clear over it, which was perfectly filled with Water, and had no Air included in it, I ſuffered it to remain, covering the Lump of Ice, till the whole was thawed, or melted into Water; and it was plain to be ſeen, that as the Ice thawed, the Blebs that were viſible in it, before the Thaw, did aſcend to the Top of the whelmed Glaſs, and then unite with one another into a conſiderable Body of Air.

THE ſecond was to ſhew that Water, though boiling hot, would yet be ponderous enough to make the Ice to ſwim and float in it. This was done by putting a Piece of Ice into a Veſſel of boiling

boiling Water: And the Ice continued to float upon it till it was all melted.

T H E Reafon of the Experiment was in order to find out the Nature of the Expanfion of freezing Water, and the true Caufe thereof; which feems to contain as many difficult Phænomena in it to be explain'd, as any other in Nature: For firft, this Body of Ice feems heterogeneous to all other Bodies; which being melted, and fuffered to cool and grow hard, are ftill condenfed and fhrunk into a leffer and leffer Room, as they grow colder; as is very obfervable in all Sorts of Metals, as Gold, Silver, Copper, Tin and Lead, every of which, when they are melted, take up more Space, or are more expanded, than when they are grown cold and hardened; as one may prefently find, by cafting any of them into a Mould, and obferving the fetting, or fhrinking of the Gitt, by which the Mould is fill'd; or by fuffering the whole Body, fo melted, to remain, and grow cold and folid in the Ladle or Crucible; for 'tis evident that the top Surface, which, when melted, is protuberant, and fwelling upwards; when cold, it is flatted, and very often concave. And fometimes alfo, in fome Metals, it is crumpled, and fhrunk into curious Figures; as is very remarkable in *Regulus Martis,* made with *Antimony,* which is therefore called *Stellatry,* for that it hath fome Refemblance to the Figure we generally make for a Star, *viz* fix Radiations from its Center. 'Tis evident alfo in Tin and Lead; Wax alfo, and fome refinous Subftances, fhrink upon hardening after the fame Manner, and Fatt, or Tallow of Animals; fo all Sorts of Vitrifications and Glaffes, and all Sorts of Oils, that will harden, and Butter, which alfo grow opaque. But Water, when it paffes from Fluidity to Solidity, proceeds very differing; *Firft,* In its inftantaneous Change.

2*dly,*

2dly, In its Expanfion, or Rarefaction. *3dly*, In its Tranfparency. *4thly*, In its Refractivenefs. *5thly*, in its Generation of Blebs, or Bubbles. *6thly*, In its Power of Expanfion: tearing and rending to Pieces the ftrongeft metalline Bodies that imprifon it; when, as yet, it leaves Room e-nough for the fmall Particles of Air to expand, if at the fame time it may not be faid to fuck it in; for I do not find that the imprifon'd Blebs are at all prefs'd, nor is their Spring at all the Caufe of this Expanfion; for by obferving the thawing of a Bleb in the Ice, I did not find the Bubble that rofe from it to be any bigger in Bulk, than the Bleb that contain'd it; whereas if the Air in the Bleb fhould be preffed with as great a Force, as the Strength of the Infide of the containing Veffel amounts unto, it muft of Neceffity reduce the Air to near a thou-fandth Part of its natural Extenfion; and confe-quently, when the Bleb comes to be thawed, and fo fet at Liberty, it muft at leaft, I fay at leaft (by rea-fon it then fuffers a greater Degree of Heat, than when it is frozen) expand itfelf into a Bulk a thou-fand Times bigger; but there is no fuch Appearance that I could obferve. Several Authors have en-deavour'd to give Solutions of this Phænomenon, as particularly the ingenious Mr. *Des Cartes*, who fuppofing the Particles of Water to be very long and limber Bodies, like fo many Eels, whilft, as it were, kept alive, and agitated by this *Mate-ria Subtilis*, are limber, and fo eafily complicate and flide one within another, and fuffer the *Mate-ria Subtilis* to have its Paffage free through them every Way; but when there is lefs Agitation of this *Materia Subtilis*, they do, as it were, die, and grow ftiff and rigid, and fo will not fo eafily comply to the Figures of each other, but grow fo-lid and hard: But then 'tis to be confider'd, that the greater Plenty there is of the *Materia Subtilis*,

the

the greater muſt be the Agitation of them; as he aſſerts in the Explication of the Particles of the Air, and conſequently the more Room muſt they take up, and ſo be more expanded when fluid, then when ſolid. Another late Author ſuppoſes, that Congelation is made by a *Sal Armoniack*, breathed, or exhaled from Animals, which, in cold, froſty Weather, is very copious in the Air, which *Sal Armoniack* does then inſinuate into the Pores of the Water, and ſo wedge up all the Pores, and widen them, and ſo make the Parts of the Water to coaleſce into a hard Body. But this I conceive to be alſo hypothetical, and not experimentally proved; for tho' there may be ſome volatile Salts in the Air, yet 'tis pretty diffi-cult to conceive there ſhould be ſo great a Quan-tity, as at once to wedge up all the Water of the Northern Part of the Earth, and yet, at the ſame Time, we ſhould not ſmell it; beſides, we do not find that the *Sal Armoniack* Spirit does perform this Effect, when it is raiſed in the Air at other Times; nor does the *Sal Armoniack* it ſelf, when mixed with Water or Ice, do it; for we find that *Sal Armoniack*, ſtrow'd on Ice, will the ſooner make it thaw, and reſolve again into Water, than make it freeze harder: Others have given differ-ing Explanations, but I have not met with any yet, that, in my Opinion, give a clear and ſatisfactory Solution of it. Nor ſhall I at preſent trouble you with Theories, or Speculations, which ſome may poſſibly have a Prejudice againſt; only ſuffer me to acquaint you with a Phænomenon or two, which, if you think any of them worth ſeeing, you may have tried, for they are very obvious, plain, and neither difficult nor chargeable Experi-ments, tho' poſſibly as inſtructive as the moſt dif-cult, chargeable, or pompous Experiments, to ſhew ſome Sorts of Expanſion.

T A K E

TAKE then a Urinal, and fit into it a Stopple of a dry Piece of Wood; then put the End of this Stopple into a Diſh of Water, and you will find, in a little Time, the Stopple will grow ſo much bigger, as to break the Urinal.

Secondly, TAKE another Urinal, and fill the ſame with Peaſe ; then filling it up with Water, ſtop the ſame with a Cork, which you may tie down faſt with a Packthread ; then let it remain ſome Time, and you will find the Peaſe will ſwell and break the Glaſs.

Thirdly, TAKE Plaiſter of *Paris,* or burnt A-labaſter, and put it into a wooden Diſh, and tem-per it with Water, till it be very ſoft and fluid, that it may be eaſily poured out ; then with this Mixture fill a Urinal or Vial top-full, ſuffer it to ſtand upright till it ſets into a ſolid Body, and you will find it ſwell and break the Glaſs.

Dr. HOOK's *Diſcourſe to the* Royal Soci-ety, May 21. 1684. *ſhewing a Way how to communicate one's Mind at great Di-ſtances.*

THAT which I now propound, is what I have ſome Years ſince diſcourſed of ; but being then laid by, the great Siege of *Vienna,* the laſt Year, by the *Turks,* did again revive in my Me-mory ; and that was a Method of diſcourſing at a Diſtance, not by Sound, but by Sight. I ſay therefore 'tis poſſible to convey Intelligence from any one high and eminent Place, to any other that lies in Sight of it, tho' 30 or 40 Miles diſtant, in as ſhort a Time almoſt, as a Man can write what he would have ſent, and as ſuddenly to re-ceive an Anſwer, as he that receives it hath a
Mind

Mind to return it, or can write it down in Paper. Nay, by the Help of three, four, or more, of ſuch eminent Places, viſible to each other, lying next it in a ſtreight Line, 'tis poſſible to convey Intelligence, almoſt in a Moment, to twice, thrice, or more Times that Diſtance, with as great a Certainty, as by Writing.

F o r the Performance of this, we muſt be beholden to a late Invention, which we do not find any of the Antients knew; that is, the Eye muſt be aſſiſted with Teleſcopes, of Lengths appropriated to the reſpective Diſtances, that whatever Characters are expoſed at one Station, may be made plain and diſtinguiſhable at the other that reſpect it.

Firſt, F o r the Stations; if they be far diſtant, it will be neceſſary that they ſhould be high, and lie expoſed to the Sky, that there be no higher Hill, or Part of the Earth beyond them, that may hinder the Diſtinctneſs of the Characters which are to appear dark, the Sky beyond them appearing white : By which Means alſo, the thick and vaporous Air, near the Ground, will be paſſed over and avoided; for it many Times happens, that the Tops of Hills are very clear and conſpicuous to each other, when as the whole interjacent Vale, or Country, lies drowned in a Fog. Next, becauſe a much greater Diſtance and Space of Ground becomes viſible, inſomuch that I have been informed by ſuch, who have been at the Top of ſome very high Mountains, as particularly at the Top of the *Pike of Teneriff,* that the Iſland of the *Grand Canaries,* which lies above 60 Miles diſtant, appears ſo clear, as if it were hard by; and I myſelf have often taken Notice of the great Difference there is between the appearing Diſtance of Objects ſeen from the Tops and Bottoms of pretty ty high Hills, the fame Objects from the Top appearing nearer and clearer by half, and more than they do when viewed from lower Stations of the Hills; and this not only when the Space between them was Land, but where it was nothing but Sea. I have taken Notice alſo of the fame Difference from the Proſpect of Places from the Top of the Column at *Fiſh-ſtreet-Hill*, where the Eye is, in good Part, raiſed above the ſmoaky Air below.

NEXT, the Height of the Stations is advantageous, upon the Account of the Refractions or Inflections of the Air; which Inflections of the Air are many and very great, ſometimes in an Air which ſeems, to the naked Eye, the moſt clear and ſerene. Infomuch that That alone does wholly confound the Diſtinctneſs of Objects appearing at a Diſtance; now the greateſt Part of theſe ariſe from Commotions of the more denſe Air that is near the Surface of the Earth, by the Rarefactions of ſome Parts of it, cauſed by Heat; which rarified Parts aſcending, do make the Objects ſeen through it, to ſeem to dance and undulate, which is in great Part avoided, if the Proſpect be from an higher Place. Beſides, the Nature of the Air itſelf, at great Heights, approaches nearer to the Nature of the *Æther*, which more powerfully propagates the Impulſes of Light.

NEXT, in chuſing of theſe Stations, Care muſt be taken, as near as may be, that there be no Hill that interpoſes between them, that is almoſt high enough to touch the viſible Ray; becauſe in ſuch Caſes, the Refraction of the Air of that Hill will be very apt to diſturb the clear Appearance of the Object, as I have often obſerv'd.

THE

The Stations being found convenient, the next Thing to be conſider'd, is, what Teleſcopes will be neceſſary for ſuch Stations. And though 'tis true in all, that the longer the Teleſcopes are, provided they are good, the better they will be for this Effect; yet ſomewhat of Limitation is re-quiſite, at leaſt, that they be not ſhorter than cer-tain Limits for ſeveral Diſtances. Theſe may be as follows: For 1 Mile, 1 Foot; for 2 Miles, 2 Foot; for 3 Miles, $3\frac{1}{4}$ Foot; for 4 Miles, $4\frac{1}{2}$ Foot; for 5 Miles, 5 Foot 10 Inch. for 6, $7\frac{1}{4}$ Foot; for for 7 Miles, 8 Foot 9 Inch. for 8, $10\frac{1}{2}$ Foot; for for 10 Miles, 13 Foot, and ſo forward. One of theſe Teleſcopes muſt be fix'd at each extreme Station, and two of them in each intermediate; ſo that a Man, for each Glaſs, ſitting and looking through them, may plainly diſcover what is done in the next adjoining Station; and, with his Pen, write down on Paper the Character there expoſed, in their due Order; ſo that there ought to be two Perſons at each extreme Station, and three at each intermediate; ſo that, at the ſame Time, Intelli-gence may be convey'd forwards and backwards.

Next, there muſt be certain Times agreed on, when the Correſpondents are to expect; or elſe there muſt be ſet at the Top of the Pole, in the Morning, the Hour appointed by either of the Correſpondents, for acting that Day; if the Hour be appointed, Pendulum Clocks may adjuſt the Moment of Expectation and Obſerving. And the ſame may ſerve for all the other intermediate Correſpondents.

Next, there muſt be a convenient *Apparatus* of Characters, whereby to communicate any Thing with great Eaſe, Diſtinctneſs and Secrecy. There muſt be therefore, at leaſt, as many diſtinct Cha-racters, as there are neceſſary Letters in the Al-phabet that is made uſe of, (as is expreſſed in *Fig.* 1.)

1) And thofe muſt be either Day Characters,
or Night Characters : If they are to be made
uſe of in the Day-time, they may all be made of
three ſlit Deals, moving in the Manner I here
ſhew, and of Bigneſs convenient for the ſeveral

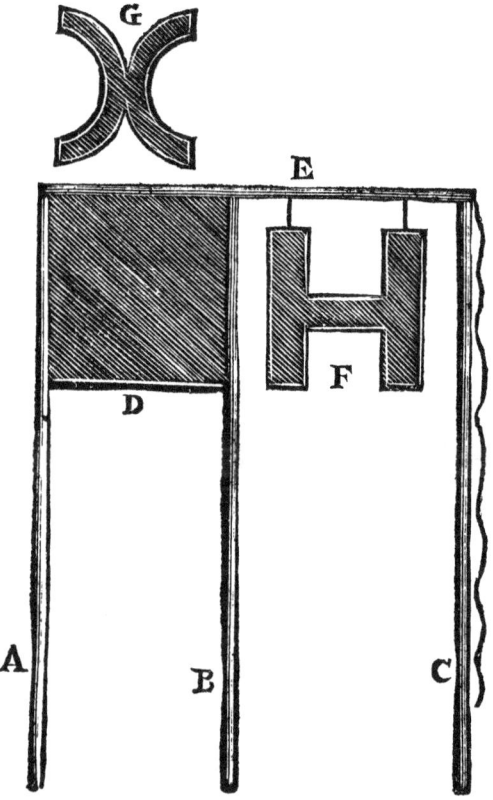

Diſtances of the Stations for which they are made,
that they may be viſible through the Teleſcope of
the next Station. Any one of which Characters
may ſignify any one Letter of the Alphabet, and
the whole Alphabet may be varied 10000 Ways ;
ſo that none but the two extreme Correſpondents
ſhall

ſhall be able to diſcover the Information convey'd; which I ſhall not now infiſt on, becauſe it doth more properly belong to Cruptography. If the Characters are for the Night, then they may be made with Links, or other Lights, diſpoſed in a certain Order, which may be veiled, or diſcovered, according to the Method of the Character agreed on; by which, all Sorts of Letters may be diſcovered clearly, and without Ambiguity.

THERE may be various Contrivances to facilitate and expedite the Way of diſplaying and expoſing theſe Characters to View, and of withdrawing, or hiding them from the Sight; but this I here ſhew, I conceive, will be as eaſy and ſimple as any: All which may be expoſed at the Top of a high Pole, and by two ſmall Lines moved at the Bottom, ſo as to repreſent any Character.

BY theſe Contrivances, the Characters may be ſhifted almoſt as faſt, as the ſame may be written; ſo that a great Quantity of Intelligence may be, in a very ſhort Time, communicated.

THERE will be alſo requiſite ſeveral other Characters, which may, for Expedition, expreſs a whole Sentence, to be continually made uſe of, whilſt the Correſpondents are attentive and communicating. The Sentences, to be expreſs'd by one Character, may be ſuch as theſe, in *Fig.* 2.

O *I am ready to communicate.*)(*I am ready to obſerve.* (*I ſhall be ready preſently.*) *I ſee plainly what you ſhew.* ‿ *Shew the laſt again.* ⌒ *Not too faſt. Shew faſter. Anſwer me preſently.* Dixi. *Make Haſte to communicate this to the next Correſpondent. I ſtay for an Anſwer;* and the like.

ALL

ALL which may be exprefs'd by feveral fingle Characters, to be expos'd on the Top of the Poles, by themfelves, in the following Manner, fo as no Confufion may be created thereby.

I COULD inftance in a hundred Ways of facilitating the Method of performing this Defign with the more Dexterity and Quicknefs, and with little Charge; but that, I think, will be needlefs at prefent, fince whenfoever fuch a Way of Correfpondence fhall be put into Practice, thofe, and many more than I can think of at prefent, will of themfelves occur; fo that I do not in the leaft doubt, but that with a little Practice thereof, all Things may be made fo convenient, that the fame Character may be feen at *Paris*, within a Minute after it hath been expofed at *London*, and the like in Proportion for greater Diftances; and that the Characters may be expofed fo quick after one another, that a Compofer fhall not much exceed the Expofer in Swiftnefs. And fo great Expedition may not only be performed at the Diftance of one Station, but of a hundred; for fuppofing all Things ready, at all thofe feveral Stations, for Obferving and Expofing, as faft as the fecond Obferver doth read the Characters of the firft Expofer; the fecond Expofer will difplay them to the Obferver or the 3d Station, whofe Expofer will likewife difplay them for the 4th Obferver, as faft as his Obferver doth name them to him, or write them down.

THERE may be many Objections brought againft this Way of Communication; and fo many the more, becaufe the Thing has not yet been put in Practice. But, I think, there can hardly be any fo great, as may not eafily be anfwered and obviated.

THERE

THERE may be many Uſes made of this Contrivance, wherein it will exceed any Thing of this Kind yet practiſed ; but I ſhall not now ſpend Time to enumerate them; only in two Caſes, it may be of ineſtimable Uſe. The firſt is for Cities or Towns beſieged ; and the ſecond for Ships upon the Sea ; in both which Caſes, it may be practiſed with great Certainty, Security, and Expedition.

A farther Explication of the Figures.

LET ABC *(Fig. 1.)* repreſent three very long Maſts or Poles erected. E the Top-piece, that joins them all together D, a Screen, behind which, all the Deal-board Characters hang upon certain Rods or Lines, and may (by the Help of ſmall Lines coming down from the Bottom of each of them) be expoſed at F, or drawn back again behind D, as Occaſion ſhall be. G is the Character for a Sentence agreed on, *&c.*

The Letters of the Alphabet in Characters, Fig. 2.

ALL the Alphabet, or requifite Characters, may be diftinctly, and without Ambiguity, expref-fed. Such a Difpofition as this, which I have here defcrib'd, I think, will be fufficient.

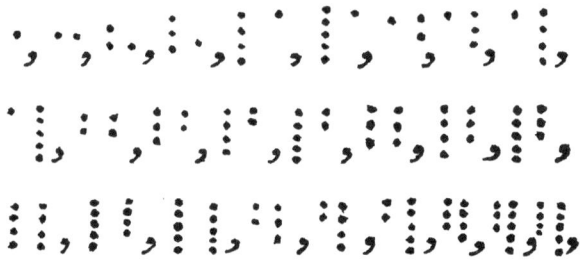

Dr. H o o k's *Difcourfe of Carriages before the* Royal Society, *on* Feb. 25. 1684-5. *with a Defcription of* Stevin's *Sailing Chariot, made for the* Prince *of* Orange.

THE Occafion of this following Difcourfe was from the Module of a *Waggon,* fhew'd to the *Royal Society*; upon which Dr. *Hook* difcourf-ed largely on the various Ways of Conveyance. Among all which he faith, But that which excel-led any, that has hitherto been done of that Kind, was the Sailing Chariot, made by *Simon Stevin,* for the Prince of *Orange,* which, in two Hour's Time, ran upon the Sand, on the Sea-Shoar, by the Strength of the Wind, forty two Miles, car-rying in it no lefs than 28 Men, with Safety and Security : Of which I have feen the Defcription, and have had the full Account. But this being only accommodated for fuch fmooth Ways, as the Sand on the Sea-Shoar, could not be made for common Ufe, and has therefore been laid afide and difus'd. However, fince there is a Poffibility of fuch

fuch a Performance, it may, perhaps, be worthy Confideration and further Enquiry, whether it may not be poffible to contrive, and make fome other Kind of Chariot, or Carriage, which may perform as much in any other paffable Ways; which, I conceive, would be of vaft Benefit to Mankind.

T h e Principal Matter, wherein it differ'd from all other Sorts of Land-Carriage, was this, That inftead of making Ufe of the *Strength* of *Men*, or of any Sort of *Animal*, he made Ufe only of the *Strength* of the *Wind*, and that after the fame Manner as it was then made Ufe of, for the moving of Veffels upon the Water ; namely, by having Mafts, Sails, and other convenient Rigging, as Shrouds, Stays, Sheets, Booms, and all other Rigging, as was neceffary for the Management of thofe Sails. Then, for guiding this Engine, he fo order'd his Contrivance, that he could, by turning the Axle-tree of the Wheels, make it go this Way, or that Way, at Pleafure, with as much Eafe and Certainty, nay, very much more than 'tis poffible to fteer a Ship, or any other Veffel upon the Water. To keep it fafe and fecure from overturning, though on fo fmooth a Plain as that paffed over, there was little Danger from the rifing of the Wheels on one Side ; yet, in the firft Attempts it being better to overdo, in making Provifions againft any Thing of Danger, he placed the Wheels at a great Diftance, or Breadth, one from another ; and, as I judge by the Draught, very near half the Length of the whole Carriage ; by which Means there could be no Manner of Danger in over-fetting ; and ftill the lefs, the more the Carriage was loaden, if the Danger of over-fetting were to be feared from the Strength of a fide Wind upon the Sails ; for the Wheels being placed at a pretty Diftance without
the

the Body of the Carriage, all the Weight of the Carriage, together with the Weight of two of the Wheels, and all the Weight of the Men muſt be lifted up, and reſt upon the two Wheels on the Leeward Side, which neither Sails nor Maſts would be able to do.

T h e Way of ſteering or guiding this Carriage, was much the ſame with that which is, and has always been practiſed in Carriages upon 4 Wheels ; namely, an Helm, or Pole, ſo faſtened to the *Axis.* that by the Means thereof, as by a Leaver, the Axis could be ſwaſhed either this Way, or that Way, upon a Center-Pin, as is now in Uſe in Coaches and Waggons, for the turning or ſwaſhing of the Fore-Axis ; only, whereas the Pole, in theſe, is turned and extended Outwards, before the Carriage, in this, it was turned Inwards. The Wheels are about a middle Size, between the uſual Size of the fore and hind Wheels of a Coach, and were made very ſtrong and ſubſtantial ; and what was peculiar in them, was, that the Rims of them were 18 Inches, or 2 Foot broad, and the Spokes were made to ſtrengthen the whole Breadth; the Reaſon of which I ſuppoſe was, that they might thereby be the better able to reſt upon the ſandy Shoar, without ſinking, or making Rotes in it, which would have made it move very much heavier, the Wheels being thereby always in a riſing Motion ; for the Weight of the whole Carriage, and the Weight within it (which muſt be very conſiderable, there being 28 Perſons in it) reſting only upon the four Points of the Wheels ; if they had been made with narrow Rims, muſt neceſſarily have ſunk pretty deep into the Sand, but being broad, and the Sand very ſmooth, as it is generally left by the Sea, a ſmall ſinking of the touching Line of the Breadth of the Wheel, doth preſently make a very broad Footing, to reſt upon the Sand.   T h e r e

T H E R E were two of thefe Chariots made, the one a larger, of about 30 Foot long, and the other a fmaller, about 10 or 12 Foot long: The larger had two Mafts and two Sails, proportionable to the Sails of a Boat, much about the fame Bignefs. The leffer had only one Maft and one Sail, proportioned likewife to its Bignefs. Each of the Sails had two Yards, the one at the Top, and the other at the Bottom, with proper Rigging to work them. The Bottom Yard, I conceive, was put upon a double Account, Firft, to keep the Sail more flat and plain, that it might, when the Carriage was to fail near a Wind, be kept more fharp and trim; the great Advantages of which I endeavour to prove upon another Occafion. And Secondly, That the Sails might be the eafier managed, and tacked, as Occafions fhould require. And though I cannot find, whether this Engine was ever tried, or made Ufe of, for Sailing by a Wind; yet, I doubt not, but that it would have far exceeded any Veffel whatfoever, that fails upon the Sea, in going near a Wind; becaufe, that in this, there could be no falling to *Leeward*, (which the beft Veffels on the Sea do more or lefs) the Wheels, in this, keeping it directly in the Line, or Plain of the Wheels.

T H E greater Carriage was guided, or fteered, by moving the hinder Wheels by a Pole, like the Helm in a Ship, and the End of it had Tackles to bend it towards this or that Side; and the Rule of Steering was the fame as in a Ship. The leffer Carriage was fteered by moving or turning the Axis of the fore Wheels; the Pole or Helm being turned backward into the Carriage, and the Rule of Moving it was alfo the fame as the former.

T H E laft Thing to be confidered in thefe Carriages, is the great Swiftnefs of their Courfe, which was fo confiderable, that no Horfes, in their full Speed,

Speed, could long keep Pace with them; and Veffels on the Sea, failing the fame Way, feem to be
carried backwards very fwiftly. This, had it not
been attefted by Teftimonies of undoubted Credit,
would have feem'd very difficult to be affented to.
But, on the other Side, if we confider the advantageous Circumftances for its Promotion, and
fpeeding forward, and the fmall Impediments for
the hindering thefe Carriages had, beyond any
other, we fhall find much lefs Reafon to doubt
the Hiftory of it: For, if we compare it with Veffels failing upon the Sea, we fhall find that this
Carriage has firft a plain, hard and even Surface
of the Shoar to pafs over, without any Rub or
Impediment; fo that it is moved in a Plain
without rifing or falling, without any unequal
Impediment, fave only fome fmall Matter in the
rubbing of the Ends of the Axes in the Naves of
the Wheels, which, being well oiled, will be very
little; whereas a Ship at Sea, when there blows
a ftiff Gale (which is abfolutely neceffary, when
much Speed is defired) is firft clogg'd in its Motion by the Lentor and Difficulty of yeilding in the
Medium of Water; by the unequal Stoppings of
the rifing Waves, which create an undulating
and unfteady Motion Upwards and Downwards,
as well as Side-ways; befides the Slope falling and
fliding away to Leeward, which muft be allow'd
for in all Side-Winds, by fteering fome Point
nearer the Wind, than the direct Way; and confequently the Length, paffed by the Veffel, will be
as much longer than the direct Diftance, as the
Secant of fuch an Angle is than the Radius. On
the other Side, if we compare its Motion with
that of a Carriage drawn by Horfes, or other living Creatures, it plainly appears that thefe were
moved by an unwearied Strength, whereas the
Horfes were not long able to hold that Pace. So
that

that upon the whole, it feems to be the fwifteft Carriage yet known, for fo great a Burthen, and fo long a Way.

Bᴜᴛ the great Objection againft this Invention is, that it is hardly practicable in any other Place, and even there but at certain Times, which poffibly have been the Reafons, why it has been fo long difufed, and almoft forgotten. To which I anfwer, That fcarce any other Invention for Carriage is practicable in all Places: Land Carriage cannot be practifed at Sea, nor Sea Carriage by Land; Carts and Coaches cannot be ufed in fome Places, by reafon of the Inconvenience of the Ways, as in *Cornwall.* But this Invention, I conceive, is not to be thought confined only to the fmooth Sands on the Sea Shoar; for I doubt not, but that if Trial were made (as I hope it will fhortly be) it might be much more practicable upon the plain Downs of *England,* than where it was ufed, by Reafon they are much more expofed to the Wind, and alfo much more hard, fo that the Wheels need not be of fo great a Breadth. I conceive farther, that the Carriage may be improved much in its Lightnefs, and alfo in the Eafinefs of moving. If fuch a Chariot were made for *Salisbury Plains, Banftead Downs, Winchefter Downs, Newmarket Row,* or fome fuch fmooth Plains, and the Wheels, (which need be but three) were moved upon fmall Steel *Pevots* or *Gudgeons,* in *Bell-Metal Sockets,* well oiled, inftead of being moved upon the large End of an *Axle-tree.* Next, if inftead of 4 Wheels, 3 only were made ufe of, placed in the Form of a Triangle, the fteering Wheel being that which went foremoft, and the Place of the Maft in the Center of the Triangle, the Weight carried, to be all placed behind the Maft, to which I would alfo have added a Contrivance to retard and ftop its

Motion,

Motion, whenever there shall be Occasion, which is easily to be done; somewhat after the same Manner as Windmills are stay'd, when there is Need. By such a Contrivance, I doubt not, but a Chariot may be made to out-run even the swiftest Race-Horse, especially where the Course is long and plain; and with a Side Wind may be carried back again to the Place from whence it set out; and both forward and backward may be carried with as great a Swiftness, even as the Wind moves, which will not be unpleasant to such as have suitable Conveniences near their Habitations; with which may be tried as many Experiments of sailing near a Wind, as can be tried upon the Sea; the Contrivance of the Wheels making the Motion as easy, as the Water of the Sea or Rivers in others; and to a very swift Motion, having much less of Impediment, especially if the Wheels be order'd to the best Advantage, all Manner of rubbing or sliding being thereby taken off, and even the Inequality of the Ways themselves may be in a great Measure removed. I have been the more particular in describing this Carriage, because it was the swiftest that has possibly yet been made, and therefore, on this Occasion, deserved more than a transient Mention, tho' I do not look upon it as an Invention of the highest Perfection, for this Effect; but may be as much exceeded, as that exceeded a Man that leisurely walks. Who it was that first invented the Wheel, is not recorded in History, it having been long before any History extant (except that of the Bible) and the first Mention we find of it there, is *Pharaoh's Chariot*, in which *Joseph* was exalted to ride: Of whose Form we know nothing but the Name, tho' it had, in Probability, been known long before that Time; which, notwithstanding, long preceded any Heathen Writings now extant.

Hyginus

Hyginus relates, in his 2d Book, where he treats *De Ophiucho,* that *Ceres* invented an One-Wheel'd Chariot, which *Triptolemus* (whose Nurse she was) first made use of, for to make Speed, to inform the World of her Bounty. *Ceres cum sua beneficia largiretur hominibus, Triptolemum cujus ipsa fuerat nutrix (qui primus hominum una rota dicitur usus ne cursum moraretur) jussit omnium nationum agros circumeuntem semine partiri. In Glossis Isiodori, Vehiculum unius rotæ,* is called *Pabo.* But how this One-Wheel'd Chariot was contrived, or used, is not to be found in History; Mention there is, of other Chariots, with more Wheels, in the ancient Authors; so that 'tis clear, it was known and practised long before any Histories of Heathen Writers were publish'd. An Invention of so great Use, that it seems impossible ever to be lost by Mankind, after it be once known: Which Consideration makes me very much wonder whence those Men came, that inhabited *America,* before the *Spaniards* over-running and conquering of it; since it seems probable, that if they, or their Ancestors, had sprung from any People here, on this Side of the World, *viz.* from *Europe, Asia* or *Africa,* they must needs have carried along with them the useful Invention of the Wheel; but it has been observed, that they knew nothing at all concerning it, nor any the least Use of it, throughout all *America,* before the *Europeans* came among them. So that we must conclude, either that they were made Inhabitants before the Invention of the Wheels was found, or that they never had any Origination from any Generation of Men in those Parts of the World, at least not from the *Tartars,* who, of all People, do most frequently use them; but this by the by. The first, and most simple of *Carriages* by Land, was this Invention with one Wheel, and may possibly be most accommodate, for attaining the End we are now inquiring after, which is Swiftnefs, it having the leaft Impediment to its Motion, and the leaft Incumbrance of any other; and may therefore, in the next Place, deferve to be confidered, and poffibly be brought into common Ufe, at leaft to be experimented, as was that of the Sailing Chariot.

Bu t before I come to the more particular Defcription thereof, I think it will not be impertinent to examine the Contrivance of the Wheel, as it is applied to Carriages, for the facilitating of their Motion. One of the greateft Obftruction to Swiftnefs of Motion being the Inequality of the Ways, and the rubbing or grating of thofe Ways againft thofe Bodies or Weights, that are drawn or flid upon them.

The Wheel being then a round Body, and moving forwards, only by its Rollings, doth not at all rub, grate, or flide upon the Way; and fo hath no Impediment at all to its Motion forward, where the Way is even, plain and horizontal, or level, there being no Impediment, or very little, from the Medium of the Air it paffes through, and fo hath no Impediment to be moved with the fwifteft Motion, like that of the Refiftence of Water to Veffels moving through it: So that the only Impediment feems to be that of its own Bulk, (of which I fhall fpeak hereafter) becaufe the outward Rim of the Wheel, in its rolling Motion, doth uniformly apply its Parts to the Parts of the Plain, by defcending down, and rifing up from them perpendicularly; and the touching Part is always quiefcent upon the Plain, and moves not either forward or backward; and confequently all Impediment from rubbing upon the Ground or Way is wholly taken off, as 'twill be evident to any one who fhall examine the Motion of any one Point of the Verge of the Wheel; for he will
find

find that every Point of this Verge doth, by the compounding the circular and progreſſive Motions together, move itſelf in a true *Cycloidal* Line, and that, in the Point of touching it, reſteth or ſtandeth ſtill in the Boundary between two ſuch Lines. So that where the Plain and the Wheel is perfectly hard and ſmooth, the Wheel receives no Impediment to its compounded Motion ; but it may be thought that the circular Motion of the Wheel is an Impediment to the progreſſive Motion, becauſe by Means of this Compoſition, the Parts of the Wheel do, in ſeveral Poſitions therein, receive ſeveral Degrees of progreſſive Motion, and ſo ſeem to go, as it were, by Starts, for that the Points, whereby they touch, have no progreſſive Motion at all ; and when they are at the Top, or at the greateſt Diſtance from the Plain, they have a double Velocity forwards, compared to that of the Center, and, in every intermediate Poſition, a differing Degree of Velocity forward. But this is no Impediment at all to the progreſſive Motion of the Whole, each Motion being ſeverally uniform, equal, and continued. For a *Pendulum*, whoſe Weight at the End is a Globe of Lead, or any other ponderous Body, ſuſpended by a String, receives the ſame *impetus* from the Power of Gravity, (which is the ſame in both Caſes) whether this Globe, ſo ſuſpended, be ſuffer'd to vibrate, whilſt it be ſwiftly whirled round upon its Center, or whether it be not ſo whirled at all, the compounding of Motions not at all intermeddling with one another ; but every one keeping its diſtinct *impetus*, as may be eaſily found by Experiment, if Trial be made in the Way I propoſe. Whence I conceive alſo, that the periodical Motion of the Earth, or any other Planet about the Sun, would be the ſame, whether the Body of any of them were gyrated round their own Centers, or not,

and

and whether the *Axis* of that Gyration were at right Angles with the Plain, in which they are mov'd or not, the Motion or Influence of the one not at all interfering, or disturbing that of the other. But this only by the by. However, I think it may be pertinent to be consider'd in the Examination of an *Hypothesis* of *Gravity*, propounded by the learned Dr. *Vossius*, in his lately publish'd *Miscellaneous Treatise*, wherein he lays great Stress upon the Position of the *Axis*, in respect of the Plain of its circular, or direct Motion.

Next, we are to consider, what Impediment to its Motion, a Wheel, thus roll'd upon a Floor, receives from that Floor. There may be two impediments then, that a Wheel, so roll'd, may receive from a Floor according to the Nature thereof. The first and chiefest, is the yielding, or opening of that Floor, by the Weight of the Wheel so rolling and pressing ; and the second, is the sticking and adhering of the Parts of it to the Wheel ; to which two may be referr'd all others, all of which proceed from the yielding or giving Way of the Parts of the Floor, and the not returning again to their bended Posture ; for, if the Floor be perfectly hard (as also the Parts of the Wheel) tho' it be very unequal, yet is there little or no Loss, or considerable Impediment to be accounted for ; for whatever Force is lost, in raising or making a Wheel pass over a Rub, is gain'd again by the Wheel's descending from that Rub, in the same Nature as a Ship on the Sea is promoted by the descending down of a Wave, as much as impeded by its ascending, or a *Pendulum* is promoted by its Descent, as much as impeded by its Ascent.

Nor is the yielding of the Floor any Impediment, if it returns and rises against the Wheel, for the same Reason ; but the yielding, or sinking of the Floor, and its not returning again, is the
great

great Impediment from the Floor ; for fo much
of Motion is loft thereby,   as there is Force requifite to fink fuch a Rut into the faid Floor by
any other Means ; whether by Weight, Preffure
or thrufting direΦly down, or any Ways ob-
liquely.

And it may alfo be calculated, by drawing on
the Wheel, whofe Weight, at the mean Time, finks
the Floor it rolls over.   Either Way it will be eafy
to bring it under Calculation, which is the Defign
of this Difcourfe.

The Second Impediment it receives from a
Floor, or Way, is the fticking and adhering of the
Parts of the Way to it ; for by that Means, there
is a new Force requifite to pull it off, or raife the
hinder Part of the Wheel from the Floor, or Way,
to which it fticks, which is moft confiderable in
moift clayie Ways, and in a broad rimm'd Wheel.
For in fuch Ways, the Wheel doth not only lofe
a Part of its Motion, by the yielding and prefling of
the Clay againft the fore Parts of the Wheel, but
by the cleaving to, and holding of it to the hinder Parts, which makes all Carriages move very
fluggifhly and heavily in fuch Ways.

Thus much I thought neceffary to confider, as
to the Goodnefs or Badnefs of the Floor, or Ways
over which Carriages are to pafs, whereof, in the
general, this may be affirm'd, that the harder the
Ways are, the lefs Impediment they give to the
Motion of Carriages over them ; and the more e-
ven they are, the more equal is the Motion.

Hitherto I have confider'd the Wheel only
as free, and, of itfelf, burthen'd only by its own
Weight. I fhall next confider it as burthen'd by another Weight. There are two Ways then of burthening a Wheel.   The firft is, by laying the Weight
at the Top of it ; the fecond is, by laying it upon
the Center, or Axis of it.

THE

T h e first Way was possibly the first invented, being of great Use for transporting of very great Weights some short Way, and is generally practised for removing of Obelisks, Columns, great Stones, or Great Beams of Timber ; and, for that Use, the Rollers, or Wheels, are generally solid Pieces of hard Timber, cut or turn'd round ; and are very long or broad, call'd Rollers ; this, of all Ways, is the easiest for removing such Weights ; but then they must be continually chang'd by being remov'd from behind the Weight, and plac'd before ; for as they roll forwards upon the Floor, so they roll backwards under the Weight, or rather promote the same with a double Velocity to that of their own upon the Floor. By the Way, it seems very strange, that the *West-Indians*, tho' in their Buildings they made use of such vast Stones, and dragged them on the Ground for so great a Distance, yet that they should not understand the Use of these Wheels, or Rollers, which, Histories say, they did not, they performing those Transportations only, by the main Strength of Men pulling at the Ends of a great Number of Ropes. By this Way, a vast Weight may be moved by a very small Strength, if all Things be hard and smooth, approaching much to the moving of a Bulk upon the Water ; but this being more proper to be inlarged upon under the Head of Strength, and not so adapted for Speed, I shall leave at present, till I speak of that Part.

T h e second Way then of burthening Wheels, is, by resting such Weight upon the Axis, or Center of them ; This may be, and has been practised also two Ways ; that is, either first, by making the Wheel move round upon the Axis fixed to the Carriage ; or, secondly, by fixing the Axis to the Wheel, and making the Axis to turn round in a Socket of the Carriage ; the first of
these

these Ways is new, and has always been the Way
of using Wheels for Chariots, Carts, Waggons, and
such other Kinds of Carriages; the second, is used
in Wheel-Barrows, and such other Carriages and
Uses, where the Wheel runs within the Frame.
Of these two Ways, the last (where it can be applied) is much the best; for that the Axis can be
much better fixed in the Wheel, so as to make it
run true in a Plain; and next, for that the Axis
may be kept more firm and steady to that Motion,
by having the two Ends of the Axis, by Means of
its Gudgeons, kept in the Sockets fitted for it; and
thirdly, because the Gudgeons, halving the
Weight, may be made very much smaller, and so
will not cause a tenth Part of the Friction which
is necessary in the other Way. This second Way,
therefore, is much better accommodated for Speed
than the former, and may also be well enough
contriv'd, to be made applicable to several Sorts of
Carriages fit for that Purpose, of which I shall
hereafter speak.

T H E next Thing to be consider'd, is the Make
of the Wheel itself; which has been several Ways
contriv'd, and made use of in differing Ages of the
World, and for differing Occasions. The first and
most simple, was that which was made of a round
Piece of Timber for Rollers, as I noted before, in
which there seem'd to be little of Art, but only
sawing it off with a Saw; these were of the
smallest Sort, and are still used for Truckles and
smaller Carriages.

T H E Second, was that of a somewhat bigger
Sort, and that was either cut out of a whole
Plank, where it cou'd be procur'd broad enough,
or else was made of two or more Planks join'd together, and fasten'd by two or more, cross Ledges,
and that was call'd *Tympanum,* and the same is
still used for the Carriages of Guns at Sea. The
third

third Way, was of bending a Piece of pliable Timber, as we now do for Hoops, and thereby making the Rim of the Wheel all of one Piece, and fixing the Spokes to it, which were also fix'd into a Nave in the Middle; which Nave was also turn'd and bor'd, as the Naves, we now use, are.

THE laſt, and moſt practicable of all, was that we now uſe, whereof the Rim was made with ſeveral Fellows join'd and yok'd together with Pins, and ſometimes with Joints, and ſtrengthen'd alſo by the Sides with Irons, and, after all, bound round with Iron Streaks and Nails. This Way is uſed for all Sort of Carriages, whether heavier or lighter; and Wheels, thus made, are differenced only by being made either bigger or leſs in Compaſs, or ſtronger and weaker in Subſtance or Bulk; whence they become alſo thicker or thinner, in Breadth or Thickneſs, and alſo heavier or lighter, according to the various Deſigns and Uſes they are apply'd unto; the Circumſtances and Accidents, that concomitate their deſign'd Uſe, beſt directing the Artiſt in the Contrivance of their Form and Make.

I SHALL not now inſiſt upon explaining, which Sort is moſt proper for every of theſe Deſigns, becauſe I ſhall do that under each proper Head; but ſhall only conſider at preſent, which Kind of theſe are beſt for Speed and Celerity, that being the Head I am now explaining.

FOR making of Speed then, thoſe Sorts of Wheels are beſt which are the biggeſt in Circumference or Diameter, becauſe firſt, a much greater Part of the Rim doth bear at once, than in a Wheel of a leſs Circumference; for the Way being always more or leſs yielding, the bigger Wheel ſinks in ſo much leſs to come to its bearing, than the leſſer Wheel, by how much the greater Circle approaches nearer to a ſtreight Line, or the Tangent

gent of the Floor.   Secondly, Becaufe the greater
the Arch, the more eafy is the Rife of the Wheel
over any Irregularity, or Rub in the Way, and
the eafier the Fall, and thereby approaches nearer
to the evening and plaining of the Way, and makes
lefs Inequality in the Draught.   On the Contrary,
the fmaller the Wheel, the worfe, for that it in-
troduces all the contrary Inconveniences.   Thirdly, The larger the Wheel is in Circumference, the
lefs is the Impediment of the rubbing and
wearing.

F o r *Firſt*, the Leaver of the Spoke is fo much
the longer, and fo the Nave will turn fo much the
eafier upon the End of the Axle; the Weight
born, in both Cafes, being the fame, and confequently the Bignefs, both of the one and the other,
needing not to be differing.

Secondly, T h e lighter the Wheel be (provided
it be made ftrong enough to perform the Bufinefs
it is defign'd for) the better it is ; and therefore
all Manner of Contrivance that tends to the making the Wheel ftrong, and yet large and light, is
to be made ufe of, for that thereby a lefs Weight
is neceffary to be moved, and confequently the
fame Strength will have the greater Effect.

Thirdly, T h e lefs rubbing there be of the Axle,
the better it is for this Effect; upon which Account, Steel Axes, and Bell-Metal Sockets, are
much better than Wood, clamped, or fhod with
Iron ; and Gudgeons of hardened Steel, running
in Bell-Metal Sockets, yet much better, if there be
Provifion made to keep out Duft and Dirt, and
conftantly to fupply and feed them with Oil, to
keep them from eating one another; but the beft
Way of all is, to make the Gudgeons run on large
Truckles, which wholly prevents gnawing, rubbing, and fretting.

T h e s e

These are fome of the good Qualifications of Wheels, prepared and adapted for the Defign of Speed, which I am now difcourfing of: There are fome other Qualifications that yet exceed thefe, of which I fhall treat fome other Time, where I fhall have Occafion to apply them.

Having thus far confider'd of the Properties and Qualifications of Wheels, fit for fuch Carriage, I fhall next confider what Kind of Carriage is beft for this Purpofe, and what Number of Wheels are fitteft to be applied.

Firft, For the Properties of the Carriage. That which is of the fmalleft Bulk, and of the lighteft Weight, and of the fimpleft, plaineft, and yet ftrongeft and moft durable Structure, is the beft; provided ftill, that, in every Particular, it be fufficient for performing what is required of it. That Carriage, which is only defign'd for carrying a fingle Man, fhould not be made either large enough, or ftrong enough, or heavy enough, to carry two; that, which can be born by one or two Wheels, fhould not be loaden, or clogged, with two, three, or four. So that upon the whole Matter of the Inftrument, fit for Conveyance of one fingle Perfon, I fee none can be better than a certain Carriage or Chariot, and for the convenient Reception of one Man, and refting or moving upon one fingle Wheel. I do not find this to be in Practice any where, but in *China*, of which there is a fhort Account in *Martinius* his *Atlas Sinicus*. But this is not fo well adapted for Swiftnefs, being moved by the Strength of Men, and, for the moft Part, by one, and fo is only a Chair, or Sedan, with one Man and a Wheel, inftead of a fecond Man; but might be contrived much better, both for Eafe and Speed, if there were two Men made ufe of with one fingle Wheel, which I fhall elfewhere defcribe;

but

but ftill it will come fhort, as to Speed, in Comparifon to one, wherein the Strength of Horfes, or
fome fuch fwift and powerful Mover, is applied
for its Acceleration.

The next Thing then to be confidered, in an
Engine for Speed, is the Application of Strength
for the moving thereof, which is the Life of the
whole ; and without which, all the reft is motion-
lefs. This I fhall difcourfe of the next Time.

[*I do not find any Account, among* Dr. Hook's
Papers, of the Matters here promifed.]

WILLIAM DERHAM.

The Number of Houſes paying Chimney-Money in every County of England *and* Wales, *in the Year* 1685.

Bedfordſhire - - -	12170	Nottingham - -	17554
Berks - - - - -	16906	Oxford - - - - -	19007
Bucks - - - - -	18390	Rutland - - - -	3263
Cambridge - - -	17347	Salop - - - - -	23284
Cheſhire - - - -	24054	Somerſet - - - -	44686
Cornwall - - - -	25374	Suffolk - - - - -	34422
Cumberland - - -	14825	Surrey - - - - -	14273
Derbyſhire - -	21155	Suſſex - - - - -	21537
Devonſhire - - -	56310	Stafford - - - -	23747
Dorſetſhire - - -	21944	Warwick - - - -	21973
Durham - - - -	15984	Wilts - - - - -	27093
Eſſex - - - - - -	34819	Worceſter - - - -	20634
Glouceſterſhire - -	26764	Weſtmorland - -	6501
Hampſhire - - -	26851	York - - - - -	106151
Hertfordſhire - -	16569		
Herefordſhire - -	15006		986765
Huntington - - -	8217		
Kent - - - - - -	29242	Wales - - - - -	42565
Lancaſhire - - -	40202	London - - - - -	30997
Leiceſter - - - -	18702	Middleſex - - -	54287
Lincoln - - - - -	40590	Weſtminſter - - -	14852
Monmouth - - - -	6490	Southwark - - -	19945
Northampton - -	24808	Briſtol - - - - -	5122
Norfolk - - - -	47180		
Northumberland -	22741	Total - - - -	1154533

Experiments

Experiments and Observations for the Improvement of the Barometer, *by* Dr. Hook, *read before the* Royal Society, Feb. 3. 1685-6.

THE Experiments I have now fhewn, are noWays pompous and furprifing. Such poffibly may better fuit a Stage or Theatre, for vulgar Spectators to admire and gaze at, who are moft taken with Shew. But thefe are plain and obvious, and only valuable, as they difcover fome Truth, that may be either ufeful of itfelf to be known, or has a Tendency to the making fome farther Difcovery, or of being ufeful, as preparatory to fome other Experiment or Invention, which may be made or founded thereupon. And indeed the greateft Part of Experiments, if they be not made for fome fuch Defign; and the material Circumftances, ufeful thereunto, diligently enquired after, and ftrictly obferv'd, and brought to a Calculation for that Purpofe, do ferve for little elfe than to hint an Experiment to fome other to try, who may have fome Ufe or Application for it.

THE Experiments, as they have been made, do exhibit the fpecifick Weight of the fluid Bodies; together with their comparative Weight with Water: That thefe three Fluids are in fpecifick Gravity to one another, as follows.

Water, 5997.
So Water to *Mercury*, as 1 to 15.

Spirit of Wine, 5102.
Oil of *Turpentine* to *Mercury*, as 1 to 17¼.

Oil of *Turpentine*, 5209.
Spirit of Wine to *Mercury*, as 1 to 17.

FURTHER

FURTHER Obfervables are,

Firft, THE great Lightnefs of Spirit of Wine, and Oil of *Turpentine,* they being, Spirit of Wine but as 51. Oil of *Turpentine,* 52, whereas common Water is 60; that is, almoft a fixth Part lighter than Water.

Secondly, THE Nearnefs of their fpecifick Gravity to one another, which may be yet made as much nearer, as fhall be requifite, or defired, by the intermingling Water, or Flegm, with the Spirit of Wine; for the Spirit of Wine being lighter, and the Oil of Turpentine heavier, fome Mixture of Water, with the Spirit of Wine, will bring the Spirit of Wine to be as near of the fame Weight, with the Oil of Turpentine, as fhall be required.

Thirdly, THE differing Nature of thefe fo feemingly fimilar Liquors.

Firft, IN that they will not mix with each other, but will float the one upon the other.

Secondly, IN that they will not eafily receive the fame Tincture, but differing; the Spirit of Wine readily imbibing a Red, from Cocheneel, which that, and the Spirit of Turpentine, a Green.

THE Ufe, or Application of thefe Experiments, is in Order to the Solution of this following mechanical *Problem.*

How to make a *Barometer,* or Inftrument, to try and find the Weight of the Air, at all Times, which fhall rife and fall fteadily, and without jumping or ftarting, otherwife than as influenced by the Air, and the hitherto unknown Alterations thereof; whofe Limits, between the greateft and the leaft Height, fhall be 10, 20, 30, 40, 50, or more Feet in Perpendicular; and the Motion, in every

Inch

Inch of the faid Height, as plainly vifible, as the Rifing and Falling of an Inch in the common fingle *Barometer*.

It is about 7 or 8 Years fince I propounded fuch a *Barometer* to this Society; and I cannot expect that many fuch will be made; however, poffibly it might not be amifs, that this Society, or fome curious obferving Perfon, would make one, and diligently remark the Changes and Motions thereof. For it might poffibly difcover fuch Changes and Motions of the Air, as we have hitherto no Notion or Conjecture of; for I did once obferve, that the Wheel-*Barometer*, a little before a great Storm of Thunder, Lightening and Rain, did appear to have a tremulous Motion, as if the Room, or Poft it hung upon, had fhook, when yet the Clouds were but gathering, and were far enough off from this Place, where I obferved it; of which I have, long fince, acquainted this Society, and, I conceive, it may be found in the Journal. But there are many other Changes in the Air, that none of the Inftruments, we yet have, will detect; and therefore there may be Scope enough for Inventions, of other Kinds, to detect them, which may give a farther Light to the Difcovery of that moft fignificant, and moft ufeful, Body of the Air. And tho' poffibly the Invention of a mechanical Inftrument may be looked upon as a trivial Thing, yet, as it may be contrived and applied, it may furnifh us with a new Senfe, by which we may be able to know fome Properties of Bodies, of which we have now no more Notion, than one born blind has of Colours, or one deaf of mufical Sounds; or than the whole World hath ever had, of the differing Gravitation of the Air, before the *Barometer* was invented and obferved.

T H E Reafon of my contriving this Inftrument, was, that I might fhew a Way how the Examination, or weighing of the Air's Preffure, might be carried to the Extreams, or as far as could well be defired; for fo it may be, by this Method, if any one will be at the Charge of making it.

A N D indeed if we confider, and a little more ftrictly examine into the Nature of Things, we fhall find, that moft of the Operations of Nature are out of the Reach of our Senfes, and cannot be plainly, if at all, difcover'd by them, and we are left to guefs at the Confultations and Defigns of the Privy Council of Nature, only by the publick Acts and Effects that are produced thereby; whereas, if we could by Senfe be informed of the Agents, and of the Method or Way of acting, ufed by thofe Agents, we fhould be much better able to give a right Judgment of the Effects.

N o w there is no Method of Information fo certain and infallible, as that of Senfe, if rightly and judicioufly made ufe of. And though the Senfes themfelves are limited in their Power and Extent, when confidered barely in themfelves, as naturally conftituted, yet their Power may be much enlarged, and their Limits much farther extended, by the Helps that Art may afford, and, moft efpecially, by Mechanicks; by Means of which, not only each of them may be made more Powerful in the Difcovery of the proper Objects of thofe feveral Senfes; but each of them may be made a *Genus*, as it were, of new Sorts of Senfe, comprifed under them, of which we have yet no Notion, nor any Senfe or Method of Difcovery; at leaft they are yet unheeded. I might inftance, in the Body of the Air itfelf, but I fhall referve it to another Opportunity.

In

In Air, 13 ʒ, ½ ʒ.
In Water, ⅛ ʒ. *gr.* 83.
In Spirit of Wine, 2 ⅛ ʒ, 28 *gr.*
In Spirit of *Turpentine* 2 ʒ, 2 ʒ, 41 *gr.*

    Air 105 ⅛.
    Water   5 ½,   3 *gr.* ——— 100 ʒ — 3 *gr.*
Spirit of Wine   20,   28 *gr.* ——— 85 ʒ + 2 *gr.*
  Ole. Tereb.   18 ⅛   11 *gr.* ——— 86 ʒ + 4 *gr.*

W H E R E F O R E I find that Spirit of Wine may
eafily be made to be 16 Times lighter than Mercury ; if then the Spirit of Wine be made of this
fpecifick Weight, by intermingling Water with it,
and the Height of the Pipes, or the Cylinder of
Spirit of Wine be defigned to play 32 Foot perpendicular ; then muft the mercurial be 2 Foot
more in Height, than the common *Barometer;*
which I have found fometimes (as particularly on
Wednefday laft) to be 30,6; and confequently the
mercurial Cylinder to counterpoife the Gravity of
the Air, and the Gravity of a Cylinder of 32 Foot
in Height of Spirit of Wine, of fuch a Rectificati-
on as I have fpecified. Now, the Cylinder of the
Spirit of Wine being always the fame, that is,
32 Foot, the Counterpoife to it of Mercury will
be always the fame 2 Foot ; and the Cylinder of
the Air only altering the Cylinder of the Mercury
alfo, that counterbalances that alfo, will only be
alter'd, and that the fame, as in the common *Barometer.* Now if the Oil of Turpentine be $\frac{1}{96}$ Part
lighter than that, then a Cylinder of Mercury $\frac{1}{97}$
fhorter than two Foot, will counterpoife it ; which
is but one Quarter of an Inch Difference in the
counterpoifing Cylinders.

Although

*ALthough I find, by the Minutes of the Royal So-
ciety, that the learned Dr. Slare had, long before the Year* 1677, *ſhewed a* Phoſphorus ; *yet it
being chiefly about this Time, that moſt of the Accounts of the* Phoſphori *were ſent, I therefore chuſe
to inſert here ſuch Preparations as I have of them.
And firſt of the*

W. DERHAM.

Bolognian Phoſphorus.

THIS Stone is found in three Places near the City
of *Bologna*; the firſt is called *Pradalbino*; the
ſecond is a ſmall Brook near the Village *Roncaria*;
the third is call'd *Monte Paterno*, and is moſt noted
for theſe Stones ; not only as having the greateſt
Quantity, but a Sort moſt eaſy to be prepared.
The Ground thereabouts is barren, yielding Pieces
of yellow *Marcaſite* of the Bigneſs of a Nut.

THE propereſt Time to gather it, is after Rain,
when the Surface of the Ground is a little waſh'd
away. It's known by a Glittering (like that of
burniſh'd Silver) which ſurprizes the Eye.

IT was firſt found out by one of that City,
call'd *Vincenzo Caſciarolo*, a Cobler, but ingenious, and a Lover of Chymiſtry; who, trying ſeveral Experiments with theſe Stones, by Chance happened on this Way of preparing them, ſo as to
make them ſhine in the Dark, after they had been
ſome Time expoſed to the Sun.

IT has no certain Figure, ſome being cylindrical, others round or lenticular ; and theſe laſt are
often the beſt, as being moſt ſhining and tranſparent.

IT's

I T's ufually no bigger than an Orange ; and tho' *Licetus* affirms, there never was any greater than that in *Androvandus's Mufæum*, weighing about two Pound and half; yet the Author hath had of five Pound.

I T's very heavy, confidering the Bulk, as being probably compounded of feveral mineral Subftances.

T H E Colour is various, as Afh, Rufty, Sky, Yellow, Earthy and White; but the beft for Ufe are Sky-colour and White.

W H E N it's well prepared, it leaves a Luftre in the Superficies, and is enlightened, not only by the Sun, but the Moon, and a Fire; but by thefe not fo ftrongly, as the Sun.

T H E Light, tho' it appear like a Coal, yet is not fufficient to read with, unlefs applied clofe to the Word.

I T will not retain the Light very long, at one Time, nor its Vertue above five or fix Years.

T H E Preparation is thus: Take a Cylinder, whofe Circumference is about two *Roman* Architect Palms, and $\frac{7}{12}$ (of our Meafure, almoft two Feet) the Height about $\frac{7}{12}$; fpread the Infide of the Cylinder with ftiff Clay, till the Diameter of the Aperture come to be but $\frac{7}{12}$; on the Top of the Cylinder make four equidiftant Notches, about $\frac{1}{12}$ deep, and $\frac{7}{12}$ broad: This being done, take another Cylinder of equal Dimenfions with the former, or fomething taller; at the Bottom, make two Port-holes, oppofite to one another, and capable to receive a Hand; make a Bottom of the fame Clay, which may reflect the Heat. This Veffel being cover'd with a thick Wire Grate, that the Air may eafily pafs through, and the other Part of the Furnace placed upon it; lay upon the Grate fome lighted Charcoal, and then other not lighted, but well charr'd, and free from Earth,

Stones,

Stones, and other fulphurous Matters, breaking the
Coal into Pieces no bigger than a Nut ; when you
have made your Bed, as high as the Notches, put
upon it your Stones, to be calcined, fo clofe, that
they muft touch ; but firft beat fome of the Stones
to Powder, and fearce it in a fine Hair Searce,
that it may come out very fine; when you have
wet your Stones, that are to be calcined, in good
ftrong *Aqua Vitæ*, roll them in that Powder, and
lay them, as before, on the Charcoal, and make
another Bed of Charcoal over them, to the Top
of the Furnace, which you cover with a round
clofe Head   When the Coals are fpent, and the
Stones cool, take the Cruft away from them, and
wrap them in Silk, putting them in a clofe Box,
till you make ufe of them.

I f you would make Figures and Reprefentati-
ons with this Light, as is often done, take the
Cruft, which comes off the Stones, and beat it
fmall, fearcing it as before ; then when you have
made your Figure, or Image, wet it with the
White of an Egg, and fprinkle upon it your fine
Powder, which will fhine like the Stone.

T h i s Sort of Furnace is not abfolutely neceffary, but convenient, as well in determining the
Time, as the Degree of Heat; which, if more,
might diffufe that Luftre which is in the Superficies of the Stones; if lefs, not raife it.

T h e Author, occafionally fpeaking of fhining
Woods, delivers this Rule, for the fure finding of
them. That an Apple-Tree is the beft Wood;
that it muft be very dry, or rotten; that being
fo, and lying under Ground, that Part under
Ground will partake of a fhining Quality, which
will not laft above three Days, nor be recover'd
again, when loft.

Phosphorus Liquidus.

SUme salem alcali *v. g.* cinerum clavellatorum bene purificatum per diversas solutiones & filtrationes, & ab omnibus impuritatibus in unum ; deinde in crucibulo novo ad salem albissimum calcinetur, tum in mortario polito & calido in minutissimas partes teratur ; deinde indatur retortæ vitreæ cum spiritu urinæ rectificatissimo imbibitus, cui applicetur recipiens bene agglutinatus; tum ignis per gradus admoveatur : hic operatione factâ debet pluries cohobari, addito semper novo spiritu urinæ in unaquaque cohobatione, atque sic tandem sal alcali cum spiritu urinæ transit in recipiens in forma butyri antimonii.

Nullius est saporis, lucet tamen scintillatione continua instar luminis stellaris, & est ultra modum volatile ac fortis odoris, quasi sulphuris accensi; ideo conservari debet in vase vitreo clauso, infusa aqua communi desuper, atque tum radios emittit per aquam, & fulgura, quæ totum occupant vitrum quando agitatur ; si enim sit extra aquam in aere libero, evanescit, tantæ extensionis est capax ut lentis magnitudine sufficiat ad illinendum totum corpus, quod luminosum apparebit, quasi igne & flammis circumdatum, absque minima erosione ; nihil aliud accendere potest quantum hucusque scitur nisi pulverem pyrium.

Phosphoros Metallorum.

TAKE *Lapis Smaragdi Mineralis* (such as is found in the Mines of *Saxony*); beat it into a very fine Powder.

I f you ſtrew this, very fine, on a Plate, of any Metal, and in any Figure, and ſet the Plate on hot Coals; in a ſhort Time you will perceive, in the Dark, a Light to ſhine; which will (ſaith my Author) laſt as long as you continue the hot Coals: And if you beat out the Fire, it may do again, for once or twice; but then the Vertue will fade.

Phoſphoros Elementaris, *by* Dr. Brandt *of* Hamburgh.

TAKE a Quantity of Urine (not leſs for one Experiment than 50 or 60 Pails full); let it lie ſteeping in one or more Tubs, or an Hogſhead of oaken Wood, till it putrify and breed Worms, as it will do in 14 or 15 Days. Then, in a large Kettle, ſet ſome of it to boil on a ſtrong Fire, and, as it conſumes and evaporates, pour in more, and ſo on, till, at laſt, the whole Quantity be reduced to a Paſte, or rather a hard Coal, or Cruſt, which it will reſemble; and this may be done in two or three Days, if the Fire be well tended, but elſe it may be doing a Fortnight or more. Then take the ſaid Paſte, or Coal; powder it, and add thereto ſome fair Water, about 15 Fingers high, or four Times as high as the Powder, and boil them together for ₄ of an Hour. Then ſtrain the Liquor and all through a woollen Cloth; that which ſticks behind, may be thrown away, but the Liquor that paſſes, muſt be taken and boil'd till it come to a Salt, which it will be in a few Hours. Then take off the *Caput Mortuum* (which you have at any Apothecary's, being the Remainder of *Aqua Fortis* from Vitriol and Salt of Niter) and add a Pound thereof to half a Pound of the ſaid Salt, both

both of them being firft finely pulverized. And then for 24 Hours fteep'd in the moft rectify'd Spirit of Wine, two or three Fingers high, fo as it will become a Kind of Pap.

THEN evaporate all in warm Sand, and there will remain a red, or reddifh, Salt. Take this Salt, put it into a Retort, and, for the firft Hour, begin with a fmall Fire ; more the next, a greater the 3d, and more the 4th ; and then continue it, as high as you can, for 24 Hours. Sometimes, by the Force of the Fire, 12 Hours proves enough; for when you fee the Recipient white, and fhining with the Fire, and that there are no more Flafhes, or, as it were, Blafts of Wind, coming from Time to Time from the Retort, then the Work is finifhed. And you may, with a Feather, gather the Fire together, or fcrape it off with a Knife, where it fticks.

THE Fire is beft preferved in a Veffel of Lead, clofed up from the Air : But to be feen, 'tis alfo put into a Glafs, in Water, where it will fhine in the Dark, but muft be clofe ftopp'd. Some of this Fire, placed in the Beams of the Sun, will kindle Gun-powder: I faw fome of it, prefs'd with a Quill that was cut, and it fired Gun-powder about it. Mr. *Concle* writ alfo with it on Paper, and the Letters all fhined in the Dark, and when they decayed, the rubbing the Paper, with the Fingers, revived it again, and this after two Days.

MY Author fays, he had once wrapp'd up a Knob in Wax, at *Hanover*, and it being in his Pocket, and he bufy near the Fire, the very Heat fet it in Flame, and burn'd all his Cloaths, and his Fingers alfo; for though he rubbed them in the Dirt, nothing would quench it, unlefs he had had Water ; he was ill for 15 Days, and the Skin came off. You may write herewith on Paper, a Wall, or any Wood, &c.

N. B.

N. B. THAT to make this Fire join in Knobs, you muſt, after gathering it from the Recipient, put it into a Glaſs (like a Urinal) and putting it *in Balneo*, or warm Sand, there will evaporate ſome Humidity that lies within it, and thereupon it will ſtick the better together.

N. B. THE Retort muſt be very well luted, to reſiſt the continued Heat : Take therefore, to 50 Pound of fat Clay, as much white Tartar, as much fine Sand, waſh'd and dry'd, and 1 Pound of Cow's Hair; all theſe, mix'd and beat together, will cloſe it Hermetically.

N. B. THAT, when the Operation is done, you muſt take off the Retort, and ſtop it with ſcme of the ſame Clay, well warmed, immediately, that the Air enter not ; for in Caſe you ſhould leave all to cool, with the Retort on, the Fire, deſired, would retire thereinto.

N. B. THAT ſome do give a little Vent to the Retort, or Recipient, becauſe of the violent Heat in the Operation, but he never does it.

Phoſphoros Baldwini.

REc. Spiritus nitri optimi, qui quodammodo ad flavedinem inclinat. q. pl. hunc mitiga cum dimidia parte *Aq. Fortis*; poſtmodum ſolve in hoc cretam optimam albiſſimam & ſicciſſimam, & quidem tantum quantum hic liquor admittit: unde tandem acquirit odorem ſuavem, fere inſtar olei amygdalarum. Hoc ſolutum filtra, filtratum infunde in cucurbitam, & igne leniſſimo abſtrahe phlegma : fortiore dein urge, ut bene fluat, & quaſi ebulliat : hoc facto, ſine ut ignis extinguatur, exime nitri diſtillati caput mortuum & in aere ſolve; ſolurum in loco calido exſicca, & habebis p. ſe. ſplendens

dens quidpiam. Vel fi vis ut fplendeat in quodam fracto fictili (.˙. pfrrbus ˙.˙) tunc accipe Verdig. & [hanc materiam] pone fuper fruftum fictilis cujuf-dam in fornacem probatoriam, aut fub veteri olla. Da vehementiffimum ignem, ut bene fluat; exime & verte feu move fruftum in omnes partes, ut liquor fluens ubivis fictili adhæreat. Reponas in loco quodam, ubi ab aere fit immune, & habebis quod quæris.

I Shall here infert the preceding Recipe, *as I met with it in* Englifh, *by reafon it contains feveral remarkable Things that are not in the* Latin.

W. Derham.

Baldwyn's Modus præparandi Phofphori Hermetici.

Take *Spiritus Nitri,* about a Pound; put it into a Glafs Body, and put into the fame, as much as you can take up, with the Point of a Knife, of the common powder'd *Creta Alba,* then it will begin to ferment, or hifs; and when it has done Hiffing, put fome more of the fame powder'd *Creta,* and continue to do fo, till it be fatiated; hereupon the faid fermented Spirit, by reafon of precipitating many *Fœces,* is to be filtrated *per Chartam Bibulam,* and afterward diftilled off, by a Retort in Sand, untill it coagulate itfelf, *in Fundo,* into a white Salt : Which muft be kept carefully from the Air, becaufe otherwife it very ea-fily runs into an Oil. Afterwards, when you would prepare it for the *Phofphorus,* there muft be a Proof-Furnace, with a Muffel, well heated, till it be red-hot.

N. B. In the Government of the Fire lies the main Bufinefs; for if the Proof-Furnace be not hot

hot enough, then the Salt flows, or afcends, not orderly high enough; but if it be too hot, then the Sulphurous Niter evaporates; then there is put, of the aforefaid Salt, two Lote (an Ounce) in Proportion of the Space, into a Proof-Pot, (or Crucible, wherein they make Ore to boil) and fet it again into the Proof-Furnace, under the Muffel, and then the Salt doth prefently run into a Water, but foon hardens again, and then runs and mounts up again, that the whole Proof-Pot, in the mean Time, is cover'd; but foon after that, the Gold will more and more confume it felf, that only in the Midft of the Crucible, the Powder 1, 7, 5, 19, 2, remains only with a little Moifture, wherein it muft be well obferv'd, that as foon as the Border of the Crucible is dry, though in the Middle there appear fome Moifture, the fame Crucible be fuddenly taken off, and let cool of itfelf. If the Work fucceeds well, then the Brim will be yellow altogether; which (Firft,) *Ex Aere* attracts the Fire, and in the Dark cafts it off again. (2dly,) In the Night, when you hold it to your warm Body, in your Bed, it fhines. And, (3dly,) When in the Evening you ftrike it with a Brufh, or Feather, or fmall Piece of Wood, fomewhat hard, it caufes very bright, fiery Sparks. But the fame Crucibles will not laft long, becaufe they attract fo much of Air and Moifture, *Magnetice*, and moulder at laft: Therefore, at the Beginning, I fet it in a Pewter Box, covered with Glafs, half the Body cut off from the Neck, and well luted, the fame to make it keep the longer. But if you would have the *Phofphorus* in the Figure of a Star, then you muft not only have the *Sal*, but many Crucibles; and when the fame are prepared, as formerly, then only that, which is yellow and fhining at Top, muft be fcraped off, upon white

Paper,

Paper, till there be a pretty Quantity of the said Powder together, according to the Proportion of the Star intended: Hereupon one takes a small round Looking-Glass, whose Foil is not made of *Mercury*, but *Lead*, in which cut therein a Star; then, after the Powder is mixed with a little white Wax, melted and heated in a Silver Spoon, over the Coal Fire, well stirred, with a little Stick; then this Mixture, while yet melted, is poured on the back Side, or hollow Side of the Glass, (which also must be warm'd, lest it break). Now as soon as the *Phosphorus* is prepared, in Manner aforesaid, then it is to be put into a Pewter, or Silver Box, and the Edges of the Looking-Glass are well secured with Sealing-Wax dropp'd upon it, round about; and then the Wax must be made handsome, and smoothed, and covered with Paper, either blue or gilded.

De Germinatione Metalli.

R EQUIRITUR ad germinationem metalli. 1. Terra apta, in qua fiat germinatio, quæ est regulus stellatus, vel etiam regulus simplex. 2. Color. 3. Humor, quo fit imbibitio.

REGULUS conficitur ex antimonio, nitro, sale communi & tartaro, æqualibus partibus, toties repetita fusione, donec regulus fiat albissimus, instar lunæ. Regulo sive terra philosophica habita, itur postmodum ad praxin sequenti modo. 1. Fiat amalgama terræ philosophicæ & mercurii, qui est humor, ad germinationem metallicam pertinens; in hac unione proportio talis est observanda. Si vis germinationem *solis*, recipe *solis* ʒj, terræ philosophicæ ʒx, fundantur simul & uniantur. Eadem dosis est *martis* germinandi. Argenti vero dosis

ʒj.

ʒj, terræ philofophicæ ʒv, eandemque dofin obtinent *faturnus, jupiter & venus.*

HAC unione facta fequitur cum ea unio mercurialis hoc modo.   *Rec.* Fruftulum terræ philofophicæ, idque craffiufcule contunde, nunquam enim uniretur, fi redigeretur in pulverem.   Huic terræ greffo modo fic contufæ adjicias tantundem mercurii, mifceafq; optime in mortario æneo tamdiu, donec totum fit unitum.   Dehinc accipe vas vitreum oblongi colli fed ventre inftar pilæ rotundo, in fui parte fuperiore recurvum, non autem in fui collo dilatatum, quia ad germinationem requiritur circulatio, non autem fublimatio. In hac pila tumulabis materialia prædicta; inque pila aperta humidum mercurii excrementitium five fuperfluum evaporabis. Facta evaporatione, pila hermetice figilletur; dein ponatur in furno claufo, hypocaufti calore inftructo, inque eo per menfem relinquatur: tum videbis metallum ramufculos furfum emittere jucundos, cavitatem pilæ occupantes. Germinatione facta frangatur vas, & ramufculi e fua terra eradicentur, inque ignem denuo exponantur, ac denique cum aquis cordialibus abluantur, ficcentur, & in vitreo vafe ad ufum ferventur.

PRO regulo etiam fumere potes Electrum, quod fit hoc modo.   *Rec.* Solis ʒij, lunæ ʒiiij, martis ʒiij, veneris ʒiiij, jovis ʒviij, faturni ʒxvj. Primo fundatur ♄, 2 ♃, 3 ♀, 4 ☽, 5 ♂.   Sed adverte, chalybem limatum prius effe debere, & mixtum cum mercurio fublimato & nitro, alias cum reliquis non uniretur: tandem & fol funditur. Atque hoc ex omnibus mixtum conficitur electrum.

HIC pulvis blande admodum purgatus, obftructiones contumaces domat, & vifcera roborat, ideoque in affixo hypochondriaco, hydrope & fimilibus morbis prodeft. Dofis eft a gr. 2. ad 4. in fyrupo, conferva, & aqua appropriata.   Eadem eft dofis electri.

                                      Toge-

Together with the Preparation of Baldwyn's *Phof-phoros, I find that of making what we call* Tin, *or* Latten-Plates; *which probably was communicated by* Baldwyn.

W. DERHAM.

The *Way of making* Latten-Plates.

TAKE tough Iron, that will bear the Hammer well; and having hammer'd it thin, ply it into the Size you would have cut your *Latten* ; then put this Iron into a Mixture of Clay and Water, of a pretty Confiftence, and let it ftand two or three Days ; then take it out and hammer it again, as thin as you will have it for your Purpofe; the aforefaid Mixture, that fticketh between the Iron Leaves, keeping them from being beaten into one another ; then cut thofe iron Leaves afunder, with ftrong Sheers, and throw by the Cuttings, as ufelefs ; then put thefe Iron Leaves into a Mixture of Rye-Meal, coarfly ground, and common Water, pretty thick, the Clay being firft rubbed off, and let them fteep therein four Days ; then take them out, and dip them into a Kettle of melted Tin, but draw them quickly out again ; then put thefe tinn'd Leaves between the Wires of an Iron Bar, made with Wires fit for this Purpofe, that the fuperfluous Tin may run off, into a Pan to receive it underneath. And becaufe the Tin will grow cold at the lower End, and fo thicker, in an Iron, an Inch deep, filled with melted Tin, dip the thicker Ends of your Leaves, one after another, and the hot Tin will melt down the Excefs of Thicknefs, but you muft take them out again quickly ; and, with a woollen Cloath, between

your

your two Fingers, wipe them off beneath; which you will fee to have been done, in all *Latten-Plates,* by certain Strokes appearing at one End. Thefe are made fhining, by rubbing them all over with woollen Rags.

In Dr. Hook's *Diary,* Dec. 26. 1673. *I find this Remark,* viz. *Mr.* Yarrington, *who had feen the* Latten-*making Works, near* Leipfick, *faid, many Plates are beat under the Hammer, at once, like Leaf-Gold, or* Tin-Foil. *The great Difficulty is, how to turn them under the Hammer quick enough.*

W. DERHAM.

The Genuine Recept for making Orvietano.

REc. Fol. Dictamni cretenfis recentior. herb. Cardui benedict. Pulegii regalis Hyperici & Scordii; radi. Ariftolo. long. & rotund. Biftortæ, Tormentillæ, Gentianæ. Imperator. Carlin. Scorzoner. Afclepiad, contrayervæ Valerianæ, Angelicæ veræ, petafitidis, bacc. Lauri & feminis Petrofelini & Dauci cretenfis ana partes æquales; & unicuique lib. pulveris, adde theriacæ Andro veteris & mithridati veri an ʒij Poftea reducatur in electuarium molle cum Extracto Juniperi baccar. vino albo parat. & in mellis cocti confiftentiam reducatur, redacti addendo fub finem pro quaq; libræ electuarii femidrach. vitrioli cyprei in pulv. tenuiff. triti; & carnis viperarum exficcatæ, pro quaque libra, ʒj. Hoc electuarium quotidie bis movere debes, per integrum menfem; deinde ad ufum repone.

This I tranflate out of the Paris Mountebank's *Paper, in* French *more at large.*

N. LE FEBURE.

THIS

THIS is the Secret of *Orvietano*, and it is made by the Heirs of *Heronimo Ferranti*, who was the firſt Inventor of this rare *Recipe*. It is now come, by the Marriage of a Daughter, to the *Contugi*, the famous Mountebank, at preſent, at *Paris*; but it was given by *John Vitrario*, the Succeſſor of *Ferranti*, to the Great Duke of *Tuſcany*, for a Sum of Money; by whom it was ſent, fairly written, and put into a great Box, unto the late Monſieur *de Guiſe*; and by him, as a great Curioſity, to the Duke of *Bouillon*; from whoſe Phyſician, Monſieur *la Febure*, my good Friend and Correſpondent, that had often made Trial thereof, with great Succeſs, I receiv'd it as a choice Secret, at my laſt being in *France*, 1652.

J. EVELYN.

The Virtues.

TO expel Poiſon: Take the Quantity of a Bean, mix'd with Oil Olive, Butter-Milk, or Broth hot; drink three or four Times, till all the Venom be expell'd by Vomit.

After which, let the Patient ſup up a good Draught of Broth, very fat, with an Ounce of *Mel Roſarum*. If any be bitten with a mad Dog, or Serpent, take of *Orvietan*, as before, in Wine; then ſcarify the Bite, and draw Blood, *per cucurbitam*, to which apply *Orvietan*, keeping the Patient waking 12 Hours.

IN Agues, Fevers, Exanthems, and all Contagions, *Rec. Orvietan* in ſome Borage, or Scabious Water, the Weight of a Crown in Gold; but to a Child, in a Fever, cauſed by the Small Pox, not exceeding the Weight of a Bean, taken in White-Wine; the Child well cover'd.

IT

IT preferves from the Peftilence, taking the Quantity of a fmall Button. Taken alfo in Wine, Broth, or a Pill, in the Morning, it corroborates the natural Heat, aids Digeftion, hinders Pains in the Stomach, Difficulty of Refpiration, ftinking Breath; cures cataraćtical Vapours and Diftillati-ons, the Cholick, windy and rhenal Spleen, *Dolores Matricis* (except *in Gravidis*) kills Worms in Children.

FOR Cattle that have Swelling, and Pains in the Belly, 'tis very excellent, giving them a Drench in half a Pint of White-Wine, warm. *Orvietan* will keep 25 Years, and more, in a cold Place; or it may be referved in Powder, and put into a Confiftency, with *Mel Rofarum* at Pleafure.

Ink for the Rolling-Prefs.

THE beft Black is the *German* Black, and comes from *Frankfort*; it looks like Velvet, and ea-fily crumbles betwixt the Fingers, like Chalk. Of this there is a Counterfeit, made of Lees of Wine burn'd, which is full of Gravel, and very pernici-ous to Plates.

TAKE excellent Nutt-Oil, and put a good Quantity thereof into a large Iron Pot (which has a Cover exaćtly fitted to it) fo as to fill it within three or four Fingers Breadth of the Top; cover it, and hang the Pot, or fet it on a Trivet, over a good Fire, till it has boil'd; but have a Care that at firft it boil not over, not yet when it boils; for 'twould indanger the Houfe. Therefore, di-ligently obferve it, and frequently ftir it with an Iron *Spatula*. Then being very hot, kindle it with a Piece of Paper, lighted. Having thus taken Fire, remove it from the Trivet, into the
Chimney-

Chimney-Corner, continually ſtirring it, whilſt it burns; which ought to be for the Space of half an Hour, at leaſt. When you would extinguiſh the Flame, clap the Cover on it, and if it does exactly cover it, you will preſently extinguiſh it, otherwiſe you muſt put a Linnen Cloth likewiſe, that no Air may enter; then let it cool a little, and pour it into a Veſſel, wherein you will preſerve it. This they call the weaker Oil, in Compariſon of the following, which they call the ſtrong Oil.

And this is made by putting freſh and crude Oil into the ſame Pot, and ordering it juſt as you did the weak, only ſuffering it to burn a great deal longer, and ſtirring it often, till it become thick and glewy; ſo that dropping a little of it upon a cold Plate, it may, in a little Time, be drawn out into Threads, like a Syrup. Some Workmen put into it an Onion, or a Cruſt of Bread whilſt it boils, and hold that it helps to cleanſe the Greaſineſs of it.

If it hap that the Fire be too violently taken, caſt in a Quarter of a Pint of crude Oil; but to prevent all Accidents, boil it in an open Court.

This done, grind, of the aforeſaid *German* Black, on a very clean Stone and Mullar, about half a Pound, pouring on it, at ſeveral Times, more or leſs, as you ſee Occaſion, about half a Pint of the weaker Oil (for ſome Blacking will take up more than other ſome) but be extreamly careful, not to pour on too much. After you have thus groſsly ground it over, re-grind it over again, by a little and a little at a Time, till it become very fine; then put it altogether on the Stone, and add to it about the Quantity of a ſmall Hen Egg, of your thicker, or ſtrong Oil, blend them well together, and cover them very cloſe, in a well-glaz'd earthen Pot, to preſerve it well from Duſt, for your Uſe.

N. B.

N. B. FOR Plates that are worn, or not deeply graven, you need not put fo much ftrong Oil into the Ink : Likewife your Black muft be good, and well ground, elfe it will give no good Impreffion, and will quickly wear the Plate. And if the Oils be not burn'd into a due Confiftency, the Black will be left behind, in the Hatches of the Plate, and the Impreffion will be pale, and nothing worth.

J. EVELYN.

Divers curious Recepts, collected by Dr. HOOK.

To give Iron the Colour of Copper.

TAKE one Ounce of Copper Plates, cleanfed in in the Fire; three Ounces of *Aqua Fortis*; diffolve the Copper, and when 'tis cold, ufe it by wafhing your Iron with it, by the Help of a Feather; 'tis prefently cleanfed and fmooth, and will be of a Copper Colour; by much ufing or rubbing, 'twill wear off, but may be renew'd the fame Way.

A Way of gilding with Gold upon Silver.

BEAT a Ducket thin, and diffolve it in two Ounces of *Aqua Regia*; dip clean Rags in it, and let them dry; burn the Rags, and, with the Tinder thereof, rub the Silver with a little Spittle; be fure firft, that the Silver be cleanfed from Greafe.

To

To make Copper into a Metal like Gold.

Rec. Diſtill'd *Verdigreaſe* four Ounces; *Tutiæ Alexandrinæ præparatæ,* two Ounces; *Salt Petre,* one Ounce; *Borax,* half an Ounce; mix all together with Oil, till they be as thick as Pap; then melt it in a Crucible, and pour it into a Fire-Shovel, firſt well warmed.

Memorandum. My Author ſays, That this will not only appear, but work like coarſe Gold; that he ſold it as dear as Silver; that the King of *Poland* had a Service of it, only mixing 15 Ounces of Gold, to 100 Ounces of this Metal.

To whiten Copper throughout.

Take thin Plates of Copper, as thin as a Knife, heat them 6 or 7 Times, and quench them in Water; then melt them, and to each Pound add 4 Ounces of *Salt Petre,* and 4 Ounces of *Arſenick,* well powder'd and mix'd, and firſt melted apart in another Crucible, by gentle Degrees; then take them out, and powder them; then take *Venetian Borax,* and white *Tartar,* of each an Ounce and half; then melt theſe, with the former Powder, in a Crucible, and pour them out into ſome iron Receiver; it will appear as clear as Cryſtal, and is called *Cryſtallinum fixum arſenicum.* Of this clear Matter, broken into little Pieces, throw into the melted Copper (by ſmall Pieces at a Time, ſtaying 5 or 6 Minutes between each Injection) 4 Ounces; when all is thrown in, increaſe the Fire, till all be well melted together for a Quarter of an Hour; then pour it out into an Ingot.

N. B. To

N. B. To make this Matter the more malleable, add a Quarter of a Pound of Silver firſt melted, and the former Metal poured into it, and then proceed *ut ſupra,* where indeed the Cryſtalline Powder ought firſt to be prepar'd.

N. B. Also that this Procefs is not to be done in a cloſe Room, by reaſon of the poiſonous Steams of the *Arſenick.*

To make tranſparent Silver.

Rec. Refin'd Silver, one Ounce; diſſolve it in two Ounces of *Aqua Fortis,* precipitate it with a Pugil of Salt, then ſtrain it through a Paper, and the Remainder melt in a Crucible, for about half an Hour, and pour it out, and 'twill be tranſparent.

Diſſolutions. Gold is diſſolved in *Aqua Regis;* 'tis precipitated with Silver, or ſooner with Quickſilver; all other Metals are diſſolved by *Aqua Fortis;* Silver then is precipitated with Copper; Copper by Iron; Iron by Lead or Tin; Tin by Lead or common Salt. *Aqua Fortis* is made by *Niter, Vitriol* and Sand. *Aqua Regis* is made of *Aqua Fortis* and *Sal Armoniac.* *Sal Armoniac* is made of Camel's Urine, preſs'd out of the Dung; or out of Horſe Urine, preſs'd out of the Dung. Volatile Salt is extracted out of Urine, Blood, Soap, and Hartſhorn.

N. B. After the Diſſolution, there remains a black Sand, the Author ſays 'tis Gold; it may be edulcorated by Water. The firſt Water of the Diſſolution dyed the Hair of my Horſe of a Purple Colour, and Yellow and Black; if there had been more Silver, or the *Aqua Fortis* ſtronger, it had been quite Black; it is apt to burn the Skin, but then did not.   *The*

The Roman Pomade.

TAKE Apples of a good Smell: Pare and core them, and cut each into fix Pieces; then take Hog's Greafe of the Bowels, which has not been melted, wafh it in Orange and Citron Flower Water *aa* ; then add *Cloves, Cinnamon, Galinga, Ligni Santali aa* ʒj. *Ligni Rofarum, Saffafras, Violarum Radicum, Benjamin, Storax Calamita aa* ʒ j. chop all into fmall Pieces, and mingle them with the Apples and the Lard; pour over all, Rofe-water a Finger high, and let it boil on a gentle Fire, till all the Moifture be gone ; then ftrain it whilft hot through a Cloath, and afterwards mix therewith fix Ounces of white Wax melted, and well ftirred together; this muft be done in a new earthen Pot, and while you are ftirring it, yet hot, pour in one after another of Oil of Cinnamon, of Citrons, Oranges, Rofes, and Jafmine, *aa* fix Drops.

To perfume Clothes.

TAKE dry'd Red Rofes, and, to encreafe their Smell, pour on them frefh Rofe-Water, and ftill drying between in the Shade ; then take Cloves, Cinnamon, Spikenard Seed, Storax, Calamita, Benjamin, Violet Roots, Nutmegs, *aa* ʒiij. to a Pound of Rofes; beat them all into fmall Pieces, and mix them with the Rofes, and put them into perfuming Bags.

Cyprefs *Powder for the Hair.*

TAKE Red Rofe Leaves in Powder, wet them as before, add Musk 12 Grains, Civet 10 Grains, Ambergreafe 8 Grains, Cinnamon, and Storax Calamita

lamita *aa* ℈ j. Cloves, ℥ ij. of the Mofs of an Oak, one Pound, well dry'd, and powdered, and fix Times wafhed with Rofe-water as before; then add three Ounces of Violet Roots in Powder, mix all together, and pafs them through a Searce, and ufe it.

To marble a Globe Glafs.

GRIND well on a Stone, *Minium* for Red, *Turmeric*, or rather *Ceruffa Citrina*, for Yellow, *Smalt* for Blue, Verdigreafe for Green, Cerufe, or Chalk, for White. Work each in Oil feparate, and with a Hog's-Hair Pencil, fingle or mix'd as you think fit, fcatter the fame into the Glafs, and roll it, or difpofe the Colours, as you like. Then laft of all, fling a little Mead amongft them, which covers all.

FOR the magick Lanthorn, paint the Glaffes with tranfparent Colours, tempered with Oil of Spike.

To gild Carps, Crawfifh, &c.

WARM an earthen Pot, till it receive as much white Pitch as will ftick round it within ; then ftrew finely powder'd Amber over the white Pitch; when 'tis growing cold, pour into it *Oleum Lini*, three Pound ; *Oleum Terebinth*, one Pound well mixed together. Clofe up all, and boil them an Hour on a gentle Fire : This is a Varnifh. Grind fome of this on a Painter's Stone, throwing to it fine Powder of Pumice-Stone, till it be as thick as ordinary Paint; then take a live Carp, or Craw-fifh, out of the Water, and dry it well with a Linnen Cloath ; then daub it over with this Paint, it will prefently dry, before which fpread your Leaf Gold, and gently prefs it with a foft dry Cloath, and then you may let it go into the Water. For the more this Varnifh is in the Water, the harder it dries and grows, and does the Fifh no Hurt.

Many

Many fuch gilded Fifh are in the Prince of *Sila-caw*'s Garden in *Bohemia*, 18 Leagues from *Prague*; he has 200 thus gilded.

The four Elements put in a Cylindrical Glafs with a Foot.

Spirit of Wine, Oil of Tartar *per deliquium*, Spirit of Turpentine and Antimony grofsly beaten : Take of each an equal Quantity, and no two of thefe will mix.

To Foil Glafs.

TAKE a Sheet of *Mufcovy*, or other Glafs, as big as convenient, and as thin as poffibly it can be made : Get alfo fome *Tin-Foil*, and laying it upon a Sheet of very fine Paper, moft curioufly fleeked, that alfo being laid upon a Plain that is exactly plain and fmooth ; then with a clean Cloath, or Piece of Leather, make your *Tin-Foil* clean, and to lie very fmooth, that there may be no Wrinkles in it ; this done, put on a little Quickfilver, and rub it upon it, with a Cloath, or Piece of Leather, fo long, until it be all Black therewith ; then with a Cloth rub that alfo clean off ; this done, put on as much Quickfilver as will cover the *Tin-Foil* all over ; then upon that, as clofe to it as poffibly you can, flide on the *Mufcovy* Glafs, fhoving off as much of the *Mercury* as you can : This done, c.ap down the other half Sheet upon it, which muft be exceeding fine, and moft exactly polifh'd ; upon this lay a Plane, that is very fmooth, left otherwife it caufe Wrinkles ; then prefs it, fo as it may be plain, for 12 Hours; then take it out, and let it ftand, or hang upright, fo as it may fend away the loofe Quickfilver ; afterwards order it as you pleafe.

After this Manner all Sorts of Glafs are foiled.

A Dif-

A Discourse of Mr. John Caswell, *late* Savilian Professor of Oxford, *concerning the going back of the Shadow on a Sun-Dial. Read at a Meeting of the Philosophical Society, at* Oxford, June *the* 22*d.* 1686.

UPON reading the Minutes of the *Dublin* Society, of *Mar.* 1. that Mr. *Tolet* had discours'd of the Shadows going twice Forward, and twice Backward, in the same Day, in a Place of the *Torrid Zone* : It was desired by our Members, then present, that I would take it upon me, to explain, at our next Meeting, how this might be. In answer thereto, I have shewn, in the following Discourse, how the Shadow of a Stile, perpendicular to the Horizon, does go Backward in the *Torrid Zone*, but not of those Stiles that point to the Pole, as it is in Common Dials ; also how, by directing the Stile betwixt the Tropicks, the Shadow may go back on Horizontal Dials in all Latitudes, and in all other Plains, if the Sun does not leave them too soon ; together with the Calculation of the Time, and Quantity, of the Shadow's Regreſſion, according to any given Situation of the Stile and Plane.

By a Stile, I understand a streight Line insisting on a Plane, and casting a Shadow thereon.

A perpendicular Stile, I call that which is perpendicular to the Plane ; an oblique Stile, which is oblique.

WHEN I mention a Stile, without distinguishing perpendicular or oblique, it is to be understood of either.

By the Meridian of the Plane, I mean a great Circle drawn thro' the Pole of the World, and Poles of the Plane.

SUPPOSE

SUPPOSE a Circle deſcrib'd on the Plane from
the Foot of the Stile (*i. e.* the Point where it cuts
the Plane), as a Center: The Way of the Shadow
I reckon on the Circumference of this Circle: And
Note, when the Shadow goes one Way round this
Circle, without any Change, during one Day,
I ſay, 'tis wholly Direct: But if it changes its
Courſe, the firſt Motion it takes before the Change,
I call Regreſſion, or Backward; and the ſecond
Motion I call Progreſſion, or Forward; for 'tis
the firſt Motion that I conceive contrary to what
is uſual, and which I therefore call Retrograde,
rather than Direct. In this Senſe the Shadow may
be twice Retrograde, and once Direct in the ſame
Day, as ſhall be demonſtrated.

Prop. I. THE Shadow of the Stile, on the Plane,
is the common Section of the Plane, with a great
Circle drawn thro' the Sun and Stile.

Prop. II. THE Semidiameter of the Earth is
inſenſible, in reſpect of the vaſt Diſtance of the
Sun from us; therefore the Foot of the Stile,
which is really at the Surface of the Earth, may
be ſuppos'd the Center of the Earth; and conſequently the Plane of the Dial may be taken for the
Plane of a great Circle of the Sphere, parallel
thereto.

Prop. III. THE Shadow cannot go Backward
(in the ſame Day, and ſo underſtand in the following), if the Stile continu'd does cut the Plane
of the Diurnal Circle, (*i. e.* which the Sun de-
ſcribes in the Heavens, and which is otherwiſe
call'd the Sun's Parallel, or a Parallel to the Æ-
quator) in, or within its Perimeter, becauſe the
Shadow is always in a Plane, drawn thro' the Sun
and Stile, if the Point of Section is in, or within
the Diurnal Circle's Perimeter, becauſe the Sun
goe conſtantly Forward, ſo will the Shadow.

Cor. I.

Cor. I. I F the Stile be the Axis of the World, the Shadow cannot go back: For the Axis cuts all the Parallels of the Æquator in their Centers; therefore in no Latitude can a Plane and Stile be plac'd, that the Shadow, which shews the Hour, with its whole Length, may go backward; only a Stile may be so plac'd, that its Shadow may go backward, and a Nodus therein shew the Hour.

Cor. II. T H E Shadow cannot be made to go backward, on either of the Æquinoctional Days, for then the Sun's Diurnal Circle, being a great Circle, is cut by the Stile, thro' the Center.

Cor. III. T H E Shadow cannot go back, if the Stile point without the Tropicks; for then it will cut the Planes of all the Diurnal Circles within their Perimeters.

Prop. IV. I F the Stile cut the Plane of the Diurnal Circle, without its Perimeter, the Shadow will go forward and backward in 24 Hours; provided the Sun shine, a sufficient Part of the 24 Hours, on the Plane. For suppose P, the Point

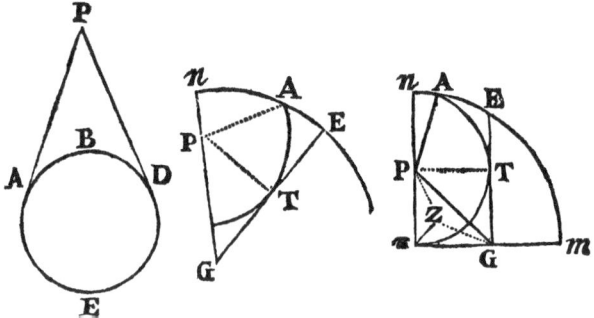

where the Stile cuts the Plane of the Diurnal Circle; from P draw two Lines touching this Circle in A D; 'tis evident the Shadow goes one Way, while

while the Sun paffes the Arc A B D; but the contrary Way, while it paffes D E A.

Cor. I ғ the Arc which the Sun defcribes, while it fhines on the Plane, be Part of **ABD**, and Part of **D E A,** the Shadow will go forward and backward.

Prop. V. I ғ from that Point in the Surface of the Globe, which reprefents the Top of the Stile, two great Circles be drawn, touching the Sun's Parallel; and if the Dial-Plane cut off an Arc of the Parallel, of which all, or part, is vifible, (*i. e.* above the Horizon); if alfo either of the Points of Contact be included within the Extreams of this vifible Arc; then will the Shadow go backward, till the Sun come to the Point of Contact; after which it will go forward, till it come to the other Point of Contact; and then the Shadow will go backward again.

Cor. I. The Shadow, in one Day's Time, in any Latitude, however the Plane and Stile be plac'd, can change its Courfe but twice, (*i. e.*) it may be Retrograde, Direct and Retrograde, but not a fecond Time Direct.

Cor. II. 'T ı s evident, there are in any Latitude innumerable Diverfities of Inclinations of the Dial-Plane to the Horizon, and of the Stile to the Plane; alfo of Declinations of both Stile and Plane from the Meridian, whereby the Shadow may be made to go backward.

Prop. VI. T ʜ ᴇ Shadow cannot go back, while the Sun is nearer the *Æquator,* than the Top of the Stile to the *Æquator.* And this holds good, whether the Sun and Top of the Stile be on the fame, or different Sides of the *Æquator.* For the Planes of all Diurnal Circles, which are nearer to the

the *Æquator*, than the Top of the Stile, are cut by the Stile within their Perimeters, becaufe the Stile paffes thro' the Center of the Sphere.

Cor. The nearer the Stile points to the *Æquator*, the more Days in the Year will the Shadow go back ; but then, in any one Day, it will go back the lefs, *cæteris paribus*.

Prop. VII. The Shadow of a Stile pointing to any one Place of the Heavens, betwixt the *Æquator* and either *Tropick*, will go back all thofe Days, wherein the Sun's Parallel is farther from the *Æquator*, than the Top of the Stile is from the *Æquator*. And this holds, whether the Sun, and the Top of the Stile, be on the fame, or different Sides of the *Æquator:* But with this *Provifo*, in both Cafes, that the Point of Contact lie in the vifible Diurnal Arc above the Plane. For Example, In our Latitude of *Oxford*, if a Stile Point as far North, as is the Beginning of the Sign *Taurus* ; then the Shadow will be Retrograde every Day, while the Sun is paffing thro' *Taurus*, *Gemini*, *Cancer*, *Leo*. But while the Sun paffes through *Virgo* and *Libra*, the Shadow is only Direct, or Forward; then in paffing thro' *Scorpio*, *Sagittarius*, *Capricorn* and *Aquarius*, 'tis Retrograde again : But with this Difference, that if the Shadow is twice Retrograde, any Day, while the Sun runs a Northern Parallel, then it will not be Retrograde once, in the Day of the Oppofite Southern Parallel. But if the Shadow is once Retrograde, in a Day of the Northern Parallel, then it will alfo be once Retrograde, in the Day of the Oppofite Parallel.

Prop. VIII. If a perpendicular Stile point any where in the *Æquator*, the Shadow cannot go back any Day in the whole Year; for all the

Points of Contact lie in the Dial-Plane, to wit, there, where 'tis crofs'd by the Diurnal Circles.

Prop. IX. I f an Oblique Stile point to the common Section of the *Æquator*, and the *Meridian* of the Plane; the Shadow will be Retrograde, during that half Year, while the Sun has Declination toward that Pole, which is elevated above the Plane. The other half Year, while the Sun is toward the deprefs'd Pole, the Shadow will be only Direct; for in the firft half Year, the Points of Contact are above the Plane; in the Second, under.

Prop. X. I f an Oblique Stile point to any Place of the *Æquator* in the Heavens, except there, where the *Meridian* of the Plane crofses the *Æquator*; if withal, a great Circle defcrib'd from the Top of the Stile on the Globe, as a Pole, does crofs the long Diurnal Arcs; (*i. e.*) of thofe Parallels which are toward that Pole of the World, which is elevated above the Plane; I fay, if the faid Circle crofs the long Arcs any where above the Plane, then will the Shadow be Retrograde, thofe Days, in which the Sun defcribes thofe Arcs, and on no other Days; for where the faid Circle crofses the Arcs, are the Points of Contact. For Example; In all Direct South-reclining Planes, above which the North Pole is elevated, the Shadow goes back all our Summer half Year; but it is only Direct all the Winter half Year. But in other Planes, the too great Declination, or Inclination of the Plane, may caufe the Points of Contact to fall under the Plane; fo as only in a fmall Part of Summer, and in no Part of Winter, the Shadow may go back.

PROB.

PROBLEM.

Any Day being given, together with the Latitude of the Place, and the Situation of the Stile and Plane, to calculate the Time and Quantity of the Shadow's Regreſſion on that Day.

Suppose the Dial-Plane were *Horizontal*, and the Stile perpendicular thereto ; becauſe the Stile muſt point within the Tropicks, then this Caſe can happen only in the *Torrid Zone*, yet not in the Æquator (*by Prop.* 8.) ; nor can the Shadow go back, when the Sun is on one Side of the *Æquator*, and the Stile on the other, tho' at leſs Diſtance, becauſe the Points of Contact are both under the *Horizon* ; but then, to recompence this, the Shadow is twice Retrograde every Day ; which to calculate, ſuppoſe Z the Zenith of the Place 10 Degrees North of the Æquator, which is now ſuppos'd the Pole of the Plane, and the Top of the Stile in the Heavens. Suppoſe the Sun in the Tropick of *Cancer*, and P the Pole of the World, T the Point of Contact (*i. e.*) where the great Circle, Z T, drawn from the Top of the Stile, touches the Tropick. Then in the right-angled Triangle ZPT, beſide the Right-angle T, there is given PT, the Sun's Co-declination, and P Z the Colatitude of the Place ; thence may be found the Angle ZPT, (as in *Fig.* 3.) the Sun's Diſtance from the Meridian, when the Shadow begins to change its Courſe ; alſo the Angle P Z T, from which, if you take the Sun's Azimuth from the North, at Riſing, the Reſidue is the Quantity of the Shadow's Regreſſion on the Circle of the Dial-Plane.

What has been ſaid of an *Horizontal* Plane, in the *Torrid Zone*, holds true for any Latitude out of the Torrid Zone, if you incline the former Plane,

till

till its perpendicular Stile point in the Meridian
10 Degrees above the *Æquator*, (*i. e.*) where it
did in the Torrid Zone.

At 5 Hours 42 Minutes in the Morning, the
Sun rifes on this Plane ; at 7 Hours 36 Minutes,
the Shadow changes its Courfe ; fo the Duration of Regreffion is 1 Hour 54 Minutes; the Quantity of Regreffion is 4 Degrees 26 Minutes. And
fo much, and at the fame Diftance from the Meridian, is the Retrogradation in the Afternoon.
Thus it is at the Summer Solftice on this Plane,
but the Regreffion will grow every Day lefs, as
the Sun, in his Diurnal Courfe, comes nearer the
Top of the Stile ; fo as when the Sun runs over
the Stile, then the Regreffion ceafes.

EXAMPLE II.

Suppose an *Horizontal Plane*, in the Latitude
of *Oxford*, and the Sun in *Cancer*, the Stile pointing
to G, (in *Fig.2.*) 10 Degrees in the Meridian, above
the Æquator: From G draw a great Circle, touching the Tropick of *Cancer* in T ; fuppofe this Circle,
(continu'd) to cut the Horizon in E P, to be the
Pole of the World, and that the Circle GP (continued) cuts the Horizon in *n*, and that the Tropick
cuts the Horizon in A. In the right-angled Triangle
GTP, GP and PT are known; thence may be found
the Angle TPG, the Sun's Diftance from the Meridian, when the Shadow changes its Way, and the
Angle PGT ; then in the right-angled Triangle,
P *n* A, P *n* the Height of the Pole, and P A, the
Sun's Co-declination, are given ; thence *n* A may be
found, and in the right-angled Triangle G *n* E,
G *n*, and the Angle E G *n* = P G T, are known,
thence E *n* may be found ; but *n* E — *n* A = A E,
which is the Quantity of the Shadow's Regreffion;
viz. The Shadow, in this Example, begins to
change

change at 7 Hours 36 Minutes, in the Morning, and the Sun rifes at 3 Hours 46 Minutes; therefore the Duration of Regreffion is 3 Hours 50 Minutes, and the Quantity of Regreffion 12 Degrees, 25 Minutes. And fo much, and at the fame Diftance from the Meridian, 'tis again in the Afternoon.

E X A M P L E III.

A PLANE having any given Pofition, fuppofe 14 Degrees Declination, or Azimuth, Weftward, and 71 Degrees Reclination, (*i.e.*) its Pole ϖ (in *Fig.* 3.) 19 Degrees from the Zenith Z, fuppofe of *Oxford*, which is Diftant from P, the Pole of the World, 38 Degrees 14 Minutes; alfo a Stile being any Ways inclin'd to this Plane, yet fo as to point be-twixt the Tropicks; as fuppofe G the Top of the Stile, or Gnomon of the Globe, has 23 Degrees Azimuth, Eaftward, and 42 Degrees Diftance from the Zenith, on any Day propos'd; fuppofe at the Summer Solftice, when the Sun is farther from the Æquator, than G from the Æquator; to find when the Sun fhall begin, and ceafe to fhine on the Plane, and whether the Shadow fhall at all be Retrograde in the Morning; and if fo, how much, and when it fhall be? (Like to which is the Cal-culation for the Evening).

FROM G draw a great Circle, touching the Tropick in T, and cutting the Dial-Circle in E; produce the great Circles ϖ P, ϖ G, till they cut the Dial-Circle in N, M. Suppofe A the Point, where the Tropick cuts the Dial-Circle. Firft inthe Tri-angle GZP, we have GZ, ZP and the Angle GZP, thence we may find the Angle G P Z, and Z G P, and P G; then in the right-angled Triangle P T G, we have P G, and P T, the Sun's Co-declination; thence we may have the Angle G P T, and the An-gle PGT. Then the Angle ZPG + GPT = ZPT,

the

the Diſtance of the Sun from the Meridian, when the Shadow ceaſes to be Retrograde, or firſt changes its Way. In the Triangle ☉ Z G, we have ☉ Z, Z G, and the Angle ☉ Z G ; thence we may find ☉ G, the Angle ☉ G Z, and the Angle G ☉ Z. In the Triangle P ☉ Z, we have Z P, Z ☉, and the Angle ☉ Z P ; thence we may have P ☉, the Angle P ☉ Z, and ☉ P Z ; then the Quadrant ☉ n — ☉ P = P n ; and the Quadrant ☉ m — ☉ G = G m ; and the Angle P ☉ Z + G ☉ Z = P ☉ G = n m. Then the two Right-angles — Z G ☉ — Z G P — P G T = E G m. Then in the right-angled Triangle E G m, we have the Angle E G m, and G m ; thence we may find m E. Then in the right-angled Triangle P n A, we know P n, and P A, the Sun's Co-declination ; thence we may get N A, and the Angle n P A. Then the two right Angles, — ☉ P Z — n P A = Z P A, which is the Time before Noon, at which the Sun begins to ſhine on the Plane. Then Z P A — Z P T = A P T, is the Duration of the Shadow's Regreſſion. If Z P A is not bigger than Z P T, the the Shadow will not be Retrograde at all. Laſtly, m n — m E — n A = A E, the Quantity of Regreſſion.

Dr. H o o k's *Way to find expeditiously and certainly, the true Meridian ; being some-what different from the Method in his Posthumous Works,* pag 361.

P R O V I D E a short Telefcope of 1 Foot, or 18 Inches in Length, fitted with a Glafs-Plate in the Focus ; upon which proper Circles muft be drawn, with the Point of a Diamond, for the Pole Star, and two other Stars not far diftant from the Pole, which is fuppofed to be in the Center of this Glafs. This Telefcope muft be fitted with two Plumb-Lines. Now by this Inftrument, in any fair Night, tho' the Moon fhine, it will be very eafy to difcover the proper Stars, thro' this Tele-fcope, and to fee that each of them be in its pro-per Circle, about the Polar Point: At which Time, the Axis of the Glafs will be in the true Meridian, and, if fitted with the Quadrant, give the Altitude; and the Plumb-Lines being in the Meridian, there may be a Compafs fufpended by them, which will alfo fhew the Variation eafily and certainly. This Inftrument is fufficiently intelligible, without any Scheme, which is therefore omitted.

A N O T H E R Way is wholly new, and the Ob-fervations are made without an Inftrument, and the Refractions of the Air do no ways influence either the Obfervations or Deductions. And that is, by obferving, with Plumb-Lines, or other proper Inftruments, either both at the fame Time, if it may be, or one at one Time, and the other at an-other, with a true Account of the interpofed Time, two Azimuth Lines, in each of which are found two confiderable Stars. By the Help of which two Obfervations, and a true Projection of
the

the Sphere of the Stars, it will be eaſy and obvious, to any Navigator, to find the Latitude of the Place, the Meridian Line, and the Azimuths of the Stars.

> *Theſe two Ways were propoſed to the* Royal Society, April 27. 1687.

W. DERHAM.

An Experiment ſhewn before the Royal Society, Jan. 26. 1689. *by* Dr. HOOK, *of the* Penetration of Dimenſions *in the Mixture of* Vitriol *and fair Water.*

THO' ſeveral Experiments have been made of the diſſolving of ſeveral differing Sorts of Salts, ſucceſſively in the ſame Water, after it has been ſatiated with one particular Salt, ſo as to diſ-ſolve, or take into it no more of that Salt ; yet, in all theſe Experiments, there ſeems not to be any real Penetration of Dimenſions ; nor do I know of any other Experiment of the like Nature, that has been made by any Perſon.   But, I conceive, it is very conſiderable in this, that Water, which has not (by the greateſt Force which has been yet applied to it) been compreſs'd into leſſer Dimenſions, ſhould yet admit a thicker, cloſer, and more ponderous Liquor to penetrate its Dimenſions, without any Preſſure or Force put to aſſiſt the Operation ; and that two Liquors, ſo differing in other Qualities, ſhould ſo readily, and harmoniouſly join and incorporate together.   They differ firſt in Weight ; for I find that the Oil of Vitriol, to the Water, is very near as 9 to 5 ; they differ in the Taſte, the one being the greateſt Acid, we

know,

know, and the other perfectly inſipid; the one very ſluggiſh, and not riſing in Fumes, but with violent Heat; the other evaporating very eaſily. It were too long to mention many other differing Qualifications and Effects; but this is worthy noting, that the Mixture of thoſe two Liquors, both actually cold, produces a very ſtrong actual Heat, and thereby cauſes a Riſing of many ſmall Bubbles out of the Water, and alſo an Expanſion of both, for a Time, as plainly appears; for that as the Mixture grows cold, ſo it retires and ſhrinks into leſſer Dimenſions, as is viſible to the Eye.

Now that I might give a more exact Account of the Succeſs I had, and what was likely to be expected upon another Trial; here I tried the Experiment with all the Care I could. Firſt then we weighed the Bolt-head, and found its Weight, empty, 2085½ Grains. Then we filled it almoſt to the Top of the Neck, with common Water, and found its Weight to be 8775 Grains; from which, taking the Weight of the Bolt-head, we found the Water to weigh 6689½ Grains; then making a Mark on the Neck, at the Top of the Water, we poured out ſo much as filled a ſmall Glaſs Cane, and ſet a Mark at the Top of the remaining Water, and found it 18 Inches and a half below the firſt Mark; the Bolt-head, and Water, now weighed 8255 Grains; whence the Weight of the Water, taken out, was 520 Grains. Then pouring off the Water, in the Cane, we filled it with Oil of Vitriol, and pouring it into the Bolt-head, we found it not to fill the former Space, and to make a conſiderable Heat in the Water, and many ſmall Bubbles to riſe: We then weighed it again, and found the Bolt-head, and Mixture, to weigh 9210 Grains; whence we found the Weight of the Oil of Vitriol to be 945 Grains: We let the Mixture ſtand about Half an Hour, by which
Time

Time we found they were fo condenfed, that 5 Inches and half, of the 18 Inches and half, of the Neck, were left empty, which is near a third Part of the Dimenfions of the Oil of Vitriol, that was poured therein; then we filled up the Vacuity, and found it to contain 138 Grains; which compared to the whole Bulk of Water, that fill'd the Bolt-head, is between a 48th and a 49th Part; for as 138 to $6689\frac{1}{7}$, fo 1, to $48\frac{11}{176}$.

FROM which Obfervations I deduce, that in this Experiment there is fomewhat more than a bare Mixture of Fluid with Fluid, as of Water with Water; where tho' they may intimately mix, and temper together, and become one uniform Fluid, yet each of them, and every Part of each, keeps its former Dimenfions and fpecifick Gravities; or of Water with Wine, Ale, or the like infpiffated Liquors; or with faline Solutions, as of Salt, Niter, Allum, Vitriol, &c. In all which, I conceive, that there is nothing but a mixing, tempering, or dilating, as in the Mixture of two Liquors of the fame Kind. Now, as I formerly hinted, I do not at all doubt, but that there may be found many other Liquors which may have the like Effects, one upon the other, upon Mixture; fo that there may be alfo found Inftances of a differing Nature, where the Mixture fhall increafe the Dimenfions of the Particulars, and diminifh the fpecifick Gravity, either of one, or both. But I think there have not yet been produced any Inftances of thefe, or the other Kind, at leaft, I think, they have not yet been proved fuch.

Mr. Waller *recommended the Trial of this Experiment to Mr.* Hawkesbee, *and if the Reader hath a Mind to fee the Succefs thereof, he may find it in the* Philof. Tranf. *of* 1711. N° 331.

WILLIAM DERHAM.

An

An Account of the Plant, call'd Bangue, *before the* Royal Society, Dec. 18. 1689.

I T is a certain Plant which grows very common in *India*, and the Vertues, or Quality thereof, are there very well known ; and the Ufe thereof (tho' the Effects are very ftrange, and, at firft hearing, frightful enough) is very general and frequent ; and the Perfon, from whom I receiv'd it, hath made very many Trials of it, on himfelf, with very good Effect. 'Tis call'd, by the *Moors*, *Gange* ; by the *Chingalefe*, *Comfa*, and by the *Portugals*, *Bangue*. The Dofe of it is about as much as may fill a common Tobacco-Pipe, the Leaves and Seeds being dried firft, and pretty finely powdered. This Powder being chewed and fwallowed, or wafhed down, by a fmall Cup of Water, doth, in a fhort Time, quite take away the Memory and Underftanding ; fo that the Patient underftands not, nor remembereth any Thing that he feeth, heareth, or doth, in that Extafie, but becomes, as it were, a mere Natural, being unable to fpeak a Word of Senfe ; yet is he very merry, and laughs, and fings, and fpeaks Words without any Coherence, not knowing what he faith or doth ; yet is he not giddy, or drunk, but walks and dances, and fheweth many odd Tricks ; after a little Time he falls afleep, and fleepeth very foundly and quietly ; and when he wakes, he finds himfelf mightily refrefh'd, and exceeding hungry. And that which troubled his Stomach, or Head, before he took it, is perfectly carried off without leaving any Ill Symptom, as Giddinefs, Pain in the Head or Stomach, or Defect of Memory of any Thing (befides of what happened) during the Time of its Operation. And

he

he affures me, that he hath often taken it, when he
has found himfelf out of Order, either by drinking
bad Water, or eating of fome Things which have
not agreed with him. He faith, moreover, that
'tis commonly made Ufe of, by the Heathen
Priefts, or rambling Mendicant Heathen Friars,
who will many of them meet together, and every
of them dofe themfelves with this Medicine, and
then ramble feveral Ways, talking they know not
what, pretending after that, they were infpired.
The Plant is fo like to Hemp, in all its Parts, both
Seed, Leaves, Stalk, and Flower, that it may be
faid to be only *Indian* Hemp. Here are divers of
the Seeds, which I intend to try this Spring, to
fee if the Plant can be here produced, and to examine, if it can be raifed, whether it will have the
fame Vertues. Several Trials have been lately
made with fome of this, which I here produce,
but it hath loft its Vertue, producing none of the
Effects before-mentioned; nor had it any other
Operation, good or bad, fince I receiv'd it with
this Account I have related; imagining I had met
with fomewhat like it in *Linfcotten's* Voyages,
which the Reader may perufe at his Leifure.

I HAVE formerly given an Account of the Effects of the Roots of *Hemlock*, accidentally eaten
by fome young Children, which, at firft, had an
Operation on them much of the like Nature with
this Vegetable; and poffibly the laft Effects might
not have been much differing, if they had not
made Ufe of Medicines, to recover them out of the
Trance, before the Period of its Operation, tho'
that be uncertain, and wants Experiences to afcertain it. Whereas this I have here produced, is fo
well known and experimented by Thoufands; and
the Perfon that brought it has fo often experimented it himfelf, that there is no Caufe of Fear, tho'
poffibly there may be of Laughter. It may therefore,

fore, if it can be here produced, poffibly prove as confiderable a Medicine in Drugs, as any that is brought from the *Indies*; and may poffibly be of confiderable Ufe for Lunaticks, or for other Diftempers of the Head and Stomach, for that it feemeth to put a Man into a Dream, or make him afleep, whilft yet he feems to be awake, but at laft ends in a profound Sleep, which rectifies all; whereas Lunaticks are much in the fame Eftate, bnt cannot obtain that, which fhould, and in all Probability would, cure them, and that is a profound and quiet Sleep.

Obfervations about Gems, *and other* valuble Commodities, *extracted by* Dr HOOK, Dec. 15. 1690. *from Captain* Knox's *Journal*; *which I think worth publifhing, by reafon they are Rules obferved at this Day.*

Directions for Knowledge of Bezoar *Stones.*

THE Monky *Bezoars*, which are long, are the beft; thofe, that are rough, prove commonly faulty, breaking with Stones in the Middle; others in Form of Tuns, fomewhat flat, which break in fmaller Stones in the Middle, are better than the rough ones. *Bezoar* is tried fundry Ways, as the rubbing Chalk upon a Paper, then rubbing the Stone upon the Chalk; if it leave an Olive Colour, it is good; alfo touch any with a red-hot Iron, which you fufpect, becaufe their Colour is lighter than ordinarily they ufe to be; and if they fry, like Rofin or Wax, they are naught. Sometimes they are tried by putting them into clear Water; and if
there

there arife upon them fmall white Bubbles, they are good, if none, they are doubtful; the Ufe of the hot Iron is efteemed infallible.

I T is beft to buy *Musk* in the Cod, for fo it will be preferved; that which openeth with a bright Musk Colour, is the beft. When taken out of the Cod, if a little chewed, and rubbed with a Knife on thin Paper, it look fmooth, bright, or yellowifh, 'tis probably good; but if of a Colour, as 'twere mixed with Gravel, 'tis bad; the Good-nefs is beft difcovered by the Scent.

Ambergriece, the beft is Grey. For Trial, if a little be chewed, and yield an odoriferous Flavour, feel-ing, in Subftance, like Bees-Wax, 'tis good, elfe not.

The Names of Precious Stones.

Diamond, Ruby, Saphir, Emerald, Topas, Hy-acinth, Amethyft, Garnat, Chryfolite, Turcois, Agate, Splen, Jafper, Lapis Lazuli, Opal, Vermi-lion, Clyftropic, Cornelian, Onyx, Bezoar.

T H E *Diamond* is the hardeft, and, when cut, the moft beautiful of all Stones: In Knowledge whereof, there is great Difficulty, having a Cruft on them before they are cut; therefore Caution to be ufed in buying them is before-hand, to make a Pattern in Lead; their Waters are White, Brown, Yellow, Blue, Green, and Reddifh; whereof take Notice, rating them according to their Waters; in our Climate, the perfect white Water is moft efteem'd. *Brut-Stone,* or rough and un-cut Stones, are in Va-lue half the Price of cut, or polifh'd Stones; nei-ther too thick, nor too thin in Subftance, is beft. A thick Stone, which is high and narrow, fable, not making a Shew anfwerable to its Weight, muft be valued at lefs than that which is well fpread, hath its Corners perfect, and a pure white Water, with-out Spots or Foulnefs, is called a *Paragon*-Stone,

and

and in full Perfection. Un-cut Stones are diſtin-
guiſhed into two Sorts, thick or pointed, which
are called *Naif*-Stones, and flat Stones; the flat
Stones are to be cut into Roſes, or thin Stones;
the *Naif* into thick Stones; and thoſe rough
Stones, which will bear a good Shape, with leaſt
diminiſhing in cutting, are in beſt Eſteem.

The Names of rough Stones, according to their
Form and Subſtance: The rough Diamonds, that
ſeem greeneſt, prove of a good Water, when cut; thoſe
that ſeem white, when rough, grow bluiſh often,
when cut. ⬦ A Point. △ A half Point. ⬡
A thick Stone. ⬭ A half ground Stone. ⊂⊃
A thin Stone. ⋈ A Roſe Stone, if round; if
long, a *Foſeel.* ⚷ A *Naif.* Care is alſo to be
taken in Choice of rough Diamond, to avoid thoſe
that have Veins; for they will never cut well, but
ſeem as filed with a rough File. For vending,
Stones of ſix Grains, or under, to one Grain and
half, are beſt. *For Trial of a Diamond*; Take a
pointed Diamond, ſuch as *Glaziers* uſe, try it on
any Stone but a Diamond, and it will cut it. The
Diamond that is of a ſandy, or hath any Foulneſs in
it, or is of a blue, brown, or yellow Water, is not
worth half the Price of a perfect Stone, of a white
Water.

For cutting of Diamonds: You muſt never
mould any of them in Sand, or Cuttle Bone, but
you muſt uſe the ſecond Lead to make a Pattern
of, becauſe the firſt will come ſomewhat leſs than
the other; never caſt it off, but of the perfect
Lead; then make a Pattern of it; but firſt weigh
the Lead, and ſet down the Weight; after, form
the Lead to the beſt and moſt advantageous Shape,
for the Stone, then re-weigh the Lead again, and
ſet down the Weight; by which you may ſee
what the Stone will loſe, by cutting to that Shape;
the

the Lead is three times the Weight of the Stone : This is a fure Rule, commonly it lofeth about $\frac{1}{3}$ Part in cutting.

To make Diamonds clean : If you fee a thick Table-Diamond in a Ring, a Jewel, or in a Locket for a Jewel, you muft firft make it clean, either with a little Pumice-Stone, or with a few hot Afhes, or with a little Oil, and boil it, 'twill make it very clean.

Valuation of Diamonds : There is a Rule accurately to be obferv'd; which is, a Stone of one Carack is worth 10 *l.* One of two Caracks is worth $2 \times 2 \times 10$ *l.* $= 40$ *l.* One of three Caracks is $3 \times 3 \times 10$ *l.* $= 90$ *l.* This, for even Caracks, comes neareft the true Value ; but for $\frac{1}{2}$, or $\frac{1}{4}$ of a Carack, tho' a Stone of two Caracks be worth 40 *l.* yet, in this Rule of Reckoning (meaning $\frac{1}{2}$ a Carack fo valued) it is valued but at $\frac{1}{4}$ of a Carack, which is 50 *s.* and one of $\frac{1}{4}$ of a Carack but at $\frac{1}{4}$ of 50 *s.* tho' a fingle Stone, one Quarter of a Grain, or $\frac{1}{2}$, worth 30 *s.* as for Example. You would know what a Stone of fix Grains is worth ; fix Grains is $3\frac{1}{2}$ Caracks ; 3 times 3 is 9, and 9 times 50 *s.* is 22 *l.* 10 *s.* which is the Value of the Stone. So of five Grains, 5 times 5 is 25, and 25 times 12 *s.* 6 *d.* is 15 *l.* 12 *s.* 6 *d.*

To make a Foil for a Diamond. A Foil, to be fet under a thick Table-Diamond, is made with black Ivory and Maftick, picked, and made very clear, with a little Oil of Maftick, to incorporate them. Black Ivory and Turpentine, heated on the Fire, is good, but the former is better. For a thin Table, black Ivory, fcraped very fine, is good; or take a Coal of the faid Ivory, with a little Oil of Maftick, and dry the fame ; or Ivory, with a little Gum ; fair Water, is alfo good.

I f you fell a Diamond, that hath high Biffals, then you may fet it upon full-fcraped Ivory, which graceth the Play of them.

A

A Rose Diamond, that is very thick; it's good to ſet it cloſe upon the Ivory, and it will play very well; or black Velvet is good, under a thin Table Diamond, ſcraped as you do Lint.

There are four Sorts of *Oriental Rubies*; that which is hardeſt, the beſt and faireſt Colour, if it be very fair, and cut Diamond-cut, is no leſs eſteem'd than a Diamond, for the Weight; but 'tis rare to ſee ſuch an one. The ſecond Sort of Rubies is white, oriental, and hard, which alſo is of good Eſteem, if cut of a Diamond-cut, but not of ſo high a Price, as the perfect red Ruby; but yet if it be in Perfection, 'tis very rare, there being few of this Sort. The third Sort is called a *Spinell*, which is ſofter than the former, and of leſs Eſteem, being not ſo hard, nor hath it the Life of the other, nor ſo perfect a Colour: 'Tis naturally ſomewhat greaſy in cutting becauſe of its Softneſs. The fourth is called a *Ballace Ruby*, not ſo much eſteem'd as the *Spinell*, being not ſo well colour'd; 'tis alſo greaſy, and will ſcarce take a Poliſh, and looks like a *Garnet*.

There are three Sorts of *Saphirs*; one perfect blue, and very hard, which if cut of a Diamond-cut, and without *Calcedonie*, is of very good Eſteem. The ſecond is perfect white, and very hard, which if of a Diamond-cut, and without Blemiſh, is likewiſe eſteem'd. The third, call'd *Water Saphirs*, are of ſmall Eſteem, being not ſo hard as the other, and commonly of a dead wateriſh Colour.

A Copy

A Copy of the Account, which Dr. WALLIS *gave to Dr.* BERNARD, *one of the Delegates for Printing, by a Meſſenger ſent from* Oxford *for that Purpoſe, the Delegates having agreed to be determined by his Opinion in the Caſe, at* Serjeant's-Inn, *in* Fleet-ſtreet. Jan. 23. 1691.

Reverend SIR,

IN Anſwer to yours, of *June* 20. concerning the Buſineſs of *Printing,* the brief Hiſtory is this:

As to the Univerſity's Right of Printing all Manner of Books vendible, before our Charter of King *Charles* I. it is not needful to trouble you at preſent; but the Uſe of Printing was firſt brought into *England* by the Univerſity, and at their Charges, and here practiſed many Years, before there was any Printing in *London*; and we have been in the continual Poſſeſſion of it ever ſince, and long before there was any Reſtraint put upon Printing, which was not at all, till Queen *Elizabeth's* Time.

ABOUT 8 *Car.* I. (and by ſeveral Charters ſince) our ancient Right is recognized, and further granted to us; beſides which Charter, Arch-Biſhop *Laud* did procure, from the *Stationers* of *London* (by Indenture under their Seal) a Grant from them of one Copy, for the *Bodleian* Library, of all Books thenceforth to be printed in their Company, in Conſideration of a Leaſe, to them granted, of tranſcribing Copies (in that Library) of Manuſcripts there, for them to print. And Sir *Thomas Bodley* gave to the Company a Piece of Plate of 60 *l.* But this, tho', for ſome Time, whilſt

whilft Arch-Bifhop *Laud* lived, 'twas, in Part, obferved, hath fince been wholly neglected, and
they give us none upon that Account.

THERE was, at the fame Time, an Agreement between the Univerfity, and that Company,
(for three Years) in behalf of the Company, the
King's Printers, and Mr. *Norton*, (with a Cove-
nant to renew at the End of that three Years)
whereby the Univerfity agreed to forbear the Printing of certain Books, and the Company to pay
200 *l.* a Year for fuch Forbearance, which 200 *l.*
was, by Agreement among themfelves, to be raifed in a certain Proportion ; *viz.* So much by the
Company, fo much by the King's Printers, and
fo much by Mr. *Norton.* But as to this Partition
between themfelves, the Univerfity was not concerned. This 200 *l.* was paid for the firft three
Years, and the Agreement renewed, with like Co-
venants, for another three Years, and obferved by
them for fome Time ; but, the Wars coming on,
the Univerfity did ftill forbear Printing, but the
Stationers gave us no Money ; and thus it continued till about the Year 1653. nor would the
Company be prevail'd with, either to renew their
old Agreement, or enter into any new one, to
that Purpofe ; but did enjoy the Benefit of our
Forbearance, without giving us any Confideration
for it.

THE Univerfity thereupon gave Leave to their
Printers (*Litchfield* and *Hall*) to comprint with
them divers beneficial Books, which prefently
brought them to fuch Terms of Agreement, (that
being the only Means to bring them to Reafon)
tho' it was then agreed to forbear, they paying the
Rent of 120 *l.* which Fall of Rent was agreed to,
upon their Complaint of Poverty and Decay of
Trade.

AFTER

AFTER the Return of King *Charles* II. Dr. *Bailey*, when he was Vice-Chancellor, brought it up to the old Rent of 200 *l.* and fo it continued for fome While.

WHEN the Univerfity devolved their Power on Dr. *Fell*, (fince Bifhop of *Oxford*) and fome others, they continued the like Agreement, with the Company, in behalf of themfelves, and fome others concerned with them, which continued for fome Time longer.

BUT after a While, the King's Printers of Bibles, prefuming that we had not Stock enough to comprint Bibles with them, broke off their Agreement, and would pay their Proportion no longer; bidding us print Bibles, if we pleafed, they would give us nothing to forbear.

MEAN while the Company and Mr. *Norton*, being well aware that we might, with a little Stock, be able to do them a Prejudice, by printing Grammars, Almanacks, and School-Books, were willing to continue the Agreement, as to their Proportions.

WHEREUPON the Bifhop and Dr. *Yates*, continuing to pay us 200 *l.* as before, did agree with the Company and Mr. *Norton*, for fo much as their Proportion came to, but did bear the Lofs, out of their own Purfes, of that which the King's Printers were to pay ; and this for divers Years, before they could put themfelves into a Capacity of printing Bibles.

AFTER fome Years, Dr. *Yates* brought into the Stock, for Printing (as I have been told) a Stock of 4 or 5000 *l.* which did inable them to fet upon the printing of Bibles.

THEREUPON the Bifhop and he printed a Bible in *Quarto*, which the King's Printers, being aware of, did print another, juft in the fame Volume, and fold it to Lofs; and did lofe by it, as
themfelves

themfelves did acknowledge, above 500 *l.* defigning, thereby, to break our Defign in Printing, by forcing us to fell fo cheap, as to lofe by it, or elfe to have the Bibles lie upon their Hands unfold; whilft themfelves would make themfelves whole, by fetting a higher Price on Bibles in other Volumes: And thus they threatened to do, with whatever Volumes we fhould print, prefuming that we were not in a Capacity to print in all Volumes.

THE Bifhop and Dr. *Yates,* finding themfelves thus over-reached, found it neceffary to take in with them fome *London Bookfellers,* as well for the better vending of Books, which did already lie upon their Hands, as for the Increafe of their Stock, that they might be in a Capacity to print in other Volumes alfo, which did effectually counterwork that Defign.

HEREUPON they firft took in *Mofes Pitt,* and one other; but finding thefe not enough to do the Work, they further took in Mr. *Parker* and Mr. *Guy;* thofe took off all the Books which the Bifhop and Dr. *Yates* had lying upon their Hands, and did effectually fet upon printing of the Bible in feveral Volumes: With fo much Struggling it was, and with fo great Charges, before we could get into a Capacity to print Bibles, without great Lofs.

THIS Difficulty being thus mafter'd, their next Attempt upon us was by a Suit at the Council-Table, about the Year, as I remember, 1679, which put us to 2 or 300 *l.* Charges; which was born partly by the Univerfity, partly by the Bifhop, and partly by our Printers, endeavouring, thereby, to get us reftrained from printing Bibles at all, or, at leaft, confined only to fome few Sorts; in which Suit, Mr. *Pitt,* Mr. *Parker,* and Mr. *Guy,* were very induftrious, and diligent in folliciting the Bufinefs, retaining and inftructing the Council, and giving

us

us other Affiftance, which we could very ill have fpared; the prefent Bifhop of St. *Afaph*, and other Friends, were likewife affiftant to us.

T H E Iffue of this Suit going for us, their next Attempt was, the fetting a Multitude of Preffes to Work, to print vaft Numbers off, and by felling them cheap, to break our Printers; fo that now the Conteft was, whether fhould print moft, and fell cheapeft; whereby the Price of Bibles, for the Advantage of the Publick, was brought down to lefs than Half of what they were before fold at; and many hundred Thoufands of Bibles, printed and fold, more than otherwife would have been; and our own People at home, and abroad in our Plantations, furnifhed from hence, which before were wont to be furnifhed in vaft Numbers from *Holland*, where Bibles were printed, far more than in *England*, becaufe cheaper; for the King's Printers did not, now, print and fell fewer Bibles, by reafon of ourComprinting,but only they fold them cheaper.

T H E I R next Attempt on us, was a long Suit in Chancery, for two or three Years, to the Charges of a great many hundred Pounds, born as before, partly by the Univerfity, partly by the Bifhop, and partly by our Printers; wherein we thought, the Lord-Keeper *North* bore very hardly upon us (and was afterwards convinced that he had done fo) but did at length difmifs us, to a Trial at Common Law: After which, if there were Occafion, it was to return again to Chancery.

A F T E R this, they vexed us with two Suits at Common Law, which are yet depending; one in the Name of the King's Printers; the other in the Name of the Company; to which we were fain, at great Charges, to put in Pleas, and to have it agreed at the Barr divers Times; but finding the Court inclinable to do us Right, they have, by delaying Proceedings, kept it off from Judgment, and the Suits are ftill depending.   T H E Y

THEY then prevailed with the Bifhop of Oxford to feparate the Interefts; and whereas before, while Dr. *Yates* was alive, they had let the whole to our Printers, at 200 *l.* and left it to them to agree with the Company, upon the Point of Forbearance, who knew, better than we did, how to hold the Company to their Agreement: The Bifhop would let, to our Printers, the Bufinefs of printing Bibles and Common-Prayer Books at Part of that Sum, and agreed with the Stationers for another Part of that Sum, to forbear printing their Copies; and this by Agreement between the Univerfity and the Stationers, for three Years, with a Claufe of Renewal after that Time.

THE Stationers now being got free of our Printers, who knew how to keep them to Terms, (better than we) they broke with us: They paid their Rent for about one Year, but then firft delay'd, and then refus'd to pay their Rent, till there was five Quarters behind, and told us we fhould be paid all the next Term, upon the *Quo Warranto.*

FOR in the Interim of this their Delay, to pay their Rent, they had caufed a *Quo Warranto* to be brought againft the Univerfity, of which they hoped the like Iffue, as of the other *Quo Warranto's* ; towards the obtaining of which, we are told of a Plate of 500 Guineas went one Way, and a Tun of Wine another Way; and 300 *l.* allowed to *Henry Hill* upon his Account, for fecret Service; and of a Bible to be prefented to fomebody (with filver Clafps and Boffes, *&c.*) which coft 60 *l.* the Binding; but thefe being Works of Darknefs, I cannot tell what to fay to them; but this we are fure of, that the *Quo Warranto* was brought, and that 14 of the chief Men of the Company did, at once, attend at the Attorney-General's Chamber, when it was there to be argued; though they
would

would now perfuade us, that it was only *Henry Hill's* Doing.

For this Arrear of Rent we did commence a Suit, (which is, I think, yet depending) but the *Quo Warranto* being then actually brought (which they hoped fhould pay all their Debts) we were advifed, as a quicker Way (they having broken their Articles, by Non-Payment of Rent) to forbear no longer, but comprint upon them, which prefently brought them to Order; and, (notwithftanding the *Quo Warranto* depending) brought down their Money, and would have paid, not only the five Quarter's Arrears (for which we had commenced the Action) but another Quarter's Rent too, which we could not fafely receive, becaufe we had comprinted upon them; but would not pay thofe Arrears, unlefs we would take that further Rent; and fo that Arrear, and all the Rent, ever fince remains unpaid by them to this Day.

'Tis true, that *Parker* and *Guy* did then depofit, with the Vice-Chancellor, Dr. *Ironfide*, that Arrear of 240 *l.* or rather fo much Money inftead thereof, and all the growing Rent ever fince; and alfo, at their own Charge, of 200 *l.* at leaft, maintain that Suit of the *Quo Warranto*; which Kind of Law-Suits were wont to be partly born by the Univerfity, and by the Bifhop, hoping, in Time, to make themfelves whole again from the Company, but (for fome Reafons) cannot do it yet; and never meant, if they continue our Printers, to trouble the Univerfity to get in thofe Arrears, or Charges, becaufe they think they can get it in, eafier than we can, if we do not difable them.

But if we take our Power out of our own Printers Hands, whofe Intereft it is, as well as ours, to preferve it, and put it into the Hands of thofe, whofe Intereft it is to deftroy it, we fhall difable both them and ourfelves for recovering
thofe

thofe Arrears or Charges ; and whatever Agreement we make with them, we may expect (upon the firft Opportunity) to have them broken, as hitherto they have been ; and if we once let fall our Printing, we can never hope to recover it again ; for where fhall we find another Dr. *Yates,* to furnifh us with fuch another Stock, and run through the many Difficulties to re-eftablifh what we now have, and may continue, without Trouble, if we pleafe.

ON the other Hand, I do not know that *Parker* and *Guy,* who are now your Printers, have ever failed in paying you, to a Penny, whatever they promifed ; nor do I find that the Company do charge them to have ever failed in any Agreement made with them, tho' but verbal.

THIS is the Account, which, as to Matter of Fact, I can on the fudden give you, from

S I R,

Yours to ferve you,

JOHN WALLIS.

Dr.

Dr. Hook's *Defcription of fome Inftruments for Sounding the great Depths of the Sea, and bringing Accounts of feveral Kinds from the Bottom of it. Being the Subftance of fome of his Lectures, in* December, 1691.

In the *Philof. Tranfact.* N. 9. and 24. we have a Defcription of an Inftrument, to found the greateft Depths of the Sea : But there were two great Difficulties that attended it : The firft was, That it was neceffary to make the Weight, that was to fink the Ball, of a certain Size and Figure, fo proportioned to the Ball, as that the Velocity of them, downwards, when united, fhould be e-qual to the Velocity of the Ball alone, when it afcended in its Return ; in Order to which, it re-quired to be prepared with Care, and required alfo fome Charge, it being almoft neceffary to make it of Lead, of a certain Weight and Figure. The other was, the Difficulty of difcovering the Ball at the firft Moment of its Return, which was like-wife of abfolute Neceffity ; and it was likewife ne-ceffary to keep the Time moft exactly of its Stay, or Continuance, under the Surface of the Water, by the Vibrations of a Pendulum, held in one's Hand ; for I was inform'd, that, upon Trial, they have, after fome Time, perhaps difcover'd the Ball floating in a Place, where they did not at firft expect it ; and fo that Experiment became infig-nificant, tho' they were at the Charge of lofing the leaden Weight, and had ufed all Diligence to keep the Time, and to watch for the firft Appearance of the Ball.

THIS

THIS Way, which I shall now explain, is freed from all these, save only of finding and recovering the Ball, after it is returned from the Bottom; for I have no Need of proportioning my Weight, provided it be heavy enough to sink, nor of making it of this, or that Figure, or of Lead, or any other Metal, since a Stone, if big enough, of any Shape, will do; nor have I any Need of counting the Time of its being under Water, since it will do as well, if I procure the Ball an Hour after it floats; so that all the Trouble is, the fetching in the Ball, when 'tis discovered, and the letting it into the Water, when it begins to sink.

IT remains therefore only to describe the Means and Way, how this Matter is to be effected, and 'tis, in short, no other than what I then experimented, and gave an Account of, in Writing, to this Society; as, I believe, will appear by the Register of that Time, which was, as well as I can remember, in the Year 1661, or 1662; but because few here, now present, may remember it, I shall now again describe it.

IT consists then of three Parts; the first is a Stone, of a sufficient Bigness, to sink it to the Bottom, how deep soever; and the bigger the Stone be, the more Expedition doth this Messenger make to its Stage. Secondly, of a wooden Ball, well pitched, which is carried down, by the Stone, to the Bottom, which then leaving it, it returns, with Speed, to the Top, and there floats upon the Water, from whence it is to be fetched aboard. Thirdly, of a Cylinder, Cone, or Hyperbolick Trumpet, that is to bring back the Information to what Depth it hath descended; this is fastened to the Ball, in the Manner described in the Figure; and at the Bottom of this is fastened the Cock, or Crook, by which they are both pulled down to the Bottom, and then let loose, as was practised

on

on the former, defcribed in the *Philofophical Tranfactions.*

THE Cylinder, Hyperbolick Trumpet, or Cone, (*Tab.* III. *Fig.* 1.) A B C is to be hollow, made of Tin, or thin Brafs, and fo contrived, as, by a fmall Hole, to receive the Water into it, lefs or more, according to the external Preffure at the Apex A, of the Fluid it defcends in; fo that it will always, by the Quantity of Water contain'd in it, give a true Account of the Preffure of the Water, at the Bottom, which is always proportionate to the Depth of it, below the Surface; this is fhewn by the Compreffion of the Air included, whofe Dimenfions are always in reciprocal Proportion to the Preffure. This is to be found after the Ball is returned from the Bottom, by weighing the Quantity of Water, contain'd in the Cone, or other Receiver, and comparing it with as much Water, as will exactly fill it, or by a Meafure of Capacity; or thirdly, if the Receptacle be perfectly regular, by a gauging Rod fet in its Axis; but the beft, and moft fure Way, I take, to be by Weight. D D is the Ball, made of light Wood, and well pitch'd, and of fufficient Bignefs, to raife up the Cone, with its contained Water, as foon as it is difcharged from the Stone or Weight. K K, which is to be of a Weight fufficient to fink it, and then flip from it, at the Bottom, by Means of the Spring-Hook; E F G the Ring to be hung upon the Hook; F H I the Cord. There is nothing in the Contrivance, but what is eafy to be made, and the Charge will not amount to a Farthing a Trial.

Emif-

Emiſſarius ſecundus ad fundum Abyſſi, ſive
Explorator Diſtantiæ Inanimatus.

T h e Opinions, concerning the Abyſs, ſeem to
have been received, and conveyed to us, from the
firſt and moſt ancient Times of the World.   And
we find that *Ovid*, tho' he ſeems to have underſtood the Earth to be Spherical, yet he, ſpeaking
of the Creation, and firſt Production of Things,
(of which, no doubt, he received his Information
from the Writings of *Moſes*, or ſome other that
had ſeen them) makes the Water to be the loweſt
of all the Elements ——— *Circumfluus humor ultima
poſſedit ſolidumque coercuit Orbem.*   I had no further Intention, but to ſhew, that the Sea was call'd
the Abyſs, and by the Abyſs was meant a Depth,
not poſſible to be ſounded, or meaſured, by the
Power of Art: But it is more properly rendered,
by our *Engliſh* Tranſlation of the Bible, *the Deep,
or the great Deep*, (when the Depth of the Sea is
meant) than by the Abyſs in the *Vulgar* ; yet there
are ſeveral Expreſſions that do ſhew, it was underſtood to ſignify a Depth, that was beyond the Power of Man to meaſure ; and ſo it ſeems to be meant
in the firſt Chapter of *Eccleſiaſticus*, where 'tis ſaid,
*Who hath meaſured the Height of Heaven, the
Breadth of the Earth, or the Deep* ; that is, the
Profundity of the Sea.   And ſo the Expreſſion in
the 37th Chapter of *Job* ſeems to intimate: The
Expreſſions in the Scripture, relating to Phyſical
Matters, being accommodated generally to the
moſt common and believ'd Opinions of Men, concerning them.   Certain it is, that no one, yet,
hath experimentally found what the greateſt
Depth of it is, except only in ſuch Places as are
meaſurable by Lines and a Plumbet, and that, for
the moſt Part, near ſome Land.   The greateſt
that

that I have met with, of that Kind, which I can rely upon, is, what Mr. *John Greaves* relates, that he tried in the Sea. The Paffage is in the 102d Page of his *Pyramidographia. In the Longitude of* 11 *Degrees* (fays he) *and in the Latitude of* 41 *Degrees, having borrowed the Tackling of fix Ships, and, in a calm Day, founded, with a Plumbet of almoft* 20 *Pound Weight, carefully fteering the Boat, and keeping the Plumbet in a juft Perpendicular, at* 1045 Englifh *Fathoms; that is, at about an* Englifh *Mile and a Quarter, in Depth, I could find no Land or Bottom.* Thefe are his Words; but where this Place was, I cannot define, becaufe he does not declare from what Meridian he computes his Longitude: Whether it were in the Ocean, to the Weftward of *Portugal,* or in the *Streights,* about *Merfeilles;* where-ever it were, it was an excellent Place, to have tried many curious and inftructive Experiments, that might be there tried, by fuch as have Opportunity to go that Way again, if it were certainly known. But this Depth is nothing, in Comparifon to what *Hefiod* fuppofes it, or *Tartarus,* which is the fame Abyfs; namely, as deep downwards, as the Heaven is high upwards; and that he afferts to be fo high, that an iron Axe would be 10 Days in falling, before it would touch the Earth; and juft fo long would the Axe be falling, before it would arrive at *Tartarus.* But to leave thefe Poetical Fictions, certain it is, that the Sea is, in fome Parts of it, very deep, and it would afford many ufeful Informations, if Inquiries were carefully made, by Means of my *Explorators,* or *Nuntii Inanimati;* for by fome, or other of them, one might be afcertain'd of divers Things, yet, never known to Mankind. That which I defcribed, the laft Day, was, for meafuring the Diftance, which it would effectually do, were it not for one Objection or two, which

which may poſſibly render its Account diſputable.
The Power of the Spring of the Air, is moſt certainly in reciprocal Proportion to its Dimenſions,
to whatever Bulk the ſame Air be reduced to, by
Compreſſion; 'tis certain alſo, that the Compreſ-
fion, or Truſion, of a fluid Body, is always in Proportion to the Length of the Cylinder compreſ-
fing; ſo that the Power of compreſſing of any one
Cylinder, of a certain Height, being known, the
Power of any other Cylinder, whoſe Length is
given, is eaſily found. The like is to be eſtimated
concerning the Powers of the Reſiſtence of the
Air, if its Power of Expanſion, or Reſiſtence to
Condenſation, be known, for any one Expanſion,
or Dimenſion of it; the Power of Expanſion is
known, for any other Dimenſion of it given. So
that both the Principles, upon which the laſt
Day's Experiment was founded, are undoubtedly
true and genuine, and conſequently the Invention,
thereupon founded, would ſucceed; and then, the
firſt *Nuntius Inanimatus* would be a true *Explorator Diſtantiæ*. But there are two Things that
may make its Information dubious; the firſt is,
the Uncertainty of the Temper, as to Heat and
Cold, in thoſe very deep, ſub-marine Regions.
For we know that Heat does augment the Power
of Expanſion in the Air, and Cold doth diminiſh
it; and therefore it will be uncertain, whether all the Contruſion of the Air, at the Bottom,
be to be aſcribed to the Gravity of the incumbent
Cylinder of Water, or to the Coldneſs of the Water of thoſe Regions, in Part; till therefore the
Temperature of thoſe Regions be known, we cannot poſitively affirm, what Part of its Condenſation
was to be aſcribed to the incumbent cylindrick
Weight of the Water.

Explorator

Explorator Temperamenti.

To know this, I have another Meſſenger, call'd *Explorator Temperamenti,* which ſhall fetch a true Account thereof; and ſo that Objection, or Impediment, would be eaſily enough removed, if Need were.

Bᴜᴛ there is another Objection (which is alſo very material) againſt the aforeſaid Method, and that is this, That, as 'tis true, that if the Water, at the Top and the Bottom of the Sea, were all of an uniform Nature, then the Rule for its Gravitation, or Preſſing, would hold exactly according to the Rules of Proportion, I have before premiſed, and the Deductions therefrom would be indiſputable ; but if there be differing Sorts of Water, in differing Depths, as no one has yet aſcertain'd us of the Contrary, then differing Sorts of Water will give differing Degrees of Gravitation, or Preſſure; and the Proportion I have aſſign'd, for an uniform Cylinder of Water, will no longer be of Uſe; for if the Water in *Specie* be heavier, (as moſt probably it is) then a ſhorter Cylinder of it will have the ſame Power to preſs, that a longer Cylinder hath, of a Water lighter in *Specie* ; ſo that if the Water be twice as heavy, half the Weight will produce the ſame Effect ; and if thrice the Weight, then a third Part of the cylindrick Weight will be only neceſſary ; and if it ſhould be as heavy as Quickſilver, or indeed as heavy as the Stone, or Weight, that ſinks the Ball ; then the *Explorator* will not dive into it at all, but ſtay at the Top of it. It is neceſſary therefore, that we be aſcertain'd of the Nature and Condition of the Water, or Liquor, whatever it be, at the Bottom, or in thoſe lower Regions, at any aſſignable Depths.

Explorator

Explorator Subſtantiæ.

AND for theſe Purpoſes I have other *Explorators*, that ſhall bring me a certain Account,
what Kind of Water, or other Liquor, it is that
poſſeſſes ſuch, or ſuch a Depth, be it 500, or
1000, or 1500, or 2000 Fathom deep, or any o-
ther greater, or leſs, aſſignable Depth; theſe I call
Exploraſores Subſtantiæ, and of theſe I have ſeveral Kinds, according to their ſeveral Employments
and Buſineſs.   There is yet another Scruple that
muſt be removed alſo, and that is, Whether the
Gravitation, towards the Center of the Earth, do
continue the ſame, at any Depth; or whether it
do increaſe or diminiſh, according as the Body is
poſited lower and lower, beneath the Surface of
the Sea; for if Gravity do increaſe, then the Body
will move downwards, or ſink faſter, than at the
Top; and if it decreaſe, it will do the Contrary.
Now there have been many, and, among the reſt,
the incomparable *Verulam,* that have affirmed, that
Stones, *&c.* in the Bottoms of deep Mines, do
weigh much lighter, than at the Top; if ſo, why
may not that be true alſo, of the Depths in the
Sea: However it be, it is deſirable, in Philoſophy,
to be aſcertained, whether it be ſo or not; and if it
be ſo, what the Differences really are ; for which
Purpoſes I have other *Nuntii* or *Explorators,* that
ſhall certainly inform me, concerning thoſe Particulars alſo.   There are many other particular Inquiries, which one would deſire to be aſcertain'd
of, which I ſhall afterwards mention, and alſo furniſh or ſupply Meſſengers, ſufficiently accoutred,
to bring back Informations, certain and inſtructive.
But I ſhall not trouble you with them at preſent;
but if there be an Opportunity of trying theſe I
have named , and many other I could enumerate,
I ſhall

I shall be ready to give my Assistance : They are Experiments indeed, not to be tried in the Presence, or at the Meeting of this Society, but yet they are such, as it were, very desirable, that the Society had a true Account of them ; as there are also Thousands of others, which, it were to be wished, this Society would procure Informations of ; which, I conceive, is in their Power to effect, if due Means and Methods were made use of, for effecting those Ends. The Harvest is great, but the Labourers are few ; and without Hands and Heads too, little can be expected ; and to rely only upon Time and Chance, is, probably the most likely Way to have all our Hopes frustrated.

Explorator Profunditatis.

B U T to leave this Digression, I shall, at present, only describe another Messenger, who is to be *Explorator Profunditatis,* or a true Surveyor of the Distance, which is not at all liable to the Uncertainties of the last, or any other, as I conceive; for be the Heat or Cold, of that Climate, what it will, or whatever the Density or Rarity, whatever the Gravity or Levity of the Water, whatever the gravitating Power, whether the same, greater, or less, whatever the Spring of the Air be, &c. none of these, or any other, that I can think of, will be material, but the Messenger will return, with a true Account of what he was sent to inquire.

T H I S *Explorator* has divers Parts, much the same with the former ; as first, a large Ball of Wood, or (*Tab.* III. *Fig.* 2.) some other convenient Material, which may be able to rise from the Bottom, after the Weight, that sunk it, is separated from it ; this

B all

Ball is marked in the Figure by A A; this has a cylindrick Hole, B B B B, open quite through the Middle of it, that the Water may paſs freely thro' it, as it deſcends to the Bottom; in this I place two Plates, C C, C C, edge-wiſe, to the Paſſage of the Water, which have each a Center-Hole to receive; and hold the Pivots of an Axis F F, ſo as to move freely therein: Upon this Axis are faſtened 4 Vanes, in the Manner as I have formerly deſcrib'd, for Meaſuring the Way of a Ship thro' the Sea; theſe are marked with E E E; this Axis has a Screw Pinion on it at G, which every Revolution turneth one Tooth of a Wheel of Account, H, whoſe Pinion turneth I, whoſe Pinion turneth K, &c. theſe keep a certain Account, how many Revolutions the Vanes do make, in their Paſſage to the Bottom; and the Revolutions do meaſure the Body of Water, they have paſſed thro', in their whole Deſcent; but that the Riſing of the Ball may not cauſe the Vanes to return backwards, I have ſeveral Inventions; that I ſhall mention, at preſent, is very eaſy, namely, a Lid, or Cover to the cylindrick Paſſage, which is ſhut ſo ſoon as ever the Weight leaves the Ball, which I effect by the Spring M, which is kept down cloſe to the Ball, whilſt it is deſcending, but ſprings up ſo ſoon as the Weight is left, it ſhutteth the Cover N, which ſtops the cylindrick Hole.

Dr.

Dr. HOOK's *Lecture, read* Dec. 16. 1691.

IN my preceding Lectures, I have described two of my *Nuntii Inanimati*, or *Exploratores Abyssi*, whose Business it is to bring back a certain Account of the Distance, or Space, between the Top and the Bottom of the Sea, which I made Choice of, in the first Place, to equip, they being previous, and the Forerunners of all the rest. The first of these, tho' it would do well enough in moderate Depths, where there is no great Difference in the Temperature of the Water, as to Heat and Cold, and other Qualities; yet in greater, especially in very profound Depths, I conceive, it may be liable to Uncertainty, for the Causes I did the last Day mention; which to prevent, and obviate any other Cause of Doubt, which I could, or can yet think of, I did contrive the second *Explorator Distantiæ*, which I described the last Day; the Contrivance of which is such, as, I conceive, will most exactly measure the said Distance, and bring back the true Account thereof. The Way I mention'd, the last Day, was contrived only to measure the Length of its Descent; which, I conceive, will be sufficient Assurance of the Extent, or Depth, thereof. However, if any shall desire to be more ascertain'd of the Truth and Exactness thereof, I have contrived a Variation of, or Addition to, the same, which is only another Prismatick Box, or Hole, with the same Kind of Helical Vanes and Wheels of Account, as the former had, which is so adapted to the Float, and contrived, that, all the while the Weight is descending, this additional Way-wifer shall stand still; and so soon as ever the Float is freed from the Weight that sank it, and it begins to ascend, this

doth

doth then begin to move, and fo continues, till it arrive at the Top of the Water: So that as the former did meafure the Length defcended, fo this doth meafure the fame afcended; which if they be found to agree, 'twill be a double Confirmation of the Certainty of the Experiment. I know it will be objected, that this will make the *Apparatus* very chargeable and difficult; and (as feeming complicate) to be apt to be out of Order; and few will ufe the Caution and Circumfpeſtion, that fuch an Inſtrument will neceſſarily require: To which I anfwer, that I can make the whole fo eafy, and obvious, that the whole Inſtrument need not coft above a Crown; and that any one, almoft, fhall be able to make, or to mend it; and any one, that can but write and read, can be able to make Trial therewith, and keep Account thereof; nor will it eafily be fo out of Order, but that it may eafily be mended, and fet to Rights again. This, I conceive, will do; all that needs to be done, to perfect this Enquiry, which being the firft, and principal, I have been the more curious, to obviate all Objeſtions, and to reduce it to as eafy and plain a Way, as can well be defired, confidering the many Difficulties which are to be provided againſt. I have not made a Module of this third, and moft compleat Contrivance of all the three; but I have prepared a Draught, fo that thofe, who underſtood, and remember the Contrivances of the firft and fecond, will eafily comprehend the Fabrick of this.

The

The Third Explorator Diſtantiæ.

A A repreſents the Ball, or Float of Wood, through which is put B B, a Stick fixed on the Top of it, for the more notable Sign, or Signal, (by which to find it, in the Sea, after its Return) but bigger, and more ſubſtantial downwards, that it may be the more fit to hold the Staple, and Hook at the Bottom C C, and likewiſe the Croſs-Piece E E, which paſſes through a Mortice made in it, and is thereby kept at Right-angles with it ; upon the Ends of this Croſs-piece, E E, are fixed two priſmatick Boxes, F F, and G G ; F F is the Box that holds the Vanes and Way-wiſer, made after the ſame Manner, as was that of the ſecond Module, which I ſhew'd the laſt Day, with no o-ther Difference, but that in this Contrivance, the Box is ſhut by the Water, ſo ſoon as ever it begins to aſcend, without any Need of the Spring which I had made in the ſecond ; and that the Box is made to open one Side, the better to fix the Vane and Way-wiſer ; and likewiſe the Inſide of it is ſquare, the better to be kept ſteady in the Water, ſo that it ſhall not be winded, or twiſted by the Helical Vanes ; which it would be more apt to be a little, if the Hollow of it were truly cylindrick. (*Table* IV. *Fig.* 1.) G G is exactly the ſame Kind of priſmatick Box, with Vanes, and Way-wiſer, as the former, but it is perfectly inverted, with reſpect to the former ; for in the former, the Valve, or Lid to cover it, is placed, or fixed by Hinges, to the Top, ſo that the Water ſhuts it, and keeps it ſo, all the while it aſcends. In this, G G the Valve, or Lid, is placed at the Bottom, and remains ſhut all the Way it deſcends ; but ſo ſoon as it begins to aſcend, 'tis opened, and the Vanes are turned by the Boxes paſſing through the Water.

Water. The Contrivance, for the opening and
ſhutting theſe Lids, is by an equal Flat, fixed on
the Axis of each, at Right-angles with them, that
of the aſcending Way-wiſer, G G, is drawn, and
marked with H H in the Figure. Thus, I conceive, I have ſufficiently accoutred my firſt *Explorator*, who is to inform me of the Depth; and is,
indeed, to be the General Poſt that muſt fetch me
all the other Informations I deſire.

The Thermometer, *or* Explorator Tempe- raturæ.

IN the next Place, I deſire to be informed of the
Temperature of thoſe lower Regions, as to
Heat and Cold: And for this Purpoſe I have contriv'd a Thermometer, that ſhall certainly inform
me; this is nothing but a ſmall Bolt-head, filled
up with Spirit of Wine, to a convenient Height
of the Stem, with a ſmall Embolus and Valve; the
Embolus is made ſo, as to be thruſt down the
Neck, as the Spirit of Wine ſhall be contracted
by Cold; and the Valve is to let out the Spirit of
Wine, when it is again expanded with Heat, in
its Aſcent; 'tis very plain, and eaſy to be apprehended, eſpecially when that is viewed, which I
have here provided: It may, poſſibly, be thought
that the great Preſſure, of the incumbent Body of
Water, may contribute ſomewhat to the Contraction, or Shrinking, of the Spirit; but tho' I am
inclin'd to think, that That will not cauſe any ſenſible Variation, yet, to try that, I ſhall ſhew a
Means how it may be diſcovered; which Diſcovery, of it ſelf, will be a Diſcovery very conſiderable, (*Tab.* III. *Fig.*4.) becauſe none of the Ways,
that have hitherto been attempted, have proved
effectual,

effectual, for the Condenfation of any Fluid, by Preffure only, though there have been made many Experiments, by this Society, on Purpofe for fuch a Difcovery.

Explorator Gravitationis.

Next I defire to be informed, whether the Preffure of the Water do exactly keep the Proportion which I have affign'd it : And for this Purpofe, the perforated Cone, defcribed in the firſt *Explorator*, fent down, and brought back with the *Thermometer*, will give an Account thereof; for by the *Thermometer*, (*Table* III. *Fig.* 1.) we fhall be informed, what is the Degree of Cold, and confequently we fhall know, what Part of the Condenfation of the Air, in the Cone, is due to that, and confequently what Part is to be afcribed to the Preffure; and by the Way-wifer, or third *Explorator*, we are affured of the Depth, and confequently we may know, whether thofe do anfwer to each other, according to the Theory, or Propofition affigned.

This I mention, to fhew that no one of the Inftruments, I have already defcribed, or fhall, for the future, explain, are ufelefs, or fuperabundant; for that, before I leave this Subject, I fhall fhew for what peculiar Ufe each of them is principally defigned, tho' many of them will not only ferve for that one, but for the Affiftance of many others; where they will be of as neceffary a Ufe, in Concomitance with others, as they are fingly neceffary for that End, for which they were principally defigned.

I T

I T may poffibly be queried, why I make ufe of Spirit of Wine to fill my Thermometer, and not of Water, or other Liquor: To which I anfwer, That firft I found fo many Trials, which I purpofely made, to perfect that Kind of Thermometers, (of which, I believe, I made the firft that were made in *England*, from the Sight of a very fmall one, brought out of *Italy*, about 30 Years fince, by the Prefident) that this Spirit was the moft fenfible of any Liquor, I could then meet with, of the Degrees of Heat and Cold. And fecondly, becaufe this Liquor was capable of enduring the greateft Degree of Cold, I could give it, by the Means of Salt and Ice, and yet remain'd fluid, without Congelation, but did continue to fhrink to the laft. Now what the Temper of the Sea may be, at thofe vaft Depths, whither this is defign'd to be fent, no Man now living, or ever did live upon the Earth, hath experimentally known, (as I am, with good Reafon, perfuaded). But, by Conjectures, one may be induced to expect, that the Cold fhould be there very predominant, and, in Probability, fuch as would congeal, and turn to Ice, a Body of frefh Water. And 'tis, in Probability, one of the Caufes that the Sea was made to abound with Salts, by the Divine Providence, who adapted every Thing to its proper Ufe and End; for 'tis very hard to fuppofe, that the Heat of the Sun fhould communicate fo powerful an Influence from the Top, or Surface of the Sea, downwards; for the Parts of any uniform Fluid, that are warmer than the reft, are alfo lighter, and confequently will afcend upwards; but that the heated Particles, at the Top, fhould fink, or defcend, 'tis not to be fuppofed. Again, that the Light, and, poffibly, fomewhat of the Heat of the Sun, may be communicated to the Bottom, if the Water be clear, 'tis not to be denied,

denied, but then it muft be fo fmall a Part, of what we fee neceffary, to keep frefh Water from freezing here above; firft, by reafon of the Quantity reflected by the Superficies of the Water; and fecondly, by the Opacities, that muft neceffarily obftruct their Paffage, thro' fo vaft a Thicknefs, that no Part, near the Poles of the Earth, can receive fo little Benefit of thefe two Qualifications of the Sun, as thefe Parts muft needs do. It feems therefore reafonable to me to fuppofe, that where there is fuch a Defect of Heat, Nature does fupply a more copious Quantity of Salt, or fome other fuch Body, as is able to refift Congelation whether Saline or Metallick; as Quickfilver, or fuch like, Time and Experiments may inform us: Which Experiments, how they may be made, I fhall, the next Day, inform you, and furnifh you with fuch Emiffaries, as fhall bring back a true Account of what Kind of Subftance the Mafs of the Sea is compofed, at any affignable Depth, not only at the Bottom, but of any interjacent Part affigned, between the Top and Bottom.

Lecture read Dec. 23. 1691.

I HAVE, in my preceding Lectures, endeavoured to fhew by what Methods, and by what Kind of Inftruments, we may be experimentally afcertain'd of feveral defirable Informations, about the lower Regions of the Abyfs, or Great Deep. As firft, and principally, what the Depth of the Sea may be, in any Place we defire to meafure it; and this by feveral Inftruments of differing Conftruction, and upon different Principles; the laft of which, I conceive, to be fo compleat, and perfect, as to obviate any Objection that can be made

againft

againſt it; as particularly that which was objected the laſt Day, that if the Water ſhould move upwards or downwards, (tho' ſuch Kind of Motions cannot, with any Ground, or Probability, be imagin'd, or ſuppoſed, ſince the Bottom, or Ground, is a Bound to the Water below, and the Superficies, or Air, is a Bound to the Water above; ſo that unleſs there be a Vent one Way, that is downwards into, or out of the Earth, or upwards, into the Air, there can be no Reaſon given why there ſhould be ſuch a Motion) but it may be ſaid, that there may be, in ſome Places, ſome ſuch *Voragoes,* as Father *Kircher* imagines, in his *Mundus Subterraneus*; that is, ſuch ſubterraneous Paſſages, as convey the Water of the Sea from one Place to another: of which Kind he tells us of many, tho', I doubt, it will be difficult to prove any one of them. I know, indeed, that Mr. *Hacluit* hath taken a Paſſage out of *Gerrardus Mercator*'s General Map, which doth hint at ſome ſuch Extravagancies; his Words are theſe:

' Touching the Deſcription of the North Parts,
' I have taken the ſame out of the Voyage of
' *James Crogen,* of *Hartzeron Buske,* which al-
' ledgeth certain Conqueſts, of *Arthur,* King of
' *Britain;* and the moſt Part, and chiefeſt Things
' among the reſt, he learned from a certain Prieſt,
' in the King of *Norway*'s Court, in the Year
' 1364. this Prieſt was deſcended from them,
' which King *Arthur* had ſent to inhabit theſe
' Iſlands; and he reported, that in the Year 1360,
' a certain *Engliſh* Friar, a *Franciſcan,* and a
' Mathematician of *Oxford* (poſſibly he meant
' *Roger Bacon,* or ſome of his Diſciples) who
' leaving them, and paſſing further, by his Magi
' cal Art, deſcribed all thoſe Places that he ſaw,
' and took the Height of them with his Aſtrolobe,
' according to the Form that I (*Gerrard Merca-*
' *tor*)

' *tor*) have fet down in my Map, and as I have
' taken it out of the Account of the aforefaid
' *James Crogen.* He faid, that thofe four In-
' draughts were drawn into an inward Gulf, or
' Whirlpool, with fo great a Force, that the Ships,
' which once entered therein, could, by no Means,
' be driven back again, and that there is never fo
' much Wind, in thofe Parts, as to drive a Corn-
' Mill.

Geraldus Cambrenfis (who flourifhed in the Year
1210. under King *John*) in his Book of the Mi-
racles of *Ireland*, hath certain Words altogether
alike with thefe ; *viz.* ' Not far from thefe Iflands
' (namely the *Hebrides*, &c.) towards the North,
' there is a certain wonderful Whirlpool of the
' Sea, whereunto all the Waves of the Sea, from
' far, have their Courfe and Recourfe, as it were,
' without a Stop ; which (thefe conveying them
' into the fecret Receptacles of Nature) are fwal-
' lowed up, as it were, into a Bottomlefs Pit ;
' and if it chance that any Ship do pafs this Way,
' it is pufhed, and drawn with fuch Violence of
' the Waves, that eftfoones, without Remedy, the
' Force of the Whirlpool devoureth the fame.

' THE Philofophers defcribe four In-draughts of
' this Ocean Sea, in four oppofite Quarters of the
' World ; from whence many do conjecture, that
' as well the Flowing of the Sea, as the Blafts of
' the Wind, have their firft Original.' Thus far is
Mr. *Hacluit*'s Quotation of *Mercator.* Mr. *Hacluit* adds, in the Margin [*There is a notable
Whirlpool on the Coaft of* Norway, *call'd* Maleftrome, *about the Latitude* 68.] The beft Account of this *Maleftrome*, that I can learn, is, that
it is a Circulation of the Water of the Sea, caufed
by fome fubmarine Rocks. But Father *Kircher*,
who is good at Fiction, has found a fubterraneous
Paffage for it, into the End of the *Bothnick Gulf*,

and from thence another, into the *White Sea*, not far from *Archangel*. I grant such a Passage may be possible, but I should be glad to have it proved; or indeed, any one of those many, which *Kircher* has asserted, in his *Mundus Subterraneus*. So that if there be any such Place in the World, it is not yet found out, or proved: And therefore there is no great Cause of supposing many, or making that an Objection against my third *Explorator*, who will perform his Business, tho' that were actually so; that is, tho' the Motion of the Water were directly upwards, or directly downwards; and not only that, but it will also, over and above, tell you, what that Motion is. This is evident, by comparing the Ascent with the Descent, for half the Sum will be the true Depth, and half the Difference will be the Motion of the Water, whether upwards or downwards, which the Way-wisers will certainly inform you of. But this, I suppose will be needless; however, I was willing to remove the Stumbling-Block, tho' it was but a Straw.

Explorator Qualitatum.

TO proceed then, I shall next shew how to fetch a Quantity of Water from the Bottom, or from any intermediate Space, or Distance from the Top.

THIS I perform, by means of a Bucket, the same I have formerly here describ'd, and verified by Trials; or by another Contrivance not much unlike it, which I shall by and by describe. The former Contrivance will serve indifferently, both for fetching the Water from the Bottom, or from any intermediate Part; but for the intermediate Parts, there is an additional Contrivance, or Invention,

vention, for freeing the Float from the defcending Weight, or Stone, after it hath been carried down a certain Number of Fathoms, which the following plain Contrivance will effectually perform, at any determined Diftance, let it be 100, 500, 1000, 1200, 1500, 2000, or more Fathoms required, where there is firft found to be Depth, fufficient for to make fuch Experiments, which is neceffary to be firft well affured of by the third *Explorator Diftantiæ*; becaufe, if the Depth be not fufficient, that is, if the Stone, or defcending Weight, do touch the Bottom, before it hath defcended the defigned Number of Fathoms, it will detain the Float, and not difmifs the *Explorator*, to return with its Meffage. The Reafon of which, you will prefently apprehend, when I have defcrib'd the Invention for the Performance thereof; tho' yet, with a fmall additional Spring, it will ferve for both Purpofes. I make ufe of the third and laft *Explorator* for this Purpofe. I fit to it two Buckets of Wood, made, according to the Contrivance I have formerly defcrib'd*; thefe are faften'd to the lower End of the Stick, which paffeth thro' the Ball, or Float, as I fhew'd the laft Meeting, and the Buckets are fet at Right-angles, to the Bar that carries the Way-wifers, or Menfurators, as appears in the Figure which I have here defign'd, where A A reprefents the Ball, or Float; B B the Stick thruft thro' it; C C C C the Crofs-Bar, for carrying the Way-wifers; D D. D D, the two Buckets, plac'd or fix'd by their Arms E E. E E, to the faid Stick, at Right-angles to the Bar; C C. C C. F F. F F reprefent the Covers at the Top of each; and G G. G G, the Valves, or Shutters for the Bottom; (*Tab.* II.

---

* *See the Defcription of thefe Buckets in* Philofophical Tranfact. N ° *9 and* 24.

Fig. 2.)

Fig. 2.) Thefe being within the Box, or Bucket, cannot be well expreffed by Delineation, but are faintly defign'd by prick'd Lines; and the Defcription and Modules, I formerly made, do make the Defign fufficiently plain. Thefe Valves, or Shutters do ftand open and upright, all the Time that the Float defcends, and the Water paffeth freely through them, changing every Bucket's Length that the *Explorator* defcends; but fo foon as ever it begins to re-afcend, they are prefently clofed, and fhut into them their whole Capacity, fill'd with the Water in which they then are. This being then underftood, for fetching up the Water at the Bottom, how deep foever, there needs no other Contrivance than what I formerly defcrib'd; for fo foon as the Weight doth touch the Bottom, the Float, and all its Furniture, is freed from it, and fo is at Liberty to re-afcend, and carry back with it, what it was defign'd to fetch. But for fetching up the Water from any intermediate Depth, (as at 100, 200, 500, 1000, 1500, *&c.* Fathoms below the Surface) I have invented an eafy Expedient, which is to let go the Weight, that finks the *Explorator*, at any Station of Depth defign'd. I have already explained the *Way-wifer*, or *Menfurator* of the Depth defcended; one of the Wheels of which doth keep Account of every hundred Fathom defcended : Upon this Wheel I put on a fpringing round Plate, with a Hoop about the Edge of it, which hath one Notch in the Circumference, or Hoop; this Notch I can fet againft any Number of the Plate, in the fame Nature as tis common for fetting the Alarm of a Clock, to go off at a certain Time defigned; which, to effect, I have contrived a very eafy Expedient, which the third Figure doth reprefent. (*Table* II. *Fig.* 3.) Suppofe then B B, to reprefent the lower End of the Stick that hath the Way-wifers

and

and Buckets, in the End of which is fixed C *c*, which is a Staple made of a flat Iron Plate; between the Sides of this is faſtened, by a Pin *c*, the Hook *d e* by the End *d* of which, doth hang the Wire of the Weight; this Hook is kept in this Poſture, by a ſmall Piece of Wood or Iron *f g*; the End *f* is cut ſloping, to anſwer the Slope of the End *e*, of the Hook *d*. Now ſo long as the End *g*, of the Trigger (as I may call it) is detain'd within the Hoop of the Wheel of Account *b b*, of the Way-wiſer, ſo long is the Hook, *d e*, kept firm in the Poſture it is here deſigned, and ſo retains, or holds the Float and Furniture faſt to the deſcending Weight; but ſo ſoon as the Way-wiſer has meaſured the Number of Fathoms deſigned, and the Notch in the Hoop be brought to the Place, where the End of the Trigger *g* may ſlip out, the Hook has no longer any Power to hold faſt to the deſcending Weight, but preſently lets it go, and the Float returns, and the Buckets cloſe, and bring back their Bellies full of the Water of that Part; or the Temperature, if the *Thermometer* be hanged to the Stick; or the Preſſure, if the Cone, together alſo with the Degree of Gravitation.

I ѕ н а ʟ ʟ only add one more Enquiry to be reſolved of at preſent, and that is to know, what Alteration ſo great a Condenſation, or Compreſſion, as muſt neceſſarily be cauſed at ſo great a Depth, will be produced in the Body of the Air, ſo condenſed; that is, ſince the Air is but about 7 or 800 Times, at moſt, lighter than Water, and that 2200 Fathoms Preſſure will, according to our Theory, reduce it to as denſe a Body; whether, I ſay, this Condenſation will not actually reduce the whole Body of the Air, ſo condenſed, into perfect Water. This may be eaſily tried, by letting down, with the *Explorator*, a ſmall Glaſs Bolt-head, filled with Air, with the Mouth of the Stem,

Stem, or Neck, turned downwards, and contra-
ćting the End of the ſame, by a Lamp, into a ſmall
Perforation, to let in theWater thereby under the
Air, as it deſcends, and to let out the Water, if the
Air do again expand, as it re-aſcends. This is ſo
eaſy to be apprehended, that I thought it needleſs
to add any Delineation, for the further Explication thereof.

Obſervations of the Lake-Wetter *in* Swedeland, *made in the Year* 1688.

*THERE being ſome Congruity between the fol-
lowing Obſervations, and that which Dr.* Hook
*had ſaid in his Lectures, about ſounding the Sea,
I find that he took the Opportunity to entertain the
Society with the following curious Relations, by
concluding his laſt Lecture with them. But who
the Author of them was, I have not found.*

W. Derham.

' WHEREAS *Olaus Magnus*, and divers o-
' ther Authors have related wonderful
' and unuſual Matters concerning the *Lake-Wetter*
' in *Swedeland*, I thought it worth while, for en-
' quiring, more particularly, concerning the Na-
' ture of it, and the Truth of the Relations, to
' viſit the Place my ſelf, one Summer, whilſt I
' went to the *Medivian Acidulæ*; thereby to be in-
' formed, from the neighbouring Inhabitants, of
' good Repute, of what I ſhould enquire, and of
' what I could not be able to obſerve my ſelf.
' The Sum of which I have here compriſed, that
' it may appear, both whatever is there more
' ſtrange, and alſo that the Truth of Hiſtories
' may

‘ may be diftinguifh’d from Fictions. Geogra-
‘ phers have fo well defcribed the Lake, that ’twill
‘ be loft Labour to add to it.   It extendeth from
‘ *Askerfund*, on the North, to *Jonekopen* towards
‘ the South, 14 *Swedifh* Miles, each of which is
‘ fix *Englifh*, and ten of them make a Degree ;
‘ its greateft Breadth three, in fome Places only
‘ two fuch Miles.   It divides *Gothland* in two
‘ Parts ; that on the Eaft is call’d *Oftrogothia*,
‘ that on the Weft, *Weftrogothia* ; near the Bound
‘ of it is a celebrated Mountain, *Ahme*, or *Ohme*,
‘ and near it the City *Wadftein*, and its Caftle on
‘ the Eaft Side ; and oppofite to it, on the Weft
‘ Side, is the old Town *Hio* ; the Lake, by Rea-
‘ fon of Mountains and Hills that encompafs it,
‘ fome with their Cliffs, others at fome Diftance,
‘ to the Spectators always appears deprefs’d, or funk
‘ into the Earth.   The Depth of it is very differ-
‘ ing, but yet great, in fome Places but fourfcore
‘ Fathoms ; but on the Side of *Oftrogothia*, and in
‘ fome of *Weftrogothia*, no Bottom can be found,
‘ at 300 Fathom deep.   Of this I was affured, by
‘ an Experiment which Mr. *Ericus Simonius*, the
‘ Minifter and *Præpofitus* of *Wadftein*, a Perfon
‘ worthy of Honour and Credit, communicated
‘ to me (he, being by long Experience well in-
‘ formed concerning this Place, was highly affift-
‘ ant to me by his Information) he told me that
‘ not long fince, one *Benedictus Amberri*, a Citi-
‘ zen of *Wadftein*, who founding the *Wetter*, near
‘ the Shoar of the City *Grennen*, with fome hun-
‘ dred Fathoms of Line, hanging an Axe inftead
‘ of a Weight to it ; and upon pulling it up, he
‘ found his Axe loft, and, inftead thereof, a Horfe
‘ Head faft to the End of his Line, but could find
‘ no Bottom.   Such another Abyfs is near the
‘ Cliffs of the Mountain *Ohme*, call’d the *Weft-*
‘ *Wall*, which has eluded the Induftry of all
                                         ‘ that

‘ that have founded it; whence none will approach
‘ that Part, for fear of a Weft Wind, which, ri-
‘ fing fuddenly, would dafh them againft the
‘ Cliff, there being no Anchoring to hinder it.
‘ So alfo on the *Weftrogothian* Side, the Gover-
‘ nor, Count *John Oxenfterne*, defiring to found
‘ the Depth with 300 Fathom of Line, could find
‘ no Bottom ; as the Fifhermen, who made the
‘ Trials, and are yet alive, do teftify.   The Wa-
‘ ter is very clear, as well as deep, fo that a fmall
‘ Piece of Money may be feen to a great Depth.
‘ Mr. *Ericus Simonius* has feen a fmall Piece of
‘ Silver, in a calm Day, 60 Cubits deep; but the
‘ Water, at a greater Depth from the Superficies,
‘ feems tinged with a Kind of Green.   And won-
‘ derful ’tis, that notwithftanding fuch Abundance
‘ of Filth is wafh’d into it from the circumjacent
‘ Hills and Woods, yet the Water fhould not be
‘ fullied.   Tho’ this Lake exceeds moft for Am-
‘ plitude, yet ’tis free from Rocks, and has few
‘ Iflands; the chief of which is *Vifingfoe*, the Seat
‘ formerly of the Counts of *Brahe* : It lies in the
‘ Middle of the Water, between *Grennam* of *Smo-*
‘ *land*, and *Weftrogothia* ; and on the North, op-
‘ pofite to the *Acidulæ Medivienfes*, lies the Ifland
‘ *Rocknens*.   Some few other Iflands, and thofe
‘ very fmall, lie near the Shores ; but the *Wetter*
‘ lying expofed to the Winds, and being encom-
‘ paffed with Mountains, ’tis no Wonder that it
‘ lies feldom quiet, but is continually ruffled with
‘ Storms and copling Seas, which does fufficiently
‘ tofs the Veffels on it ; and this oftentimes hap-
‘ pens fo fuddenly, and unexpected, that its Sur-
‘ face, being as fmooth as a Looking-glafs, becomes
‘ to be fecretly moved, before any the leaft Breath
‘ of Air can be felt; which feems to be caufed by
‘ a Storm in fome other Part of it, that communi-
‘ cates it under Water, before it can arrive above
‘ by

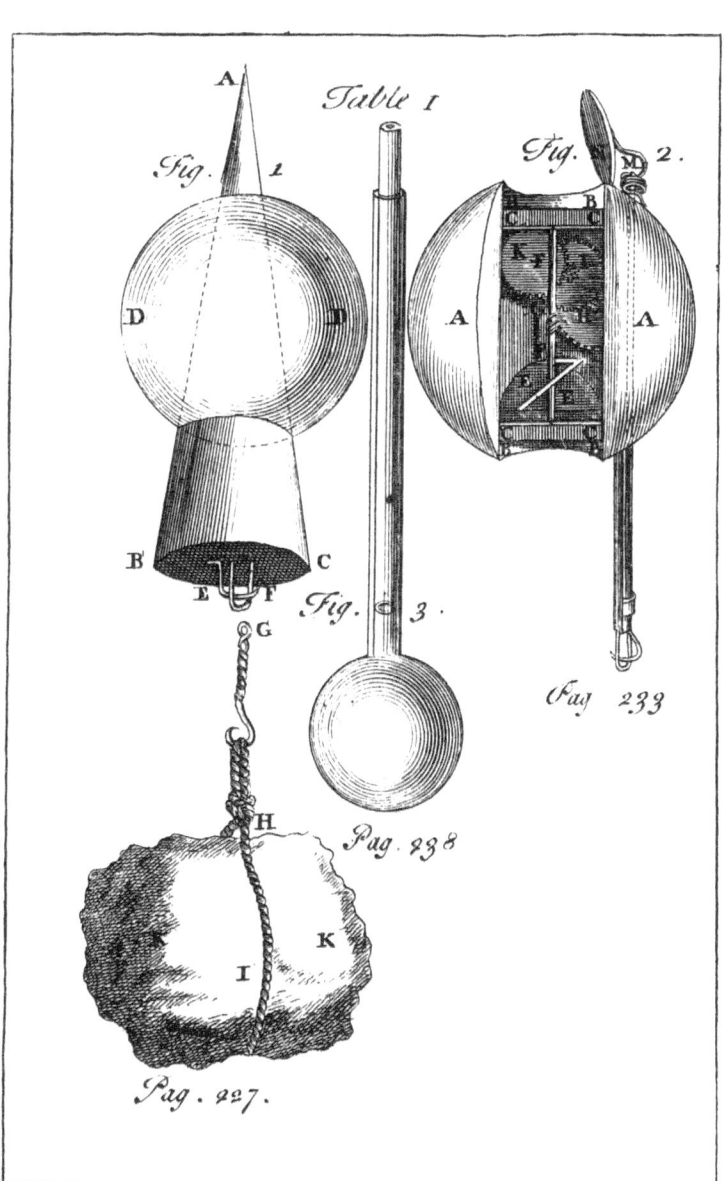

Table 1

Fig. 1

Fig. 2.

A

D D

B C
E F

Fig. 3.
G

H
I K

Pag. 227.

Pag. 238.

Pag. 233.

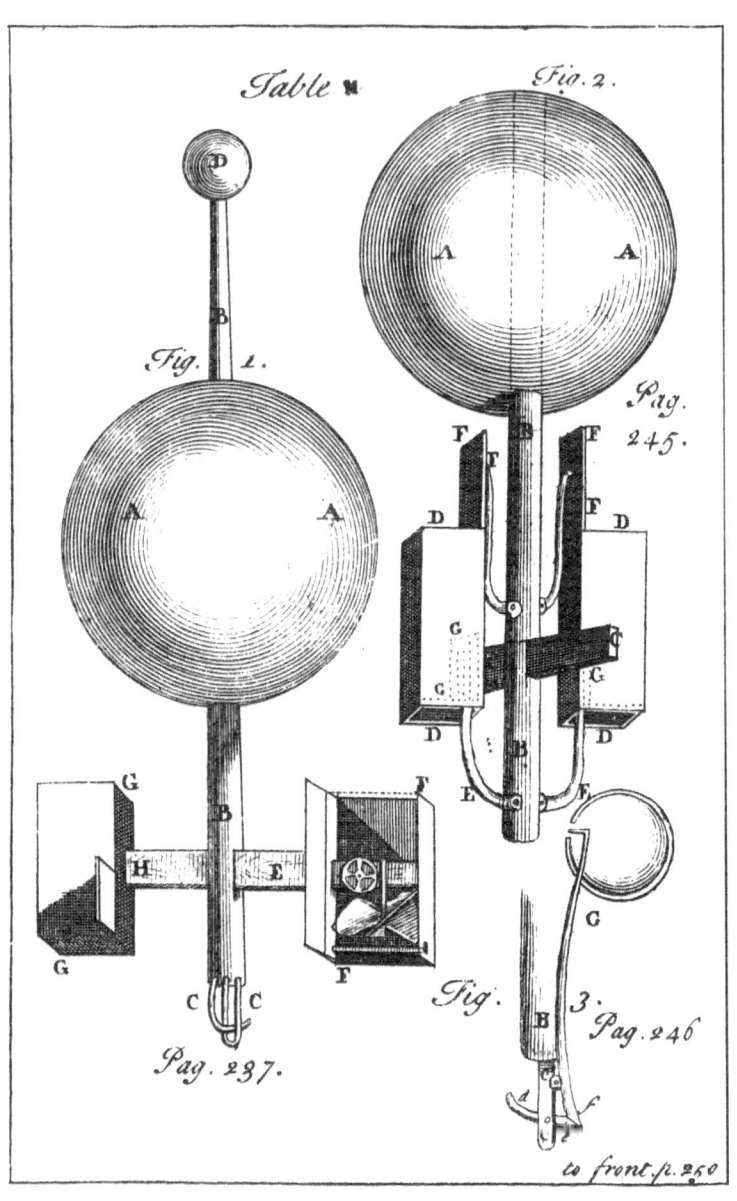

Table M

Fig. 2.

Fig. 1.

Pag.
245.

Pag. 237.

Fig. 3. Pag. 246

to front p. 250

‘ by the Air ; for it often happens, in the *Wetter*,
‘ that Ships are ruffled, and driven with Storms
‘ in one Part, whilft others, hard by, are rowing
‘ and becalm’d ; which is a plain Sign that thefe
‘ are caufed by the Eruption of fubterraneous
‘ Winds, as *Varenius* explains them, in his Gene-
‘ ral *Geography*. Divers Phænomena confirm this
‘ Sufpicion : For, upon the approaching of a
‘ Storm, and Tempeft of Rain, there is perceiv’d
‘ a Rumbling, or thundring Noife, of the Water,
‘ whilft the Air is yet ferene and calm ; which
‘ happen’d to me alfo going to the *Acidulæ* ; for
‘ I heard that Thundring, whilft the Air was moft
‘ calm; but always a whirling Storm prefently fol-
‘ lowed. This the Inhabitants of *Wifingfoe* do
‘ more plainly know ; who, lying oppofite to thofe
‘ Iflands whence the Storm comes the next Day,
‘ do hear this thundring Noife, like the Difcharge
‘ of Cannon. And when this Roaring is heard
‘ from the Eaft, the Eaft Wind rageth, with Hail
‘ and Rain. The fudden Huffing of Vapours, and
‘ Rifing and Gathering of them together, which
‘ fome have remarked in this Lake, are worthy
‘ Notice. Some fuch thing the Architect *Abraham*
‘ *Winandz* (paffing thefe Coafts with his Compa-
‘ ny) obferved, not without Admiration, that the
‘ Water being yet calm, there were darted out,
‘ as ’twere from the Bottom, certain fmall Clouds,
‘ which, coalefcing together in the Air, infefted
‘ the Travellers all Day with fmall Rain ; all
‘ which do confpire to prove thefe fubterraneous
‘ Winds.
‘ To the fame Caufe, without doubt, ’tis, that
‘ the Ice in the Spring is one Hour fo thick and
‘ ftrong, as to bear Horfes and Trahys, upon the
‘ Coming of a Storm ; the next Hour, thofe, that
‘ were fecurely carry’d in thefe Tra’ys on the Ice,
‘ may as fafely navigate the Lake in Boats, the
‘ Ice

' Ice fo fuddenly breaking and difappearing. But
' before fuch Kind of Ruptures happen, there
' is heard a Roaring of the Water, which,
' with Terror, warns the Travellers to fly off;
' though oftentimes fuch, as are far from the
' Shore, are either drowned prefently, or, with
' great Difficulty, at laft efcape on Pieces of Ice.
' Sometimes alfo the Ice fuddenly finks, when the
' Air is not in the leaft moved. Now, whether
' metallick *Halitus*'s may contribute to the Rai-
' fing thefe fubterraneous Winds, for the prefent,
' I fhall not difpute: But that fuch are not there
' wanting, the divers Mountains that encompafs
' the *Wetter* on the North, and the *Weftrogothian*
' Shores, richly furnifh'd with Iron Ore, and alfo
' with others more rich lately difcover'd, and
' others alfo, as *Antimony*, *Magnefia*, *Mica Sterill*
' but fhining, the Species of *Galæna*, *Ochre*, *Py-*
' *rites*, &c. whence have been extracted *Sulphur*,
' *Vitriol*, *Alum*, and other mineral Juices, do
' plainly prove. Nay, the Water affords great
' Quantity of *Pyrites*, and a Kind of Iron
' *Ochre*; divers Pieces of which I my felf, for
' Curiofity, collected. To thefe alfo are to be
' afcribed the *Ignes fatui*, frequently obferved not
' only upon the Shores; but, in the Night, up-
' on the Middle of the Lake, they fly to and fro,
' and confound the Fifhermen; which are gene-
' rally afcribed to an Increafe of metallick and
' fulphureous Vapours: Nor are the *Granates*,
' *Porphyries*, *Jafpers*, *Chryftals*, and divers other
' choice Stones, fuch as were heretofore collected
' by the Count *Peter Brabe*, and by Art reduced
' to fuch Luftre as to be ufed for Marriage Jewels
' at *Wifingburg*, to be believed to be generated
' without mineral Steams: For, all thefe are the
' Off-fpring of Minerals, as are alfo the *Acidulæ*
' *Medivienfes*, of which more another Time.

BUT

‘ But among many other ftrange Qualifications
‘ of our Lake, we muft not pafs over the won-
‘ drous fubmarine *Vortices*, and pertinacious Tor-
‘ rents, which caufe great Trouble to the Fifher-
‘ men, when the Wind fets againft the only Exit
‘ of this Lake ; from which venting of Rivers
‘ and Winds from below, and its unfathomable
‘ Depth, ’tis believ’d that the *Wetter* has Com-
‘ munication, by fubterraneous Paffages, with
‘ another *Swedifh* Lake, called the *Wenner*, about
‘ ten *Swedifh* Miles diftant : And the feveral *Vo-*
‘ *ragoes*, that are between thefe two, do feem to
‘ confirm the Conjecture ; two of which lying in
‘ the Parifh of *Fagren*, and called, the one the
‘ *black*, and the other the *white Vorago*, Mr. *Had-*
‘ *dorphius*, a celebrated Antiquary of *Sweden*, has
‘ endeavoured to found, but found them of un-
‘ meafurable Depth ; he obferv’d alfo an inteftine
‘ Motion in them, as if they were in a Fermenta-
‘ tion.   This Opinion alfo is augmented, by rea-
‘ fon the Water of the *Wetter* is fome Years aug-
‘ mented, and the next Years confiderably dimi-
‘ nifh’d.   Mr. *Daniel Ridelius*, the Paftor of *Mo-*
‘ *talen*, has noted, that thefe laft feven Years the
‘ Water of fome Parts of the *Wetter* has fo much
‘ wafted, that many Places were left bare which
‘ ufed to be cover’d with Water to carry Boats ;
‘ whereas the Rains have been very plentiful all
‘ about in the Years 1680, 1682, 1684, 1685 ;
‘ but, in the Year 1686, towards Autumn, the
‘ Water began again to increafe, and has conti-
‘ nued fo to this prefent Year 1688 ; but whether
‘ our Lake does obferve fuch Periods of feven
‘ Years in Increafing, and feven in Decreafing, as
‘ the *Wenner* is affertd to do, by thofe that have
‘ enquired, I cannot now pofitively affert.   It is
‘ alfo wonderful, that in a calm Air the Guns of
‘ *Stockholm*, and other Places 30 Miles diftant,
‘ are

' are plainly heard here: As, when in the Year
' 1685, the Princes were buried at *Stockholm*,
' every Shot was diſtinctly heard here at Five of
' the Clock: So alſo, the Broad-ſides at the Sea-
' Fight in the Year 1676, at about 30 Miles Di-
' ſtance, were diſtinctly remarked. But what
' *Olaus Magnus*, *Meſſenius*, and other Hiſtorians,
' relate of the Cave of *Gilbert*, in the Iſland of
' *Wiſingſoe*, I leave to their Credit: Only, this
' is true, there is a Cave, at preſent, that is fill'd
' with a Stench of Sulphur very odious, which,
' with the Conſent of the Inhabitants, has been
' collected into a Cave near the Water of the
' Lake; which, by being long pent up, it eructates
' noxious and ſulphureous Vapours, which others
' have aſcribed to other Cauſes, which I cannot
' approve: And Antiquity has diſcover'd its
' Weakneſs, in ſo eaſily giving Credit to ſuch
' Fables; tho' they relate ſtupendious Things of
' the ſaid *Gilbert*, and his *Præceptor Catillus Ru-*
' *nes*. But that there do appear divers *Spectra*
' and Phantoms in the neighbouring Parts in the
' Shape of Women, Horſes, or other Animals,
' none that are intent about theſe Matters do
' gainſay. Theſe might be evinced by Relations
' of modern, as well as antient Times; but, for
' the preſent, I omit them. But I muſt not omit
' the celebrated River *Motala* (the only Mouth of
' this Lake) which at certain Times ſeems at a
' Stay, and dried, ſo that one may go and take
' up the Fiſh that are left at the Bottom, with-
' out Impediment, as it happen'd in the Years
' 1682, and 1685, at *Chriſtmas*. And the com-
' mon Inhabitants believe, that this Stop of the
' Water never happens, but either Dearneſs of
' Corn, War, or ſome other publick Calamity is
' portended by it; as much as the *Engliſh* believe
' the Coming of a Whale into the *Thames* is omi-
' nous.

‘ nous. But, for my felf, as a Naturalift, enqui-
‘ ring only the Caufes of natural Effects, they
‘ were no ways fatisfactory to me, unlefs I found
‘ them conformable to the known Laws of Nature:
‘ I was therefore more follicitous about thofe
‘ Things which were advantageous to this Purpofe,
‘ for explaining this Phænomenon of the River,
‘ tho’ I had not the Opportunity of feeing this
‘ Stopping: And tho’ they divers Ways endea-
‘ vour to folve the Phænomena, by faying, that
‘ at that Inftant the Waters do recede from the
‘ Shores, and go to the Bottom; yet I always
‘ fufpected, that the Ice, or Snow, did fome
‘ ways obftruct the Paffage of the Water above,
‘ whilft at the fame Time the inward Water flowed
‘ out into the Sea. What hinted this Conjecture
‘ was, 1. That this Mutation never happened in
‘ the Spring, Summer, or Autumn, but always
‘ about *Chriftmas*, or in the Beginning of the
‘ Year. 2. That this only happen’d near the
‘ Bridge, where the Water is but three Ells deep,
‘ and the Heaps of Stone, on which the Bridge
‘ is founded, do impede its Courfe. And this
‘ Sufpicion the Paftor of the Church of *Motalen*,
‘ who lives hard by the Bridge-Foot, does judge
‘ very rational, from his own and others Experi-
‘ ence ; for he has noted, that divers long Plants,
‘ fuch as *Potamogiton*, *Polygonum aquaticum*, &c.
‘ do grow in the Parts near the Bridge, and that
‘ by thefe the Ice and Snow will be clodded and
‘ bound together, which, being carried by the Ri-
‘ ver to the Bridge-Foot, do in Time fo accumu-
‘ late againft it, as to make an abfolute Dam to
‘ the River. The Millers alfo that live there con-
‘ fefs, that ufually, before fuch a Stop, there are
‘ divers white Lumps flow out of the Lake, which
‘ fticking to the Bodies they meet with, like
‘ Glue, do by degrees fink there to the Bottom.
‘ Nor

' Nor is it unfrequent, that all the Water of the
' Lake ſhall be one Day quiet, and the next Day
' be ſtopped near the Bridge. Whatever it be,
' 'tis wondrous that this Retardation happens not
' in the ſharpeſt Winters, but in a more mild Sea-
' ſon, and for the moſt part about *Chriſtmas* or
' *New-Year*'s-*Tide*, when the Cold is yet intenſe
' under the Water, tho' more mild in the Air ;
' or, that the Ice, being leſs harden'd, is detain'd
' and implicated by the Weeds which cauſe theſe
' Obſtructions. Before I leave this Subject, I
' cannot paſs by the mentioning what I underſtood
' from the Reports of the ingenious Paſtor of
' *Nijen*, (where the *Acidulæ Medivienſes* are) and
' of divers others, concerning a certain Fountain
' not far from the Shore of the *Wetter*, in the Pa-
' riſh of *Nijen* not far from the Church, and Pa-
' ſtor Mr. *Jonas Frodel*'s Houſe ; to wit, That
' they call this Fountain the *Foreteller of Dearth* ;
' becauſe it is never ſo fill'd with Water, as when
' a Dearth ſucceeds the next Year. 'Tis encom-
' paſſed round with ſoft ſandy Hills, between
' which and the Fountain is a low Vale, but
' not marſhy : Out of this, by occult Paſſages,
' iſſues this Fountain ; ſingular in this, that in
' rainy Summers it waxeth dry, and in dry Sum-
' mers, when Famine is fear'd, or (by others whom
' I regard not) War, it overflows the King's
' Highways of *Wodſtein* and *Motala*, as is atteſted
' by many of the Inhabitants : Nor does it contra-
' dict this Report, this preſent dry Summer ; for it
' abounds with Water, now all the neighbouring
' Fountains are dry'd up. And though this may
' ſeem fabulous or ſuperſtitious, yet 'tis confirm'd
' by many Experiments ; which ſhews, that there
' are many internal Operations of Nature that yet
' are kept ſecret, and cover'd with a Veil, which
' we are yet unſufficient to diſcover. However, the
' fol-

' following Obfervations may fomewhat affift : 1.
' That this Scarcity of Corn is foretold to *Oftro-*
' *gothia* and the Places near the Fountain. 2. That
' in all this Region, and efpecially near the Foun-
' tain, the Plain is fandy, but in fome Places it is
' thick Clay ; which require much Water to make
' them ufeful. 3. That Corn is thin only in dry
' Years ; the contrary of which happens in *Jemtia*
' and other Northern Provinces. 4. That the
' Phænomena of Meteors are caufed by fubterra-
' neous Influences for the moft part. 5. That this
' Fountain is fupply'd by ftraining through fecret
' fandy Veins from thefe Sand-Hills. 6. That from
' fome natural Caufes, the Waters may afcend a-
' gainft a dry Seafon, and fink againft a wet Seafon.

Dr. Hook's *Difcourfe concerning* Telefcopes *and* Microfcopes; *with a fhort
Account of their Inventors, read in* February 1691-2

Of Friar Bacon, Baptifta Porta, Diggs, Metius,
Galileo, *and other Inventors of* Telefcopes.

How much the great Improvements of natural Knowledge have been owing to the Difcoveries and Improvements that have been made
in Opticks, I think few can be ignorant of, that
have inquired into the Reafons and Grounds of
the Progreffes made in this laft Century, fince it
hath been actually effected : For, though it be
evident that *Roger Bacon* did underftand fomewhat of the Grounds of it, and, in Probability,
would have further improv'd that his Knowledge,
if he had met with a Generation worthy thereof ;
yet fuch was the ill Treatment he receiv'd by falfe
' Accu-

Accusations, scandalous Reports, Imprisonment, and Loss of Places, that we hear no more concerning it, but only some Hints that he gave, of his being able to see things at a Distance as if they were near, in his Apology for himself, addressed to the then Pope, to protect him against his Persecutors. This Persecution quash'd it for that Time; and we find nothing of the Revival thereof, till the *Lyncean* Academy became founded in *Italy*; where, from the Encouragement that divers ingenious Men received, it was again started: And we find that *Johannes Baptista Porta* had made a Discovery of it, as is very plain by some Passages of his natural Magick; and our *Diggs* had done the same thing here, as is testified by his Son, who printed some of his Father's Works after his Death. These two Testimonies we have, that somewhat like the Telescope was known in the preceding Century, both the said Books being printed before the Beginning of this Century. We find nothing further concerning its Description, or Use, besides the Hint that it was then known to these two Men, some Years before *Galileo* put it in Practice. In the Beginning of the present 17th Century, *Metius*, a Spectacle-maker in *Holland*, light upon a Composition of a Convex, with a concave Glass set at due Distance in a Tube, which made a perspective Glass to see Objects at a Distance. And *Galileo*, in *Italy*, whether excited by a Hint thence received, or from *Baptista Porta*, or by his own good Genius, is uncertain, did the same thing at *Florence*: But not contented with the bare Invention, and Use for terrestrial Objects, he improved it farther, and made Use thereof for Discoveries of the Cœlestial Bodies. By this Means he detected the *Galaxia* to be an infinite *Congeries* of small Stars; as also the cloudy Stars, to be of a like Composition.
By

By the fame he difcovered the Roughnefs and Inequality of the Surface of the *Moon*, and the Phænomena of the Shadows and Lights of thofe rough and uneven Parts, and the Progrefs and Recefs of the Light of the *Sun* thereupon. By this he difcovered the four Stars about *Jupiter*, and in fome Sort adjufted their Periods, and hinted the Ufe of them, for the Difcovery of the Longitude of Places upon the Earth. By this alfo he difcover'd the unufual Figure of the Body of *Saturn*, the Waxing and Waining of the Light of *Venus*, and the Spots in the Face of the *Sun*, together with their Motions and Changes; which laft, whether it were not primarily, or at leaft at the fame Time, detected by *Scheiner*, is difputable, fince both lay Claim to it. This, I think, may truly be faid for *Scheiner*, that whoever firft detected them, he was the Man that perfected the Theory of them, fo far as it has hitherto gone; which he hath performed in that moft elaborate Work of his *Rofa Urfina*.

These Difcourfes excited the Curious of thofe Times to inquire into and improve the Knowledge of Opticks, efpecially that Part of it which had been leaft cultivated, namely, the Bufinefs of Refractions. (*Stelliola*, who was a *Lyncean*, feems to have been the firft that difcover'd the Ground of Refraction, in his Book *Il Telefcopio overo il Specillo Celefte*.) *Kepler*, in his Opticks, explain'd the Reafon of the Phænomena of *Senfes*, and the Caufes thereof; and alfo, that the fpherical Surface did not give the true Figure requifite to refract all the parallel Rays that fell upon it to one Point, but a Figure fomewhat elliptical; but made no Demonftration what the true Figure was, nor the true Proportion of Refraction. But *Defcartes*, by thefe two Helps, went through with the Demonftration, and proved both the true elliptical

cal Figure, and also moft ingenioufly and mechanically explain'd the Ground and Caufe of Refraction.

FERMAT foon after, taking a contrary Suppofition, explain'd the fame Phænomena; as did alfo *Emanuel Maignan,* in his *Perfpectiva Horaria,* by a third Suppofition; and our Countryman Mr. *Hobbs* by a fourth; but thefe two laft by Ways lefs intelligible and more improbable. Others fince have gone other Ways, but fall fhort of the firft. However, the firft Succeffes caufed it to be exceedingly cultivated by very many ingenious Men. And that not only as to the Theory, but as to the Practice alfo: Thence many Attempts have been made by divers ingenious Men, as *Defcartes, Hevelius,* Sir *Paul Neile, Divini,* Mr. *Smethwick,* and others, to make Object-Glaffes and Eye-Glaffes of elliptical Figures, but all without Succefs. However, of the fpherical Figure they made good Improvements, by making Object-Glaffes of much greater Lengths, and truer Figures, than they were at firft able to do: For, *Galileo*'s Glafs, of which he made fo good Ufe, I have been informed, was not above four or five Foot long, at the moft; and, I am apt to think, that the Glafs, *Hevelius* ufed for his *Selenography,* was not better, if, at moft, it were fo good; fince as many Particulars, as he has noted in that Book, may be made with a Glafs of three Foot. But Sir *Paul Neile* made fome of 36 Foot pretty good, and one of 50, as I have been informed, but not anfwerable. *Divini* and *Campani* made alfo Glaffes of thofe Lengths, but how good I cannot knowingly affirm: However, if we may be allowed to judge of them by the Difcoveries they made with them of the true Figure of *Saturn,* I conceive they were but ordinary, and did not exceed our 12 or 15 Foot Telefcopes; for, by one of that Length, I
plain-

plainly difcover'd the Ring and Satellite of *Saturn*, to be as Monfieur *Hugenius* doth affert in his Book ; and, with the fame Telefcope, I firft difcovered the permanent Spot in the Belt of *Jupiter*, which proved its diurnal Motion on its Axis. Since that, Mr. *Reive* firft, and then Mr. *Cox*, made fome good Glaffes of 50 and 60 Foot long, and the laft one of 100 ; but how good, I cannot affert, having not made Trial of it. And, as it hath been cultivated here, fo others, in *France* and *Italy*, have not been idle : Particularly one Mr. *Borelli*, at *Paris*, who prefented one of a confiderable Length, to this Society, which Mr. *Flamftead*, I fuppofe, has in his Keeping, Sir *Jon. Moor* having borrowed it of the Society for his Ufe. But tho' there has been fome Life left in the Grinders of Glaffes, yet the Warmth of thofe, that fhould have ufed them, has grown cool ; and little of new Difcoveries hath been made by them, befides what Mr. *Caffini* has done at *Paris*, in difcovering four new Satellites about *Saturn*, befides that of Mr. *Zulichem*.

Much the fame has been the Fate of Microfcopes, as to their Invention, Improvements, Ufe, Neglect and Slighting, which are now reduced almoft to a fingle Votary, which is Mr. *Leeuwenhoek* ; befides whom, I hear of none that make any other Ufe of that Inftrument, but for Diverfion and Paftime, and that by reafon it is become a portable Inftrument, and eafy to be carried in one's Pocket.

If we enquire into the Reafon of this Change of Humour, in Men of Learning, in fo fhort a Time, we fhall find that moft of thofe, who formerly promoted thefe Enquiries, are gone off the Stage ; and with the prefent Generation of Men the Opinion prevails, that the Subjects to be enquired into are exhaufted, and no more is to be done :

done : Befides, they pretend that all the Difcoveries that have been hitherto, or that can be made,
for the future, by thefe Inftruments will afford
no gainful Profit, and all other Notions are infipid
with them, befides fuch as bring ready Money.

But thofe, who make fuch Eftimates, may, perhaps, find themfelves very much miftaken in their
Judgment, if the Subjects were duly profecuted,
as they are capable of fo being. For, as to the
Difcoveries that may be made in both Kinds, I
conceive they are vaftly greater, both for Number
and Value, than thofe few that have been already
made; and not only for the Information of the Intellect, but what anfwers their greateft Objection,
even for the increafing their Treafure.

Having given this fhort Account of the Hiftory of Telefcopes, as alfo of the Ufe and Difcoveries that have been hitherto made with them,
which, as they have been very confiderable, as to
the Improvement of the phyfical or natural Knowledge of the Cœleftial Phænomena, I may obferve
that a further Improvement and Ufe of them, will,
in all Probability, afford much greater, and more
confiderable, not only for the perfecting and com-
pleating the Knowledge of thofe Particulars which
have been already, in Part, detected; but alfo for
making of ether new Difcoveries, which as they
are yet much further removed from the Power of
the Senfes to comprehend, fo they have been,
upon that Account, never afforded Entrance into
the Imagination and Intellect; if at leaft *Ariftotle*'s
Maxim be true, That there is nothing in the Intellect, but what was firft in the Senfe: And tho'
there are many Things that may be imagined, and
gueffed at, by Analogy, and the Uniformity of the
Proceedings and Productions of Nature; yet there
are certain Non-pareils of Nature, of which Kind,
poffibly, nothing like them have been produced in
                                                    all

all thofe Particulars, which are more common and obvious, as I might inftance in the Body of *Saturn*.   For who would ever have imagined fuch a Configuration or Fabrick, as that of the Ring of *Saturn?* what is there in all the other Celeftial Bodies, we yet know, that is analogous to it? and from the Imperfection of the firft Telefcopes, what extravagant and irrational Conceptions were form- ed thereof, as does more evidently appear, by the Defcripions and Explications of the Phænomena of it, before the more perfect Difcovery made by Monf. *Chr. Huygens*, and his ingenious Explicati- ons thereupon.   And that *Autopfia* is not only ufeful, but abfolutely neceffary, to give one a true *Idea* and Conception of many Phænomena, without which, the Imagination is very apt to rove, and go out of the true Way, as I might confirm by many Inftances, there being enough; but I fhall only mention one, namely, that of Dr. *Voffius*, his Explication of the Phænomena of the Moon, publifhed in his laft Book, upon which I did formerly read a Lecture to this Society, to fhew the Irrationality thereof, and how little Ground or Probability there was to be found in all the Phænomena of that Planet, viewed and ex- amined with a good Telefcope.   And therefore I did conclude, that that learned Man did never, himfelf, obferve the Phænomena, or if he ever did, it was certainly with a very fmall, and very imper- fect, Telefcope.   Upon which Account, *Autopfia* is not only neceffary for directing the Mind and Intellect, in its Progrefs to be made, for what is to be gone thro' with; but 'tis neceffary alfo, for the reducing it to its right Way, from which it may have been mifguided, by the falfe and erro- neous Suggeftions it hath formerly met with, ei- ther in fome famous Authors that have pofitively afferted, or defended a Falfity; or of fome other
                                                    Perfon

Person reputed eminently skilful in this, or that Part of Knowledge. With which Kind of Information, how full are the Authors that have treated of some Subjects? and that not one or two, but Hundreds, nay, Thousands, if we consider natural Philosophy and Physick, with the Arts subservient thereunto: What shall we say to the whole Generation of Astrologers, which have yet always prevailed, and possibly always will, with some especially, who have once been prepossessed or prejudiced for it: The like may be said of those who defend the four *Aristotelian Elements,* or the four *Chymical Principles,* or the three *Cartesian Materia's,* or his *Mundane Vortices,* which are, in Probability, all alike *Chimera's* which have sprung up, and got rooting in the Minds of Men, in several Ages of the World; and having once prevailed, they become prolifick, and propagate themselves in new Soils, and new Assertors and Defenders of those Doctrines do daily spring up: Among these may also be ranged the *Solid Orb Men,* the *Plastick Faculty Men,* and the *Sympathy* and *Antipathy Men,* each of which, having once embraced their respective Doctrines, will maintain and defend them to the last, against all others whatsoever. 'Twas from the first of these Sects (as I may call them, from their Division from the true Philosophy) namely, the *Solid Orb Men,* that poor *Galileo* was put into the Inquisition, and, to save his Life, was necessitated to lose his Doctrine, and to unsay what he really knew, and had discovered and asserted ; and tho' he, as well as *Copernicus,* was encouraged, at the first, by Popes, Cardinals, and Princes, yet in the Conclusion all fail'd, and their Doctrine must be condemn'd. Thus it happen'd also to *Roger Bacon,* and, I am apt to suspect, to the far greater Man, the Lord Chancellor *Bacon,* for being too prying into the then receiv'd

ceiv'd Philofophy: But notwithftanding all this, there is a real Beauty and Allurement in Truth, that will produce fome Votaries in the worft of Times ; and that will in Time prevail, and fhine out, and difpel the Clouds of Error that encompafs it. *Multi tranfibunt & augebitur Scientia,* was the prophetick Saying of *Daniel,* and ufed by the learned *Verulam.* And there is no doubt, but there is yet behind, much more to be difcovered, than what is already known, if fit Methods, and fit Inftruments be apply'd, and profecuted with Diligence. Some Ufes 1 have made of the *Telefcope,* and not without fome confiderable Succefs; as in the Difcovery of the Figure, Motions and Qualities of the *Cometical* Bodies ; as namely, of following them for near a Month after they difappeared, and finding them retrograde, in obferving their flame-like Figures and Qualifications ; in difcovering the Smallnefs, or rather Inconfiderablenefs of their Parallax, by a Way not taken notice of before, by any that I know of: And tho' Monf. *Caffini* has defcribed it in his Obfervation of the Comet in 1680, yet he hath added nothing more to it, than what I publifhed in my *Cometa* fome Years before, fave the Application of it to that Comet. By thefe I difcovered the Parallax of the Earth's Orb, and the Vifibility of the fix'd Stars, at all Times of the Day   Upon which Occafion I cannot but take Notice of a Paffage printed Page the 385th of *Ozenam*'s Mathematick Dictionary, and, by him, faid to be written by Monf. *Caffini* ; the Senfe is this; By the Means of great Telefcopes, fixed to certain Parts of the Heavens, thro' which the fix'd Stars pafs, which are the moft proper for this Obfervation, one may beft examine whether there be any Difference (of the Situation of thofe Stars, as to *Parallax*) in different Seafons of the Year ; for this Defign, in the Foundation
of

of the Royal Obſervatory, there is left an Opening thro' all the Vaults, by Means whereof one
may ſee, from the bottom of the Vaults, the
Vertical Stars, thro' Teleſcope Glaſſes of 160
Foot in Length, which will be prepared againſt
the Obſervatory is finiſhed. Notwithſtanding the
Engliſh Aſtronomers have begun to practiſe a Method like to this, we are aſſured, by an Eſſay of
Obſervations which they have made with great
Subtilty, that they have found ſome ſuch Difference, which have verified that the Diameter of the
annual Orb of the Earth hath ſome ſenſible Proportion, compared to the Diſtance of the fix'd Stars;
which, nevertheleſs, is not yet evident to us, by reaſon that the Obſervations, we have made of ſome
fix'd Stars Variations, do not agree with this Hypotheſis; for that the Variation was not found in the
Way that this Hypotheſis requires: But if the Obſervations ſhould confirm it, and be correſpondent to
the Hypotheſis, yet then we may doubt, whether
the Variation be from this Cauſe, or from ſome
conſtant Variation of ſome fix'd Stars, which hath
no Relation to the Earth's Motion; I ſuppoſe, he
here means *Mallement de Meſang*, who, to evade
the Strength of the Argument for the Earth's Motion, drawn from the ſenſible Parallax amongſt the
fix'd Stars, aſſigns every fix'd Star to move in a
ſmall *Epicycle* that will anſwer the Appearance.
(Obſerve only the Humour and Ingenuity of theſe
great Philoſophers and Aſtronomers, and judge
how likely 'tis, by any Means in the World, to
convince ſuch of any Error they ſhall once aſſert.)
Yet, be pleas'd to obſerve his Concluſion; *viz.*
But when we have found, by a great Number of
Obſervations, that a ſufficient Number of the fixed
Stars have a Variation conformable to this Hypotheſis, then we may judge that there is ſome
Foundation for it, notwithſtanding ſome Irregu-
larity

larity that has been, in Part, obferved to the contrary.   The Obfervation is extremely difficult and
long, becaufe the Period of the Variation, propos'd to be obferv'd, is of a whole Year, and requires that the Inftrument fhall be unfhakeable.
It is for this, that it can no where be better done,
than in the Royal Obfervatory.   Thus far Monf.
Caffini.   To which Mr. *Ozenam* adds, [That
the Royal Obfervatory is a haughty Building,
which the King has caufed to be built in an
eminent Place, without the Suburbs of St.
James's, for making Phyfical and Aftronomical
Obfervations; and that it is called Royal, for that
it was built by the Munificence of *Louis le Grand,*
whofe Liberality has extended to divers Perfons,
diftinguifhed for their Merit, and principally to a
certain Number of learned Men, chofen out of
the reft, who have endeavoured, with *Eclat,* to
make Sciences flourifh in this Kingdom, who compofe the Academy Royal of Sciences.]   When my
Attempt firft was publifhed, I was informed fome
of that Affembly were angry at it, for that it had
not been firft thought of by them; but I confefs I
did not believe it.   But meeting with this Paffage
does feem to make it probable enough.   However,
they needed not have regretted it, fince there
were enough befides, as confiderable to have fhewn
their Penetrancy of Spirit, and Accuratenefs of Obfervation; and tho' *England* poffibly wants thofe
Affiftants which they can boaft, yet I hope to
fhew, that weaker Means may effect many Things
that their more powerful have fail'd to perform,
if God grant me Life and Health.

I f we confider, in the next Place, the Fate of
Microfcopes, we fhall find much the like to have
attended their Performances.   The firft notable
Thing performed by it, that I have met with,
was the Figure of the Bee made by Sir *Francifco Stelluti,*

Stelluti, a *Lyncean,* and prefented to Pope *Ur-*
ban VIII, which is mention'd by *Johannes Faber,*
in *Hiftoria Plantarum & Animalium Mexicanorum,*
lib. 1. p. 757. *Tam mirabilem anatomen præbuit*
partium omnium externarum, quæ in Ape funt mi-
nuto animalculo, oculorum, inquam, linguæ, cor-
nuum, jubæ, aculei, pedis, digitorum, aliarumque,
& nuper in æs incidi commifit, atque felicitati Ur-
bani VIII *dedicavit, ut hæc omnia malim te ocu-*
lis tuis intueri quam rudi meo calamo adumbrare.
And *Fabius Columna,* upon the fame Place, fays,
it was *Impreffum a Lynceorum Academia S. D. N.*
Papæ Urbano VIII *in perpetuæ devotionis fymbo-*
lum oblatum fuit anno 1625. *Cum noftratis Apis*
imagine accuratiffime a D. Francifco Stelluto novo
quodam Microfcopio obfervata, ut qui illam viderit
in admirationem incidat; tam multas partes orga-
naque depicta difcernit, quæ ab intuentium oculis
in ipfo animalculo omnino abfconduntur. Thefe
Difcoveries were alfo highly favour'd and practis'd
by Prince *Cefius* himfelf, which greatly encouraged Obfervers, and produced many in divers
Parts of *Italy.* Accordingly we find fome Obfervations made by *Hodierna,* in *Sicilia,* about 1640,
and others recorded by *Panarolla* about the Year
1650, namely, the Poroufnefs of Man's Hair, the
red Sands in Urine of calculous Perfons, and the
Worms in Vinegar. Many others were alfo found
to make fome few Obfervations in other Countries; but, by Degrees, it is become almoft out
of Ufe and Repute: So that Mr. *Leeuwenhoek*
feems to be the principal Perfon left that culti-
vates thofe Enquiries. Which is not for Want of
confiderable Materials to be difcover'd, but for
Want of the inquifitive Genius of the prefent
Age.

Dr. Hook's *Invention of a Reflecting Telescope:*

*W*HICH *I insert after the foregoing Papers, by Reason of its Congruity therewith; because I know not the Time when this Telescope was invented, whether before, or after Mr.* Caffegrain's, *in* Phil. Tranf. N. 83. *from which it differs in some very material Matters.* W. Derham.

I Have lately made a Telefcope by Reflection, with which I look directly at the Object, and fee it very diftinct, and magnified. And this is by Planting a fmall *Lens* in the Middle of the *Object Speculum,* and Planting another fmall *Concave Speculum,* beyond the Focus of the *Object Speculum;* the Manner of which your Lordfhip will readily underftand by the annexed Scheme; where *a b* reprefents the Object Specu-

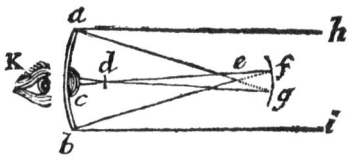

lum, *e* the Focus of that Speculum, *f g* a fmall concave Speculum, ferving to reflect the Rays to a fecond Focus *d,* where the Eye *k* fees the Object by the Help of the fmall *Lens c.* 'Tis eafy fo to contrive the Cell for the Eye, that the Rays that pafs on each fide of *f g* fhall not difturb Vifion.

We long much to hear of Monf. *Hugenius's* Opticks and Mechanicks: They are Subjects capable of vaft Improvements, and cannot be rationally expected from any more likely, than from
his

his acute Wit and excellent Pen.   But, my Lord,
I fear I have too far trefpafs'd upon your Lordfhip's Patience, and muft humbly therefore beg
your Lordfhip's Pardon, and fubfcribe my felf,

MY LORD,

Your Lordfhip's moft Faithful

and moft Humble Servant,

R. HOOK.

Mr. WALLER's *Obfervations upon* Dr.
HOOK's *Difcourfes, concerning* Telefcopes *and* Microfcopes.

NOvember the 29th 1693, Dr. *Hook* read a Difcourfe concerning Microfcopes, their Ufes
and Advantages in difcovering the Textures and
Motions of Bodies, as well animate as inanimate ;
obferving, that all Examinations by Fire, or Chymical Menftruums, deftroyed or altered the compounding Particles, or mix'd them with, and con-
founded them with heterogeneous Parts of the
Fire, or Menftruum, made Ufe of ; whereas the
Microfcope difcovers them in their natural State
and Actions.   Obferving farther, that the Motions of the Vifcera and of the Fluids, in the fmall
Veffels, are, by that Inftrument, to be feen, by
their different Colours and Refractions, through
the tranfparent Skins and Bodies of many Infects :
Natural Hiftory, hitherto, being for the moft
Part only converfant about the outward Shape and
Colour of Plants, Animals, and the like ; but the
Microfcope would afford a very large Field of En-
quiries

quiries and Obfervations, not yet much cultivated, which he recommends as one of the moft
proper Ways of difcovering the true Texture and
Mechanifm of Bodies.

In the next Place, he takes Notice of a Treatife lately publifh'd by Sig. *Bonani* in *Latin*, call'd
Micrographia Curiofa, &c. in which the Author
defcribes the feveral Sorts o Microfcopes, and
gives his Way of Grinding Glaffes for that Purpofe in a Cypher, which Dr. *Hook* thus decyphers.
The Tools are to be made of Brafs or Tin, and
of a due Form; that the Difh ought to be three
times as large as the Glafs that is to be ground in
it ; that the Difh is to be held in the Left Hand,
and the Glafs in the Right, and fo wrought and
turned every way, one to the other, till the Glafs
has acquired its due Figure ; that the Glafs ought
firft to be ground near the Figure defired, in a
larger, and then finifh'd in a fmaller Difh or Tool.
And, for its Polifhing, *Bonani* prefers the Way of
gluing a fine Paper into the Difh in which it was
laft ground, and by fpreading on the Paper fine
Powder of *Tripoli* ; work the Glafs therein till it
has acquir'd its due Polifh : And for this he recommends a Mandrill to fix the Difh on, made to
run fwift round, by Means of a large Wheel. Neverthelefs, Dr. *Hook* approves better the Ways
ufed by our Artifts by a reciprocating Motion,
and judges the bare Tool, without Paper, better
for Object-Glaffes, at leaft of Telefcopes ; tho'
for Eye-Glaffes he grants the Way by Paper and
Tripoli, fufficiently exact.

He concludes this Difcourfe with the Defcription of Stings or Thorns of the Prickly Pear, or
Indian Fig, thus ; The brown Tufts on the Prickly
Pear confift of a great Number of very fmall and
fharp-pointed Thorns, fmaller than the fineft Needle, and ftiff, fo that they eafily pierce the Skin

of

of whoever touches them; and what makes them yet more troublefome is, their being all barbed with Thorns like a Bramble, or rather a Bee's Sting, fo that they cannot eafily be got out, when they are once enter'd into the Skin. Of this he gives a Microfcopical Figure.

THE 6th of *December* following he read a Difcourfe of Telefcopes, in which he obferves, that *Galileo* firft difcover'd the fmall Stars, not vifible to the naked Eye, in the cloudy Stars, and the Galaxy, which *Ariftotle* afferted to be a Vapour, with the Figures of the Planets, at firft, indeed, not exactly true, as to fome of them, with their different Magnitudes, their Revolutions on their Axes, the Satellites of *Jupiter*, &c. He fuppofes that *Reita* was the firft that made Ufe of Convex Eye-Glaffes, taking in a larger Area than the Concave ones ufed before; and that he invented the Rete, or Menfurator, placed in the common Focus of the Glaffes; which Sir *Chriftopher Wren* perfected, and invented the angular Inftrument, confifting of two Telefcopes joined at a moveable Joint, fo as to take Angles by two Obfervers, to a Quadrant; and that himfelf had improved and recommended the Ufe of Telefcope Sights for Aftronomical Inftruments, in his Animadverfions on *Hevelius's Machina Cæleftis.* He proceeds to an Account of the Difcoveries made by feveral learned Men, as the true Figure of *Saturn,* and of its Satellite, by Monf. *Huygens*; the Satellites of *Jupiter* by Mr. *Lawrence Rook*; four
other

other Moons about *Saturn*, by *Caffini*, with the Periods of *Jupiter*'s Satellites more exactly limited by the fame. Next, he mentions his own Telefcopical Obfervations of the Comet in 1664, and 1665. thofe of the Stars in the *Pleiades*, being 80, great and fmall, which Obfervations and Figure of them, in his Micrography, he here afferts to be very exact, and made with great Care ; tho' Mr. *Caffini* and *De la Hire* have publifh'd Figures of them very different from his, both as to their Number and Situation; whence he infers there has been an Alteration in that Afterifm, as, he fays, Mr. *De la Hire* alfo believes, he having found them differing from what he had himfelf at firft obferv'd.

> *Dr.* Hook's *Lecture here mentioned, being long, and Mr.* Waller *having extracted every thing in it obfervable, at leaft it being contain'd in the preceding Paper, I have therefore chofen to publifh Mr.* Waller's *Extract.*

W. D E R H A M.

An Account of an Earthquake at Deal, *and other* Places *in* Kent, Portfmouth, *on* Sept. 8. 1692.

Deal, September 9. 1692.

YESTERDAY the People of this Place and Country were under a great Confternation, occafion'd by an Earthquake, which began precifely at two of the Clock in the Afternoon, and continued about fix Minutes; during which Time the Houfes

Houfes fhook; Pewter, Brafs, and other Kitchin-Goods, totter'd from the Shelves; empty Glafs-Bottles, where they lay, dafh'd one againft another; Beds and Tables in the Houfes fhook fo much, that People could not, for that Time, write; Some Chimnies fell, and feveral Houfes fhaken. This was at *Canterbury, Sandwich,* and many Villages thereabouts, tho' not fo violent, yet we had the fame at *Deal,* particularly at *Deal*-Caftle; altho' the Wall thereof be of a vaft Thicknefs and Strength, yet it fhook fo much, that the Inhabitants thereof thought it would have fallen on their Heads. In *Deal* Town feveral Houfes fhook, and fo all the Country over; fome Houfes ready to tumble down, others fafe, and felt nothing. In this, feveral Chimnies fell, and fome Houfes much damnified.

Portfmouth, Sept. 9. 1692.

HE RE fell much Rain Yefterday, and between two and three in the Afternoon this Town and Point, for about three Minutes, had a very fenfible Touch of an Earthquake, to the great Terror and Affrightment of many. The Tower of the Church, with many Houfes, were found to fhake confiderably; but, bleffed be God, I hear not of the leaft Damage, nor any thing more felt thereof fince.

A Con-

Tab: III.

Fig: 1.

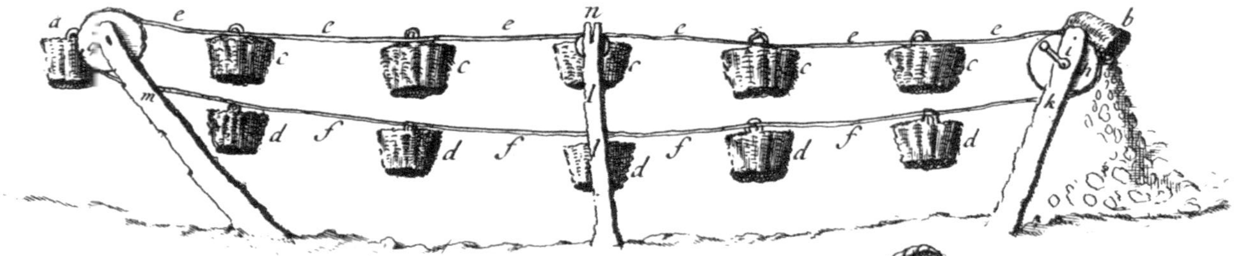

Fig: II.

Fig. III.

Fig. III.

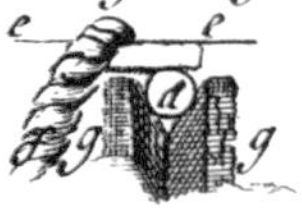

Place this at Page 275.

S. Parker Sculp.

A Contrivance which Sir Robert Southwell
faw at Brandenberg, *for ſpeedy Conveyance of Earth, and to fill up, or raiſe
Ground,* &c. *communicated to Dr.* Hook,
Sept. 9. 1692.

The Explanation of the Figures.

FIGURE I.

a. The Basket to be filled.
b. The Basket emptying itſelf, by the lower Part
of it hitting againſt the Axis of the two Pullies, *b.*
c. The filled Basket paſſing from *a* to *b,* ſupported by the Pullies, *n.*
d. The empty Baskets returning without any Support between the Extremes.
e. e. e. The Rope carrying the filled Baskets.
f. f. f. The ſame Rope returning them empty.
g. The Pulley at the filling End, ſupported by
the Poſt *m.*
b. The Pulley at the emptying End, ſupported
by the Poſt *k,* and turn'd by the Windle *i.*
l. The Poſt to ſupport the interpoſed Truckles *n.*

FIGURE II.

b. b. The two Pullies faſten'd on.
r. r. An Iron Axis to be turned by
i. q. A Winch, or Windle.
e. e. The Rope, lying in its tapering Edge, as expreſs'd in *Fig.* III.
o. o. The Basket hung by its Handles.
p. p. A Stick tied with Packthread, or girt on the
Rope *e. e. e. e.*

FIG.

FIGURE III.

g.g. The Shape of the Edge of the Pulley, exprefs'd in the firft *Fig.* by *b*, which is grooved
with a tapering Groove, that may hold the
Rope, on which is tied *e*, the Stick thruft thro'
the Handle *f.f.* of the Basket.

FIGURE IV.

a. The Rope on which is tied
b. The Stick, thruft through the Handles of the
Basket.
c.c. The Truckles Edge, hollowed with a half
round *b*, for the Rope to run on.

A fmall Module would exprefs all the Particulars, much plainer than any Draught. By this
Way 'tis eafy to tranfport Earth, Sand, &c. 1,
2, 3, 4, or 500 Yards, whether afcending, or
defcending; and, I conceive, two Men can do
more than fix in the common Way.

Tranfcript

Tranfcript of a Paper of a Quantity of Gold up the River Gambay, *in* 169 ,.

Your Importunity, together with my Gratitude to you, for your moft curious Informations and Inftructions in the Mechanicks (without which, I confefs my Labour had been in vain) has extorted that from me, which, I confefs, the Refolution I had a-new taken to the contrary, by refolving never to divulge, either for Love, or Force; to which End I expect, according to your faithful and folemn Vows of Secrecy, both of the Bufinefs itfelf, and likewife of which I would not fhould be known to the King for 10000 *l.* being content with what Proportion it hath pleafed God to affign me, as well as with the King's Revenues. Nor fhall I wholly, or fully, difcover the vaft Proportion of Gold I difcover'd there, being fo much, not fit to be communicated to Paper, as not knowing to whofe Eyes, or through whofe Hands this may come. I fhall only tell you, I was more troubled to obfcure its Abundance from my Fellows, than to bring down what I got; and, I am confident, that if your felf go upon this Defign, and follow the Directions of my Journal, and, attain your Purpofe, you your felf will be of my Opinion; for, as it is faid, *What will the whole World profit a Man, if he lofe his Soul:* So I fay, What will the Riches of both the *Indies* advantage, if thereby you forfeit your Security, Life, and Freedom? And how will you be affured of any of thefe, if thefe Things fhould come to Knowledge of fuch as have Power of you, and to command you in what they pleafe; That I do truly tell you, did I not value my own Peace and Quiet at fo high a Rate as I do, I fhould come willingly, and

mani-

manifeft it to his Sacred Majefty; though I am not fatisfied in that neither, as not knowing whether the Information may prove good or bad to the Publick; however, I conjure you a-new, that, whatever you attempt, you conceal me, fo that directly or indirectly I be not difcover'd.

I F you go on the Bufinefs, let your Boat be flat-bottom'd, for mine being fome feven Tons, or thereabout, and made after the common Fafhion, was extremely troublefome, both at Fords and at Falls, where we were forced to unlade her; and, having unladed her, to heave her, or launch her over Land: You ought alfo to have a little Boat for common Ufe, which you will find extreme ufeful. You advifed me to take 20 Pound of Quickfilver, for Trials; if you go, take at leaft 100 Pound, for fome in working will be loft, as you know, better than my felf: Your Advice alfo, for 50 Pound of Lead, is too little, take 150 Pound, much more you cannot well carry, for the peftring of your Boat.

T H E *Sal Armoniack* I ufed little of, for it I can give you no Advice: The *Borax* I ufed all, wifhed for more, if you go, carry 50 Pound; my Sand ever did me rare Service, I ufed it all, better have 10 Pound too much than too little, therefore take 40 Pound. I am confident, if I had carried the Philofophers Bellows, I had done very well; I was fo troubled with fitting the other, tho' I confefs them better when a-new placed. *Antimonia Horn* did me little Service; I believe it rather from my Ignorance, or wanting the perfect Ufe and Inftruction you gave me. Ingots I would take two, I carried but one, I wanted another for Expedition.   Wedges 12, with a Sledge or two, or Beetle; for about 12 *Englifh* Miles from the firft Fall, or fomewhat more to the Southward, in the Side of a barren Rock, looking Weftward, there

there is a Cliff in the Rock, rather

moſt rich between the Stones, almoſt half a Handful thick in ſome Places. Our Pick-axes did here ſtand us in no great ſtead, but having with us ſome Iron Tools, that we could hardly ſpare, with much ado made a ſcurvey Iron Wedge, and preſently we found the Benefit of that, for ſome 12 or 14 Days, till improvidently one of us driving the Wedge up to the Head, and not having another to relieve it, we were forced to leave it behind us, to our great Loſs and Grief. Wooden Bowls from *England*, ſix or eight, are very neceſſary, and will do better than Gourds, that I was forc'd to make uſe of; you may take Store of them, 'tis no Sore.

For the Crucibles I muſt inform, that four large melting Pots, in our large Work, will ſtead you much, and make better Diſpatch than ſix Neſts of Crucibles; though you cannot well ſpare thoſe, I was forced to make uſe of a broken Earthen Pot, that I carried along with me; I made uſe of it till it broke, had I had Crucibles, and Pots enough, I had brought ſo much Gold in Sand or Tyber.

For the ſeparating and diſſolving Waters, I uſed but little, becauſe their Uſe was troubleſome, neither had I Conveniencies to erect a Still a-ſhoar; but for the *Aqua Regis* I uſed it all, and could have done more, if I had had it; yet, in my Opinion, the Trials of Quickſilver are better, had I had it. But I carry Coals to *New Caſtle*; you know better the Operation than my ſelf. Let your Mortar be of Iron, and large; I wiſh I had follow'd your Directions in that, for my Braſs one put me to a double Trouble, and I was enforced to leave the Refining of much, till I came into *England*, for the *Mercury* got a *Spurca* from thence, which is communicated to my Gold, which no Art, I underſtand,

ftand, could free it from; in this Particular you left me lame, or my Memory much failed.

THERE is a Tree much like our Corners in *England*, but very large, which we felled, and made a Shift to make Charcoal of, which we did thus; we cut off the Boughs, for we wanted a Saw, and therefore could not meddle with the Body of the Tree, and cut them into fhort Pieces; then we digged a good large Pit, or Hole, in the Ground, about a Yard wide, and fo deep, or deeper; in the Bottom we kindled a Fire, and filled it with Wood, and, when it was well burn'd, threw Earth upon it, and damped it; and, when it was cold, we took out the Coals: You will eafily find the Place, if you obferve but the Cautions; you will come to a broad gathering together of Waters, not much inferior to *Ronnander Meer*, in the Edge of *Lancafhire*: Here we fpent a Week in fearching many Creeks and In-falls of Rivers; but we followed that which points South Eaft and by Eaft. My miferable Ignorance, in the Mathematicks, cannot direct you, neither for Longitude or Latitude. Up the buffing Stream, with fad Labour, we wrought, and fometimes could not go above two Miles in a Day. You muft pafs the firft Fall; yet there my Exceed of Gold was 47 Grains from 10 Pound of Sand. When we, or you come to the upper Fall, you will be much troubled, I believe, as well as I, to get your Boat over Land; but being up, proceed till you come to the In-fall of a fmall Stream to the South, directly thence liften, and you fhall hear a Fall of Waters; you cannot get your Boat thither, by reafon of the Smallnefs of the Brook; you will there find our Reliques on the Side of the Rock, with many of our Names, I mean, Letters of our Names, cut with our Knives. Here, tho' the Sand, by the Wafh, yield plentifully, yet do you afcend the Top of the Rock, and, pointing your

Face

Face directly Weft, you will obferve a Snug of Rocks fomewhat to the Left Hand of you; and, under that, if the Rains and Force of Weather have not wafhed away the Earth and Stones, you will difcover (they being unmoved) the Mouth of the Mine it felf; where, being provided with Materials fit for that Work, you will not defire to proceed any further, or with a richer Vein.

Take this, all along, for a conftant Rule, which I, in my Search, obferved up the River, That in the low, and woody, and fertile Country, I could never find either Metal or rich Mine, but always among barren Rocks and mountainous Countries, and commonly accompanied with a reddifh Kind of Earth. Other Inftructions I fhall not give you, being (as I conceive) a thing needlefs to you, unlefs I fhould return you your own Principal, this being but only the Intereft of what is due, befides that Obligation which tieth me unalterably to remain, *&c.*

I began my Voyage up the River, *December* the 4th, about two Hours before the Sun fet; in my Company no more than feven Men, befides my felf, all *Englifh*, and four *Blacks*, whereof one was a *Maribuck*, who, being acquainted with the *Portugal* Language, I intended for an Interpreter, if I fhould ftand in need; but the main was, to help us in our Labour againft the Stream. My Provifions were chiefly of two Sorts: For my Voyage and for Accommodation, three Barrels of Beef, ten Gammons of Bacon, two Barrels of white Salt, befides Bay Salt for Trade; alfo two Hogfheads of Bifcuit, befides Rice; half a Barrel of Gunpowder, and Shot proportionable; Strong-Water, Vinegar, Paper, Beads, Looking-Glaffes, Knives 18 *d. per* Dozen, fome Iron, little Brafs Chains, Pewter Rings, and a deal of fuch like Stuff, as

Occa-

Occafion permitted: The other Sort of Provifions were, a Pair of Goldfmiths Bellows, Crucibles four Nefts, Scarnelles two Nefts, Quickfilver, Borax, Sal-Armoniac, Aqua Regis, Aqua Fortis, a Mortar and Peftle, and Leather Skins to ftrain, Brafs Scoops and Ladles with long Handles, to take up Sand, and other Implements for my private Defign: All which had laden my Boat far deeper than I defir'd; for thereby I drew much Water, which, I was jealous, might hinder our Progrefs over the Flats, if we fhould meet with any.

December the 7th, We arrived near *Settico*, being 14 or 15 Leagues above where our Men ftay'd; but paffed one half League further up, where we anchored, the River there being broad, we always chufing the Middle, as being freeft from Difturbance, though it oft fell out otherwife; for our ugly Neighbours, I mean the Sea-Horfes and Crocodiles, (it feems) ill pleafed or unacquainted with any Co-Partners in thefe watery Regions, did often difturb us in the Night, not only with their ugly Noifes, but their Vicinity to our very Boats, which caufed us to keep Watch.

December the 23d, We were much troubled that Day with getting over a Flat, under the Wafh of a fteep and high Mountain bearing South. Here I firft put in Practice my Defign, and took up fome Sand at the firft Trial of the Ford, and, out of five Pound Weight of that Sand, got three or four Grains of Gold. I tried alfo in another Place of the fame Ford, but did get lefs. I faw neither Town, nor Houfes, nor People, fince we left *Baracunda*.

January the 14th, At a Ford between two high Mountains, I tried again; and out of ten Pound Weight of Sand, I wafhed 30 Grains of Gold. I made a Trial likewife with Mercury, and found
out

out of five Pound 47 Grains.   Here my Hopes increafed, yet refolved to try higher.

January the 27th, We were much troubled with great Trees that lay in the Water upon the Side of a Rock, on a craggy, barren Mountain adjoining.   I afcended, with three Men with me, to make Difcovery; and carrying a Pick-Axe with me, which, as we were digging up a Piece of Ore, as I conceiv'd, we were affaulted with an incredible Number of monftrous great Baboons; whom, no Oratory, but our Guns, could perfuade to let us retreat to our Boats; for, having killed two or three of them, fo incenfed the reft, that had, not the Report of our Guns terrified them, I verily believe, they would have torn us to Pieces: Having attained our Boat, I fell to try my Ore; which proved but a Sparre.

February the 6th, I made a Trial of a certain glittering Sand, which I took up from the Side of a Rock, the River here inclining Southward, with a fudden Turning like an Elbow.   The Wafh of this afforded 41 Grains from 10 Pound Weight of Sand: By other Trials, from five Pound Weight of Sand, 57 Grains.   Here I thought to make a Stand; yet, upon more ferious Advice, had refolved to proceed.

February the 15th at Night, a Sea-Horfe ftruck our Boat through with one of his Teeth, which troubled us fore, being all bad Carpenters; which caufed us to unload her on a fmall Pinnacle to mend her; and, to prevent the like Mifchief for the future, I invented this Device, To hang a Lanthorn at our Stern; and thereby we were freed from all After-Troubles of that Nature, they not daring to come within three or four Boats Length of Light fhining in the Water.

February the 24th, I tried the Ufe of *Virga Divina*, upon a high, barren and rocky Mountain:

tain: But, whether it afforded no Metal, or whether my Rod, being cut in *England*, and being dried and carried far by Sea, had loſt its Vertue; or, whether it hath no ſuch Quality (which I rather believe) I am not certain. However, my Companions laugh'd me out of the Conceit.

March the 16th, Between two mountainous Rocks iſſued a Creek; and, putting up therein, diſcover'd a Fall of Waters from the South of the River. Here, making Trial by the Way, I found 63 Grains of Gold from five Pound Weight of Sand. Other Trials, more exact, afforded very large Proportions; ſo that here we ſpent 20 Days; and, plying hard our Work, in that Time had gotten 12 Pound *Troy*, five Ounces, two Penny-weights, 15 Grains, of good Gold.

March the 31ſt, Our Materials waſting apace, I was willing to try further, here beginning our greateſt Toil; for, often in a Day, we were conſtrained to ſtrip our ſelves, and leap into the Water, with main Strength to force our Boats and the Flats. Nor was this our greateſt Affliction; for the River Water ſmells ſo ſweet and musky, that we could not drink of it, nor dreſs our Meat with it; and, as we conceive, by reaſon of the Abundance of Crocodiles, which have the ſame Scent.

April the 7th, We perceived the In-fall of a ſmall River South, the Current quick, the Land all rocky and mountainous, and, in the Silence of the Night, could hear the Noiſe, perfectly, of a great Fall of Waters; and, before the Mouth of it, anchored that Night.

In the Morning, into that we put, and came as near the Fall as we well could. Our Water failed; but our indefatigable Induſtry overcame all Difficulties; for, what I could not by Water, I did attempt by Land; Where arriving, I found the long
expected

expected End of our moſt toilſome and long Voyage; for, I believe, never any Boat, nor any Chri-
ſtians, have been ſo high in that River, as we.
Here, upon the firſt Trial I made, the Exceed of
Gold was ſo much, that I was ſurprized with Joy
and Admiration : However, here I was reſolved
to ſet down my Staff; and, to that End, the firſt
thing I did, was to go   the Boat; and, a-
bout a League and half thence, I found Wood.
Here we practiſed to turn Colliers, and laded our
ſmall Boat with as much as ſhe could well carry
back; we went and fell to Work, for which I
hope (to God alone be Praiſe) none of the
Company hath Cauſe to repent, for the great Pains
and Labour he took, tho' we choſe the worſt
Time of the Year almoſt, the Waters being then
at the very loweſt; but had we gone immediately
after the Rains, which is *June, July* and *Auguſt,*
or before the Waters were fallen ſo low, we had
been free from much of that Trouble, at Fords
and Falls, by having Water enough to carry us
over.

At the End of the Paper are theſe Words,
Tranſcribed *verbatim* from a Paper Manuſcript,
lent me by Mr. *Fr. Lodwick, Octob.* 2. 1693. by
R. Hook.

*This Paper (which I have here publiſh'd exactly as
I found it) I not long ſince lent to a Perſon of great
Quality, for the Service of the African Compa-
ny, (then ſetting out for an Expedition into thoſe
Parts) and I hope it hath, or will, prove as
much for their Benefit, as my Wiſhes are. The
Paper ſeems to have been written by one that had
gotten great Riches, in King Charles the IId's
Time, by his Progreſs up the River Gambay:
And his Deſcriptions of the Openings, and Turn-
ings of the Gambay, the Inlets of other Rivers
into*

into it, the adjacent Mountains, &c. may be a good Guide to Undertakers, how to find out the Place, where our Author met with Gold, even to Satiety. Who he was, can scarce be known, he conjuring his Friend, Mr. Lodwick, (to whom I conceive this Letter was addressed) to the greatest Secrecy, being, I suppose, afraid to be known, or talked of, lest he should be commanded away, by the King and Government, upon another Expedition, from that peaceable and satisfactory Retirement he enjoyed, after his Acquisition of sufficient Wealth.

W. Derham.

Experiments and Observations about heated Iron; communicated to the Royal Society, January 3. 1693-4.

HAving lately met with some Experiments which are not much known, tho' they are obvious, and easy enough to be observed and experimented, I thought it might not be altogether impertinent, nor unacceptable to this Assembly, to give a short, but true Account thereof; and so much the rather, because they are very pertinent for the Proof and Confirmation of a Theory which I have formerly read before this illustrious Society, and have published in the 8th Observation of my *Microg* for the explicating the Phænomena observable about the Sparks of Fire, struck from the Steel, by the Edge of a sharp and hard Flint, or some other such hard and stony Body: These I found to consist of small Globules, looking like melted Iron, or else some small Sliver cut off from the Steel, and thereby made red-hot, but not melted, but keeping the Shape it received by the

Stroke,

Stroke, or Gaſh of the Stone ; which Phænomena
I did there thus explicate. ―――― *It ſeems that
ſome of theſe Sparks,* &c. *Pag.* 45. *Line* 24, to
Pag. 46. *Line* 14. ―――― Muſcovy *Glaſs.*

THERE are two Particulars, therefore, that I
have there alledged, which, by the Experiments
I am now to mention, will receive great Confirmation. And the firſt of theſe Experiments is,
That two Smiths, taking each his ſmall Bar of
Iron, both perfectly cold, and each of them ham-
mering his Bar upon the ſame Anvil with ſmall
Hammers, in a very ſhort Time, and with not very many Strokes, reduced them both to ſo great a
Heat, that immediately laying them one upon another, and continuing to hammer them a very
ſmall Time longer, they were thereby perfectly
welded, or joined together into one Piece, as
firmly, as if they had been welded the common
Way, by being ſufficiently firſt heated in the Fire,
and then hammered together. This Experiment
I have not yet ſeen tried my ſelf, but I have been
aſſured of the Truth of it by a knowing Perſon,
who ſaw and examined all Circumſtances thereof,
inſomuch as I do no way doubt the Truth and
Certainty thereof.

THE other Experiment is this, That taking a Bar
of Iron, and heating it to a white Heat, ſo that it
ſpurts, or darts out of it every Way, very ſhining
and fiery Rays ; then immediately laying the ſame
on the Anvil, or a Tile, and blowing the glowing
Iron with a Pair of Bellows, inſtead of cooling the
ſame, as moſt would be ready to expect, the cold
freſh Air from the Bellows will make it glow and
burn much brighter and hotter, and will continue
to do ſo for a ſonſiderable Time ; and if the Bar
be ſufficiently heated at firſt, the Bellows, by ſo
blowing, will melt the ſame, as if it were Pitch or
Roſin on Fire. The laſt Part of this Experiment
I have

I have not yet verified my felf, but the former Part I have, and obferved it to burn and wafte under the Blaft of the Bellows, as if it had been a Piece of kindled Charcoal, fo blowed upon; and the Flame, or Light thereof, to be fo very ftrong and vivid, that one cannot well endure to look upon the fame, without much offending the Eyes, as if one look'd upon the very bright Face of the Sun it felf.

B y the former Experiment it is evident, that the Force of the Blow or Stroke, which is able to cut off a Sliver of hardened Steel, may not only be fufficient to heat the fame, to a Degree fufficient to fet Fire on the Tinder, but to intend it, fo far as to make it of a welding or white Heat, which having acquired, and flying off into the Air, with a very quick Motion, by the 2d Experiment, 'tis evident that the Operation of the Air is fufficient to intend the Heat yet further, fo as to melt, or vitrify the fame, and thereby to caufe it to be formed into a Globule, Ball, or Shell, as it often appears through the Microfcope. All which Effects are more eafily perform'd on fo fmall a Body, as are thofe Slivers which are ftruck or cut off from the hardened Steel. But the Globules, Balls, or Shells, that are made by the melting of the heated Iron, blown on by the Bellows, are much bigger, and more confpicuous, but of the fame Form and Subftance. Nor is this Combuftibility peculiar only to Iron, tho' therein it be very notable and confpicuous, but the other Metals have alfo their Combuftibilities in their diftinct Kinds, as Copper, Brafs, Lead, Tin and Silver; upon each of which the Menftruum of the Air will work and diffolve, or burn them when they have firft been prepared by a proper Degree of Incalefcency, as I fhall, at fome other Time, make manifeft, by plain and evident Experiments.

Dr.

Dr. Hook's *Account of Monf.* De la Hire's *Difcourfe of Froft.* 1694.

I HAVE have perus'd the Book of *Dan. Bartoli,* concerning Froft and Ice; and tho' he hath many Arguments to deftroy the Sentiment of feveral of the Moderns on that Subject, of *Valefius, Des Cartes,* Mr. *Boyle, Olaus Magnus,* Sir *Kenelm Digby,* &c. yet I do not find any other Doctrine affirmed concerning it, but that he conceives it done by a nitrous Subftance, which is of a cold and dry Nature, which operates after the fame Manner in coagulating the Water, as the Runnet doth in coagulating Milk; but, how that is done, I do not find he does explain.

HAVING therefore fail'd of my Expectation from him, I refolved to fee what Satisfaction I fhould have in perufing a Difcourfe, upon the fame Subject, of a much newer Date, namely, that of Monf. *De la Hire,* publifh'd at *Paris* in 1694. whereas that of *Bartoli* was publifh'd at *Rome* 1681. This I found to be much more concife, and plain, and pofitive in what he has deliver'd, and much more clear in explicating of his Notions and Conceptions of it; fo that tho' I could not meet with fuch an Information concerning Ice and Froft, as I could have wifhed, yet in perufing 16 Pages in *Quarto,* which is the whole Treatife, I was fatisfied that I underftood fully what he intended to communicate; whereas I was to feek, what was intended by the other, in almoft ten Times the Number of Pages.

Monf. *De la Hire* then begins his Difcourfe, by defining or explaining what he means by Cold; that is, the fenfible Quality in Froft; and this, he fays, is nothing elfe but a lefs Agitation or Motion on of the aqueous Particles, whether blended with the Air, or united in a Mafs, than of the like Particles in the Skin, or Pores and Veffels of our Body. He might, he fays, have added another Affertion, that all the Agitation of aqueous Particles proceeds from that of the fubtile Air; but he thinks it fufficient, for this Difcourfe, to fhew how all the Phænomena of Cold will be plainly folved, by the Explication he has premis'd.

N E X T, he fays, that thefe Particles are depriv'd of their Motion by certain Particles of Salt, which are very minute, long, ftiff and fharp, which, by their Motion, are eafily carried and blended with the Air, but do more eafily infert themfelves into the Particles of Water, than of any other Body, nay, even than of the Salts themfelves, from whence they proceed, which he reckons to be from common Salt a little, from Niter more, but moft of all from *Sal Armoniack*, which Salts do therefore eafily diffolve in Water; that thefe Particles do penetrate Metals, and even Glafs, but that they are moft entangled, and ftay'd by the Particles of Water, which he fuppofes, with *Des Cartes*, to be long and flexible, like Strings or Threads, and by that Means they deftroy the Motion or Fluidity of each other, which compofes a folid, hard and dry Body, which is Ice. This Ice, he fays, encreafes Extenfion, by Means of thefe Salts, and fo breaks the Veffels that contain'd the Water; and, being fo extended, is lighter, and fo floats on the Water. Hence 'tis, he fays, that Blebs are form'd in the Ice; but his Explication of this Phænomenon is not confonant to the other Suppofitions. By the bye, he explains the Expanfion of Water by Heat, and that he makes to proceed from the Expanfion of the Parts of the Air contained in it. Hence he concludes, that there is a middle State of the Water,

2

which

which is its fpecifick Expanfion, and it is then
cold, becaufe tho' it may be eafily moved, and fo
retain Motion enough to keep it fluid, yet it has
fo little Motion of its own, that it communicates
none to other Bodies. The Reafon, why Oils and
Spirits freeze not, he fays, is becaufe they have
few of thofe entangling aqueous Particles : By this
he gives a Reafon of the not freezing of other Bodies. To fortify his Hypothefis, he explains the
Experiment of the Expanfion of Spirit of Wine,
by the Application of Snow ; and freezing other
Bodies by Application of Niter and *Sal Armoniack,*
and Spirit of Wine ; and, by the Way, he tells a
pretty Method of cleaving Mill-Stones, by the
fwelling of fmall wooden Pins, drove into certain
Holes, drill'd in a Line on the Stone where 'tis to
be cloven. He takes Notice alfo, that the Refraction of Ice is lefs than that of Water, and
quotes his own Publication of it in 1693. though
it was fhew'd by me, to this Society, 30 Years before ; (but 'tis not ufual for thofe Writers to own
Difcoveries to be made by any but themfelves,
who take themfelves to have an Empire over all
the reft of the World). He alfo takes Notice of
feveral other Phænomena, and Experiments of Ice
and Froft, mentioned by Mr. *Boyle,* but without
naming him. Upon the whole, I conceive, he has
more particularly applied the *Cartefian* Notions of
Particles, Motions, Figures, *&c.* to the Explication of thefe Phænomena of Cold : But as the Suppofition of fuch qualified Particles is wholly pre-
carious, fo neither will thofe, without a great
many other fupplemental Suppofitions, fuffice to
folve the Phænomena fatisfactorily, unlefs the Particles be fuppofed to act and operate by Inftinct ;
and tho', poffibly, they might ferve to put a feemingly probable Explication of thefe Phænomena of
Cold, by fuppofing them thus, or thus, qualified
and

and adapted; yet, I very much fear, there are some Phænomena of Heat, or of other Qualities, wherein the same Particles of Bodies are necessary to be introduced as the principal Agents, tho' their Actions in those be quite contrary to their Actions in these. It would be, therefore, but a second lost Labour, to shew that these Particles are of another Nature than what they are here supposed, and to assign them other Figures, Motions, and Qualifications: Because, first, it would be almost as much Labour to demolish this Fabrick, as it was to raise it, and a third fruitless Labour to erect another. Nor can it be expected to be otherwise, till such a Structure be founded upon a natural, firm, and solid Ground, and not upon feigned and imaginary Suppositions.

An Instrument of Use to take the Draught, or Picture of any Thing. Communicated by Dr. Hook *to the* Royal Society, Dec. 19, 1694.

AMONG the Instruments that may be of Use to curious Navigators and Travellers, one is, for procuring the Pictures, Draughts, or true Forms and Shapes of such Things as are, or may be, taken Notice of by them; that is, not only of the Prospects of Countries, and Coasts, as they appear at Sea from several Distances, and several Positions; but of divers In-land Prospects of Countries, Hills, Towns, Houses, Castles, and the like; as also of any Kind of Trees, Plants, Animals, whether Birds, Beasts, Fishes, Insects; nay, of Men, Habits, Fashions, Behaviours; as also, of all Variety of Artificial Things, as, Utensils,

fils, Inftruments, Engines, Ships, Boats, Carriages, Weapons of War, and any other Thing
of which an accurate Reprefentation, and Explanation, is defirable. For, tho' a Defcription in
Words may give us fome imperfect Conception,
and Idea, of the Thing fo defcrib'd; yet no Defcription, by Words, can give us fo full a Repre-
fention of the true Form of the Thing defcrib'd,
as a Draught, or Delineation of the fame upon
Paper. Nor can we fo perfectly conceive, or
imagine, the true Colours, by Words, as by feeing the very Colour it felf imitated and compared
with the Life, or the real Thing: Whence we
find how imperfectly the Colours of Plants are reprefented by Herbals, which are wafh'd, or colour'd, only from the Defcriptions which are made
of thofe Colours in the Books.

Now, though this be not a new Defign, or a
Thing that has never been done before by any that
have given us Accounts of their Travels; yet, if
we do but confider, how the moft of thofe have
been done, it will, I conceive, make this, which
I propound for this Effect, fo much the more valuable. 'Tis well known, that the Books commonly
made for the Ufe of Seamen, (now commonly called *Wagoners*, becaufe one *Wagoner* printed a
Collection of many fuch Obfervations) that thefe
Books, I fay, are full of the Profpects of Countries, as they are faid to appear upon the Sea, at
fuch Diftances and in fuch Pofitions: And I lately
faw a Book containing the Profpects of all the
Weftern Coafts of *America*; but any one, that underftands Profpect, will eafily difcern, how rude,
imperfect, and falfe a Reprefentation, all fuch
Books contain of the Places themfelves: For,
not to mention the Impoffibilities they often reprefent, as the Over-hanging of Mountains for half
a Mile, or a Mile, which, tho' the Mountain were
made

made of cast Iron, were impossible to be sustain'd
in such a Posture : The extravagant Heights they
generally raise the Hills to, and the sudden and very decline Descents they make them have into the
Vallies, do plainly enough demonstrate them to
be no true Representations of what they are design'd for.   And, indeed, they are most made by
the Hands of the Mariners, who are, generally,
very little skill'd in the Art of Delineation ; and,
therefore, 'tis not to be expected that they should
be very exact : However, even these are of very
good Use for Navigators ; and they furnish them
with a better Idea of the Appearance to be look'd
for, than Descriptions by many Words would inform them.   Again, we find that many Relations
of foreign Countries do give us Pictures of Towns,
Prospects, People, Actions, Plants, Animals, and
the like ; and those beget in us Ideas of Things,
as they are there represented.   But, if we enquire
after the true Authors of those Representations,
for the Generality of them, we shall find them to
be nothing else but some Picture-drawer, or Engraver, here at Home, who knows no more the
Truth of the Things to be represented, than any
other Person, that can read the Story, could fancy of himself, without that Help.   Such are all
the Pictures in the Books of *Theodore de Brie*,
concerning the *East* and *West-Indies* : Such are also the greatest Part of the Pictures in Sir *Thomas
Herbert*'s Travels ; and those of Mr. *Ogylby*'s *Asia*,
Africa, and *America* ; which are Copies of the
Dutch Originals, and are, originally, nothing but
Mr. Engraver's Fancy : So that instead of giving
us a true Idea, they misguide our Imagination,
and lead us into Error, by obtruding upon us the
Imaginations of a Person, possibly, more ignorant
than our selves.

I t

I T is, therefore, the Intereſt of all ſuch, as deſire to be rightly and truly informed for the future, to promote the Uſe and Practice of ſome
ſuch Contrivance as I ſhall now deſcribe ; whereby any Perſon that can but uſe his Pen, and trace
the Profile of what he ſees ready drawn for him,
ſhall be able to give us the true Draught of whatever he ſees before him, that continues ſo long
Time in the ſame Poſture, as while he can nim-
bly run over, with his Pen, the Boundaries, or
Out-Lines of the Thing to be repreſented ; which

being once truly taken, 'twill not at all be difficult
to add the proper Shadows and Light pertinent
thereunto. By the ſame Inſtrument alſo, the
Mariner may very eaſily and truly draw the Proſpect

fpect of any Shore, and from Time to Time denote the Rifing thereof, as he does nearer and nearer approach it, and the Depreffion, or Sinking of it, as he does recede.

THE Inftrument I mean for this Purpofe, is nothing elfe but a fmall Picture-Box, much like that which I long fince fhewed the *Society* for Drawing the Picture of a Man, or the like; of the Bignefs of the Original, or of any proportionable Bignefs that fhould be defired, as well bigger as fmaller, than the Life; which, I believe, was the firft of that Kind which was ever made, or defcribed by any. And, poffibly, this may be the firft of this Kind, that has been applied to this Ufe; tho', upon the firft Inftitution of the Royal Foundation of *Chrift-Church*, I propounded it to the Governors there, for the Ufe of the Children: But Sir *Jon. More* undertaking to write an Inftitution, and having omitted it, it has not been there brought into Ufe.

A Way to meafure Heights and Diftances, &c. at Sea, Feb. 13, 1694-5.

THAT, which I fhall at prefent explain, is a Method of Meafuring the Bearing and Diftance of Objects feen at Sea, fuch as Ships, or Shores, Iflands, Promontories, Caftles, Towns, Mountains; their Heights, as well as Diftances: Alfo the Courfe, Length, Breadth, &c. of Rivers, and the like: As alfo, for knowing the Diftance from any Light, or Light-Houfe, feen in the Night. Now, tho' experienced Navigators do, by long Practice and Ufe, give pretty near Gueffes at them; yet the Way I fhall propound, I conceive, will come much nearer, and be much

more

more certain, and may eafily enough be put in
Practice ; which if the Gentleman, that defcrib'd
the Coafts of *England,* had known, or put in Practice, I conceive, he would have prevented many
Miftakes he has therein committed. However,
tho' it be now too late for that Purpofe, yet it
may be of good Ufe for fuch as may attempt the
Amendment of thofe, or any other Coaft-Maps,
or Charts, for the future. And I have the rather
mentioned it at this Time, for that fomewhat of
that Kind is fhortly defign'd to be undertaken.
And it would be, as I conceive, very much the
Intereft of all Mariners, Merchants, nay States
that are concerned in Maritime Affairs, to be at a
conftant Charge to have fuch a Defign profecuted,
till it be compleated for the whole World, at leaft
for all Coafts that are traded to, or much frequented, or which are often paffed near, or touch'd at,
in farther Voyages ; that Seamen, in Cafe of Di-
ftrefs, might know where to find convenient Har-
bouring, and alfo Accommodations of frefh Water, Wood, Victuals, *&c.* I know the Work is
great ; yet it is neceffary, and ought to be done,
fome Time or other, and therefore the fooner the
better. Somewhat of this Kind, I know, is accidentally done almoft by every Navigator, and recorded in their Journals ; but moft of thofe being
kept by themfelves, they are of little publick Benefit, and ferve only for their own future Information. But thofe who have made it their Bufinefs to collect and digeft fuch Journals, and to
print the Refults thence deduced, which the *Hol-
landers* and *Englifh* have profecuted more than any
Nation befides, have very much deferved the Ac-
knowledgments of all the reft of the World ; as
all fuch for the future will do, who fhall promote
and encourage fuch a Work.

T H E

THE Way then, which I propound, is perform'd by taking the true Bearing of an Object at
the fame Inftant from two Stations, which, the farther they are removed from each other, the more
fit they are for this Purpofe.   Now, becaufe both
thefe Stations are to be comprifed within the Ship,
or Veffel, made Ufe of, I would have them to be,
at the Extremities, of the Length of the Veffel, to
wit, at the Stern and Head, or in the Round upon the Head of the Boltfprit, which will add fomewhat to the Diftance of the two Stations; for, upon the Meafure of that depends the Meafure of
all the other Lengths or Diftances.   Now, in each
of thefe Places which are pitched upon for the Stations, I would have a fix'd Frame, or Pedeftal,
for the holding of the Inftrument to be ufed on
it, and the Inftrument fo fixed to it, as to remain
firm and fteady in any Pofture defired, and yet,
with the greateft Eafe imaginable, fo to be moved,
as to refpect directly the Object requir'd, and,
when the Obfervation is made, to be as eafily removed, and as eafy again to be fixed.   The Inftruments I would have to be Sextants of about
two Foot Radius, moft exactly graduated; on each
Side from the middle Line, that is, to 30 Degrees
on each Side, and to be fitted with Perfpective
Sights, whofe *Rete*, or Sight-Point, fhall always
be in the Center of the Inftrument, and that Center always in the Line and *Terminus* of the Diftance of the two Stations, which fhall be invaria-
ble, however the Inftruments are moved to refpect
the Objects; to which Purpofe each of the Inftruments fhall have a double Motion; one of which
fhall be exactly upon the Line of Diftance of the
Inftruments, whereby the Plane of the whole Inftrument is moved; and the other of the Sight,
upon the Plane of the Inftrument it felf, fo as to
refpect the Object, and give the Angle that the
Line

Line of the Sight makes with the former Axis of Motion, or with the middle Line of the Inſtrument ; which middle Line ought to be exactly perpendicular to the Axis of the Motion of the Plane of the Inſtrument, which is the Line of Diſtance. Next, there ſhould be two expert Obſervers placed to make Uſe of theſe Inſtruments, and each of them, at the ſame Inſtant, ſhould direct his proper Sight to the ſame Point of the Object; which, that it may be done the more exactly, I think it convenient, eſpecially in large Ships, to have a Line, Packthread, or Wire, to paſs between the two Obſervers, by which they may, at the Inſtant they deſire, advertiſe the correſponding Obſerver, of what will be neceſſary, according to the Signs or Directions they have before mutually agreed upon. By this Method, if well executed, I do not doubt, but that Heights, Diſtances, and Poſitions of Objects, ſeen on the Sea, may be eſtimated ten times more exact than any that are now made by Judgment, (as they ſay) or rather by Gueſs. And, if any one will endeavour to put it in Practice, I ſhall be very ready to explain any Part thereof more fully, and particularly, for his Information.

Dr.

Dr. Tho. Smith's *Letter to Dr.* Halley, Jun. 12. 1695 *concerning Mr.* Greave's *Observations in* Egypt.

Excerpta out of Mr. *Greave's* Note-Book.

I N his Astronomical Observations, he begins the Day with the Rising of the Sun, as seeming most natural.

Obliquitas Zodiaci, A. C. 1639. 23° 30′ 15″.

T H E Colours of the Planets not different at all from what the Antients make them, and from what we see in *England.*

A. D. 1638. Mense Decembri.
Declinatio acus magneticæ a meridiano Alexandriæ occidentem versus, e multis observationibus, iisque accuratis, 5° 45′.

J. Gravius Anglus.

T H E Altitude of the Pole at *Alexandria* 31° 10′ N. but I find, in other Places of the Book, 31° 5′ and 31° 3′. [Which of these three Observations he determined to be the most accurate and certain, I could not find.]

Posidonius, as *Cleomedes* writes, observ'd the Altitude of *Canopus,* at *Alexandria,* to be 7° ½ ; he observed it there to be but six Degrees, and almost half. *Canopus,* says *Ptolemy,* has Long. 17° 10′. Lat. Austr. 75°. *Snellius* finds the Altitude of the Æquinoctial at *Alexandria* to be 58° 58′, and so the Pole consequently 31° 2′. The Sun's Meridian Altitude taken by him,

11 *March* 1637. S. V. at *Galata,* by *Conſtanti-nople,* 49° $\frac{11}{100}$.

11 *Sept.* 1638. at R*hodes,* 53° $\frac{53}{60}$.

19 *Decemb.* 1638. at *Alexandria,* 35° $\frac{201}{100}$.

The Diameter of the Sun, taken *January* 25. S. V. 1638. 2^h $\frac{40}{100}$ *p. m.* and ſo again 4^h *p. m.*

> As 10000 to 103,
>
> So 100000 to 1030, the Tangent of 35′ 25″ the Diameter of the Sun.

Jan. 29. S. V. 1638. about 5^h *p. m.* he found the ſame Diameter.

A T the Riſing and Setting of the Sun in *Ægypt,* eſpecially about *Alexandria,* there is great Store of Vapours. At a good Diſtance from the Horizon, the Body of the Sun grows ruddy, and appears bigger than it uſually ſeems in *England.* Few Nights, and thoſe without Wind, that he could ſee the Stars near the Horizon: The Reaſon was, becauſe when the Winds blow, they raiſe Sands, which make, oftentimes, the Sky to look, as when it is hazy Weather in *England.*

H E could obſerve no Spots in the Sun, for ſeveral Weeks together, in the latter End of *January, February,* and *March.* On the 5th of *April,* S. V. 1639. three little Spots in the Sun, whereof two cloſe together.

A T this Day but four Channels, or *Oſtia,* of *Nile;* two natural, *Damiata* and R*oſetto,* which make the *Delta;* and part ſome twenty Miles below *Cairo:* Two Artificial, 1. The one on the South Side of *Alexandria,* and has its Beginning ſome 30 Miles above *Roſetto:* By this all Merchandiſe was anciently brought to *Alexandria,* which now comes from R*oſetto* by *Giermas,* with great Uncertainty, by reaſon the *Bocca* of *Nilus* is very dangerous, both becauſe of the N. W. and N. N. W. Winds, which bar in all thoſe Ships, as alſo for the Sands and

Shallows;

Shallows ; tho', at the overflowing of *Nilus*, good Ships may pafs.

2. T h e other at *Boulas*, where it falls into a *Sinus* of the Sea ; *i. e.* in the Mid-way between *Rofetto* and *Damiata*, and like to that at *Madiga*, which is in the Mid-way between *Alexandria* and *Rofetto :* Between thefe two Places, about 40 Miles *Englifh*, *Rofetto* lies from *Alexandria* Eaft and by South.

T h e Courfe of *Nilus*, allowing for the feveral Turnings S. S. E. wherefore *Memphis* and *Alexandria* cannot be in the fame Meridian, nor *Rhodus ;* for from *Rhodus* they fail S. S. E. to *Alexandria*.

Dr. H o o k's *Contrivance to augment the Divifinos of the* Barometer, *in a Difcourfe to the* Royal Society, Dec. 17. 1695.

*T*HE *following Contrivance I met with in a fmall Script of Paper, and find it was a Part of a larger Difcourfe on the Subject, which never came to my Hands.*

W. D e r h a m.

T h e other by a Counterpoife and Wheel, whereby I could make an Index point the Divifions of a long fpiral Line, not only of one Revolution of that Line, but many whole Revolutions in a fpiral Line : So that if one Round of the Spiral were fix Foot Compafs, and fo eafily fufceptible of 1000 Divifions, I could eafily make it move fix or eight Revolutions, each of which fhould be equally capable of the like Number of plain and very vifible Divifions, which maketh the Difference of two Inches in the common, to become 40 or 50 Foot in this, and confequently capable of

eight

eight or ten Thousand Divisions, as sensible and plain to be seen, as the half Decimals of an Inch; and the Contrivance is such, that there is no Manner of Stiffness or Rubbing in the Contrivance, but each of these Divisions will be as exactly pointed to by the Index, as the Index, in the common single Barometer, can be pointed to by the Surface of the Mercury; which, since it is usually comprised within 40 Decimals, or Parts of an Inch, or two Inches, and this Way it may be made 40 or 50 Foot; it follows, that consequently the Alterations will be 200, or 250 Times more visible and discoverable, than by the common Barometer.

AND having brought it to this Pass, that I could, by these Methods, be able to make the smallest Alterations, (that have yet been imagined) to be sensible and measurable, I desisted from improving this Subject, by further Contrivances upon these Principles. However, I may, in Time, shew some other Instruments for Discovery of the Weather, that may, come to be of as good Use.

Dr.

Dr. Hook's *Conjectures about the odd Phæ-nomena observable in the Shell-Fish call-ed the* Nautilus. *Read to the* Royal So-ciety *Dec.* 2, 1696.

FOR *the right Understanding of this Matter, I shall give a brief Account of this Animal from* Ariſtotle, Pliny, Oppian, Ælian, Bellonius, *and their Tranſcribers,* Geſner, Aldrovand, *and* Jon-ſon, *viz.* That *the* Nautilus *is an Inhabitant of the Deep :* That *it hath three Motions, viz. a Power to raiſe it ſelf up from the Bottom to the Surface of the Sea ; that it can ſail thereon ; and again ſink itſelf to the Bottom :* That *its Shell is made very commodiouſly for theſe three Motions, with divers Cells :* That *it can erect its Shell edge-ways for Sail-ing :* That *it hath two (ſome ſay three) Arms, or Claws, with a thin and light, but ſtrong Mem-brane between them, like that of Palmiped Birds :* That *this it hoiſts up and ſpreads like a Sail, and is driven thereby on the Surface of the Sea : Be-ſides which, that it hath alſo other Parts on each Side of it, that it lets down to ſteer and guide its Courſe, as with a Rudder, ſo long as no Danger is nigh : But, if it perceives any Danger from the more powerful Animals, or Storms, that then it fills its Shell with Water, and ſuddenly ſinks itſelf to the Bottom.*

BUT *for the Reader's Diverſion, if he hath a Mind to ſee* Oppian *the Poet's elegant Deſcription of this Inhabitant of the Waters, as tranſlated by* Lippius, *he may find it thus in* Aldrovand. de Te-ſtaceis, *l.* 3. *c.* 5.

———— ——— ——— Quem dicunt nomine vero
Nautilon, inſignem ponto ſua gloria fecit,

Per

Per freta dum cautus fub Navis imagine ludit.
In fabulo domus eft, fumma defertur in unda
Pronus, neu pontum capiat, plenufque gravatus,
Cum nando vehitur, per fluctus Amphitrites,
Extemplo verfus tumidam per marmoris undam
Labitur, ut nandi doctus, puppifque peritus.
Atque pedes geminos tendit, de more Rudentum,
Quos inter medios tenuis membrana tumefcit
Extenta, atque pedes contingunt æquora fubter,
Themoni affimiles, navem, pifcemque domumque
Deducunt.   Si forte malum fupereminet ullum,
Abforbet fluctus intus, lymphifque gravatus,
A tumidis trahitur cum pondere fluctibus unda.

Hinc (faith *Aldrovand)* *homines navigia inve-
nerunt,* & *ex eodem Oppiano citat Lilius Gregorius
Gyraldus.*

W. D e r h a m.

The Account which 'Dr. Hook *gives is thus :*

T h e Structure of the Shell of the *Nautilus,*
which as it is very curious, and indeed very
wonderful, fo it is not lefs inftructive to one that
fhall contemplate on it ; and to me, as yet, it appears to be the only Inftance of a Contrivance
truly wonderful ; for that I do not know any thing
like it in the whole Genus of Fifhes, tho' there
are fome Inftances that tend that Way.   It is, in
fhort, this, The Creature, it feems, to whom this
Shell is adapted, by Accounts we have of it, is an
Inhabitant of the Abyfs, or Great Deep ; which
how deep it is none yet knows, nor will know, till
fome of my *Nuntii ad Abyffum* (which I have formerly acquainted you with) be fent thither, and
bring back Tidings concerning it ; or, till this our
prefent *Nuncius* can find a Way to manifeft, how far
he has afcended to come up to the Day, or how far

he

he defcends to go to his Refting-place at the Bottom of the Sea. For thefe Progreffes he is faid to make, befides his Voyage, when he fails on the Top of the Ocean. Now being conftituted by Nature to perform thefe, and yet to be without Wings or Fins, to help himfelf by Labour to move in any of thefe three Ways; it is wonderful to confider, by what a plain and eafy Contrivance the All-wife Creator has endowed him with fufficient Faculties to perform the fame, with very little or no Fatigue at all, but to be carry'd in his Chariot, or rather Ship, from Place to Place, as he has Occafion to change his Refidence.

T h e Manner of which (if I am not miftaken in my Conjecture) is this : Nature has furnifhed him with a curious Shell, dividing it into many diftinct Cells or Cavities, by certain Valves, Diaphragms or Partitions, which have no Communication with each other, but only by Means of a Gut or *Ductus*, which paffes through them all from the Bowels or Body of the Creature, placed in the Cavity of the Mouth of the Shell to the very End of the Spiral Cone, or conical fhaped Shell, which ends in the very Center or Beginning of the proportional *Spira*, and has there a *Spiramentum* or Vent, which I have formerly difcover'd, by examining more curioufly one of that Kind, by opening it, though it has not hitherto been taken Notice of by any Author that I have met with. The Axis, or middle Line of this Cone, or conically-fhap'd Body is fpiraled round exactly in a Plane, and not helicated on a conical Surface, as in almoft all the Shells of other the conchylious Fifhes, it is obfervable. Now this admirable Structure feems to me not a mere *Lufus Naturæ*, or a Form by Chance, to exprefs, a Variety, but an Emanation of that infinite Wifdom, that appears in the Shapes and Structure of all other created Beings, which

is

Is to endow them with fufficient Abilities to perform thofe Actions, which are made neceffary to
their Well-being. Now, the Relations of Hiftories of this Creature inform us, that it has three
Kinds of Motions through the Water, that is, afcending, defcending, and progreffive ; and fince
there is one Pofture of the Shell, that is moft proper to perform each of thefe, therefore it is, as I
conceive, that the Shell is fo contriv'd, as to be
put, and kept in that Pofition, whilft it performs
that Motion : The Shell then is contriv'd to be
all a Cavity, and to have no other Part or Bowel
of the Creature within the firft Cavity, but only a
fmall String, Gut, or *Ductus*, which paffeth from
the Body of the Creature, placed in the Mouth of
the Shell, to the End of the conical Cavity. Now
by this I conceive, that when this Cavity is fill'd
with Water, the whole Bulk becomes heavier than
the Water, and fo muft fink to the Bottom of
the Sea : But when the Cavity is fill'd with Air,
then the Whole will be boyant, and lighter than
the Water, and fo rife to the Top, and float on
its Surface : Thefe Powers it would have had,
fuppofing the Cavity of the Shell had had no other
but the firft or greateft Diaphragm, and the reft
had been one entire Cavity : But this would not
have difpofed the Shell to all thofe Motions, it is
to perform, into the moft convenient Poftures ; for
that Pofture, that is fitteft for its rifing, would not
be fo for its finking, nor for its failing, nor poffibly for its Progreffion at the Bottom, (if fuch a
Motion it does perform, as to me it feems rational enough to fuppofe) for that every one of them
will need a different Pofture. We find, therefore,
this Cavity all fubdivided by internal Diaphragms
or Partitions, into a great Number of diftinct Cells,
(I have found 40 in fome Shells) and every one
of thefe penetrated by this Gut or *Ductus*, fo that

by

by Means thereof, I conceive, the Animal has a
Power to fill or empty each of thofe Cavities with
Water, as fhall fuffice to poife and trim the Pofture
of his Veffel, or Shell, fitteft for that Navigation or
Voyage he is to make ; or if he be to rife, then
he can empty thofe Cavities of Water, or fill them
with Air which lie toward that Side, that part the
Shell, that beft penetrate the Water: If he be to
defcend, he can fill thofe with Water, and empty
the oppofite ; if to fail on the Top, he can evacuate thofe Cavities that will trim his Shell fit to
fail with the Mouth of it upwards, that he may
there expand his Sails and ufe his Rudders ; and
if to move at the Bottom, he can fill thofe, and
empty the oppofite, fo as that the Mouth may be
downwards, to refpect the Ground or Bottom over
which he paffes, fo to difcover his proper Nutriment
or other Convenience, and to defcend to it when
he finds it. Now it may be imagined, and objected, that thefe Operations may be too notional
and fanciful, and fo feem to have more of Defign
and Counfel, than the Creature feems to be capable of : To which I anfwer, that it is no more, nor,
may be, fo much, as moft other Creatures are endow'd with, and conftantly perform : For whoever
confiders what Defign and Contrivance there is
for the Performance of all mufcular Motion, where
this or that Mufcle is to be ftrained, and that
or the other Mufcle is to be relaxed, and prefently the quite contrary Effects are to be effected,
and all thefe to proceed from the Will, or Intention
of the Creature that moves himfelf thereby, which
Way it pleafeth, will not think it fo ftrange to
conceive, that this Creature may have implanted
in it a Faculty, to make ufe of the Organs for Motions, as well as any other : There needs no Inftitution of a Bird to make ufe of his Wings to fly,
or of his Tail, to poife or guide him in his Flight ;

no,

no, Nature, or the infinitely wife God of Nature hath taken Care to give him an Inftinct or Impulfe, which enables him to do thofe Things, that are neceffary to be done, for the producing the defired Effect. Now, though the fhaping, and triming, and fteering of an artificial Ship, doth require the Underftanding of the Men that are to act in that Ship, to know, and accordingly to difpofe of all Things, for the effecting what is neceffary or defired ; yet 'tis' not thence to be argued, that the Operations of animal Motions muft be perform'd by the Operations of Reafoning. No Man can tell how, or by what Means, he moves his Finger, or any one Mufcle of his Body ; no, Nature hath fet all Things in Order, and endow'd us with a Power to perform what is neceffary, though we know not how, nor by what Means ; nor is the Notion, I have hinted, fo extravagant, or fo much beyond the other Contrivances, for the effecting of various Motions in other Animals, as fome may imagine, fince, when I come to treat of that Subject, I fhall fhew, and prove feveral Contrivances, that are actually made Ufe of, that are abundantly more wonderful.

[*On* Dec. 16. *following,* Dr. Hook *refum'd his Confiderations of the* Nautilus, *and having taken Notice of feveral Tranfmutations, as particularly of Water into the folid Parts of Vegetables, as alfo into Earth or Ice ; he then proceeds, and faith,*]

W. D e r h a m.

B u t this Metamorphofis, or Tranfmutation of Elements, I take Notice of here, only by the by, as it may be of fome Ufe for the Explication of another Metamorphofis of a contrary Nature, and that is, of Water into Air, which is by Rarefaction,
for

for such an Operation Nature seems to have;
and somewhat of this Kind is producible by Art,
as has been prov'd to this Society by many Experiments, heretofore made, for the Production of
artificial Air; which, though under that Notion it
seem'd not to be regarded, yet, as such another,
published a good While after all those Experiments, as his own, not owning at all he had been
inform'd of them, by some of the Members of this
Society: But to pass by that at present (because
there are Abundance of Instances of the like Nature that have been given, which I may on some
other Occasions manifest) I had a further Prospect
in the Success of those Trials than what was, for
the like Reasons, then spoken of; one of which
was, for the Solution of such a Phænomenon as this,
of the floating and sinking of the *Nautilus*, which
I discoursed of the last Meeting but one. It seem'd,
indeed, very strange, how that Creature could so,
at his Will fill, and empty, the Cavities of his Shell,
with Water; it was easy to conceive, how he could
fill his Shell with Water, and so sink himself to the
Bottom; but then how (when there, at such a Distance, from the Air) he could evacuate the Water,
and fill the Cavities with Air, that was difficult to
comprehend, especially being under so great a Pressure of Water: But if Nature had furnish'd him with
a Faculty of producing an artificial Air, then the
Riddle would quickly be unfolded. I found, therefore, that by Art it was feasable to produce such
an artificial Air, and that it was endued with a
very great Power of Expansion, so that it would
not only make itself Room to expand, notwithstanding the incumbent Pressure of the Air on all
Sides; but, if sealed up in strong Glasses, it would
break out the Sides there of, which might have as
much Power of Expansion as might counterpoise,
nay,

·nay, out-power both the Preſſure of the Air, and alſo the Water too, though 100 Times greater than that of the Air. It will be, I confeſs, a difficult Matter for me to prove, that the *Nautili* have ſuch a Power, for that I could never yet get a Sight of that Fiſh that inhabits thoſe Shells, nor do I find that any of the Authors, that pretend to deſcribe it, have, nor has any of them given a Deſcription of it that can give one any true Idea of it : Yet, methinks, it might be procured from ſome ingenious Perſon, that has an Opportunity of viſiting the *Barbadoes*, and ſome of the other Leeward Iſlands, where there are found great Plenty of a ſmaller Sort of them, which though of a differing Shape, in the Coil of the conical Body, yet they agree with all the other Kinds oᶠ hem in having the Diaphragms, and a *Duttus*, oɾ Vᴜſſel paſſing through them all, from the Baſis to the Apex of the coiled Cone, and the Axis of that Cone is alſo coiled in a Plane, as are all the other Kinds of the *Nautili* ; of which I have one here to ſhew, given me by one who had a whole Box full of them, which he had there collected, and brought with him to *England*.

Some farther Obſervations relating to the Nautilus, *and other Shell-Fiſh. Read* Dec. 23, 1696.

W. DERHAM.

I Explain'd, the laſt Day, the Fabrick and Structure of a Creature, which, as Authors inform us, is an Inhabitant of the Abyſs or Great Deep, which does often perform a Voyage from thence to this ſuperior Region of the Air ; and, after the Diſpatch of his Buſineſs here, returns again to his own Habitation. I explain'd alſo, by what Method he per-

form'd

form'd these Voyages, as I conceiv'd, from the Consideration of the Structure of the Shell, and the Effects perform'd by it. I cannot be positive in it, as not having ever had an Opportunity to see the Creature itself: But by considering of the Contrivance of other Fishes, to help them to float in the Water, or at least to buoy them up, or counterpoise them with the Water, by the Help of the *Swim*, as 'tis call'd, or Bladders blown up by Air, or Vapours, I think there is great Probability in the Conjecture.

F o r the *Nautilus* is not the only Inhabitant of the Deep, or of the Bottom of the Sea; no, questionless, there are a Multitude of other Sorts of Animals that are there bred, and do there reside; for we do not only find Oisters, Scalops, Cockles, Periwinkles, and most other Kinds of Shell-Fish, but most Sorts of crustaceous Animals, as various Sorts of Lobsters, various Sorts of Crabs, and various Sorts of Prawns or Shrimps, and such like; nay, we find there also several Sorts of Fishes, not furnish'd either with Shells, or Crusts, which the Fisher-men always find and catch, near the Bottom of the Water, where they fish for them: And I myself have proved, that the best Place, to lay the Bait to catch Whitings, Grundells, Place, Flounders, Beards, is, at within a Fathom of the Ground, where the Depth of the Sea was about 25 Fathoms, or 150 Foot; and, from as great a Depth, I have known Lobsters and Crabs to have been taken by the same Fish-hooks, which were baited for the catching those other Sort of Fishes: And, indeed, most Part of the Lobsters, Crabs, and Prawns, are taken, in Fish-Pots, or Fish-Cages, laid at the Bottom of the Sea, when there has been found a Place frequented by them: As also Scates, Thornbacks, Monk-Fish, Dog-Fish, and the like, which are catched by baited Fish-hooks, laid at the Bottom

of

of the Sea, they being all ty'd by ſtrong ſhort Lines, ty'd to a Rope, there extended between two Stones, which there keep it extended. So that moſt Fiſh, of all Kinds, do, for the moſt Part, there reſide, and thence it is probable to conjecture, that there they find the greateſt Part of their Food and Nouriſhment, and that there do likewiſe grow abundance of diſtinct Sorts of Vegetables, which may be uſeful for that End ; for we find, in Seas that are not very deep, that divers Sorts of Algas, Sea-phans, Sponges, Cotulli, and the like, are there produc'd ; and why then may there not be Multitudes of others ? Nature, we find, does accommodate every thing it produces with all Conveniencies, neceſſary for its Support and Well-being, and fit every Thing neceſſary for the Carrying on and Perfection of its Deſigns ; ſo that I ſee no Reaſon to doubt, that theſe Sub-marine Regions are as well ſtock'd with Variety of Animals and Vegetables, as the Surface of the Earth, which is only Sub-aerial, only we are leſs knowing of them, becauſe they are out of our Element, and we want *Nuntii* or Meſſengers, to ſend thither to bring us back Information, and alſo the Productions and Commodities that this *Terra incognita,* or unknown World, does afford. I have heretofore produced ſome ſuch *Nuntii*, for this or that particular Deſign, but when there may be an Opportunity of ſending them, I ſhall be able to produce divers others, for other Purpoſes, if God ſpare my Life ſo long as to ſee the Seas again free from Rovers, and that the Study of Arts does ſucceed the Study of Arms. It is now above thirty Years ſince I try'd many Experiments, for this very End, to know under how great a Preſſure a terreſtrial or aerial Animal could live, and conſequently a Man ; and I ſhew'd a Way alſo how to ſupply him with freſh Air from above, to whatever Depth he ſhould

be

be able to defcend, without prejudicing his Health or Life : I fhew'd alfo how to accommodate him for feeing with Spectacles, and acting freely in the Water as he could do in the Air, by Means of other Accoutrements, whenever he was able to endure the Preffure. And I have many other Experiments, which would be not only inftructive, but ufeful for thefe and other Defigns, but I want an Apparatus and Affiftance to perform them. And, probably, moft People will treat me as *Columbus* was, when he pretended the Difcovery of a New World to the Weftward : But I have been accuftomed to fuch Kind of Treatments, and fo the better fitted to bear them. However, I think, that fuch Objections as moft will be apt to make, that Animals and Vegetables cannot be rationally fuppofed to live and grow under fo great a Preffure, fo great a Cold, and at fo great a Diftance from the Air, as many Parts at the Bottom of very deep Seas are liable and fubject to ; I fay, I think that thefe Objections may be eafily anfwer'd, by fhewing, that they all proceed from wrong Notions that Men have entertain'd, from the fmall Experience they have had of the Effects, and Powers, and Methods of Nature, and a few Trials will eafily convince them of the Erroneoufnefs of them. We have had Inftances enough of the Fallacioufnefs of fuch immature and hafty Conclufions. The Torrid and Frigid *Zones* were once concluded uninhabitable ; and to affert *Antipodes* was thought atheiftical, heretical, and damnable ; but Time has difcover'd the Falfity and Narrownefs of thofe hafty Conclufions.

Dr.

Dr. Hook's *Difcourfes to the* Royal Society, *in the Beginning of* 1697, *concerning* Amber.

The Sum of Dr. Hook's *Opinion, in thefe Difcourfes, Mr.* Waller *gives in this following Preface, viz.*

HAVING met with a Treatife concerning *Amber*, publifh'd by ——————— (of which he gave an Account) he proceeds from feveral Obfervations therein mentioned, and fome of his own, to give his own Sentiments, *viz.* That *Amber* being found almoft all over *Pruffia*, as well in the Inland Parts, as in the Sea, on the Shore, in the Caverns, Clifts, and under the Hills, by digging, and this in a Sort of *Minera arenaria* ; which, by the Subftances found in it, fuch as Shells petrify'd, and the like, Dr. *Hook* judges to be a certain Layer, or Bed of Sea-Sand, the Remains of the Bottom of fome Sea that formerly covered the whole Country, which, in Procefs of Time, has been raifed above the Level of the prefent Sea ; but, at a certain Depth, all that fandy Bottom yet remains, containing fuch Subftances as were there depofited, whilft it was in that State ; at leaft, fuch of them as have not been rotted and confumed by Time, fuch as petrified Shells, Wood, Bones, with Vitriol, Alum, Niter, and Sea-Salt, together with Lumps of *Amber*, are frequently now found in digging into this Sand, for Wells, or the like. Here he has Recourfe to his Hypothefis, formerly difcourfed of, for the Solution of thefe Appearances, *viz.* That not only the Vales, and lower Parts of the Land, have been fome Time the Bottom of the Sea, but even the Tops of Hills and Mountains ; as the feveral

veral

veral Subſtances now found thereon make evident. *Amber* then being thus found, either at the Bottom of the Sea adjoining, or in theſe Layers of Sand, the Queſtion is, How it came there? and from whence? To anſwer this Inquiry, tho' the Author of the Treatiſe is of another Opinion, yet, from ſeveral Obſervations therein mentioned, Dr. *Hook* judges it to have been the Gum of a certain Tree petrified, and altered to the preſent State and Appearance it has. Thus far Mr. *Waller*; next follows,

Dr. Hook's *Diſcourſe of* Feb. 24, 1696-7.

I HAVE lately ventured to aſſert my Opinion, That *Amber* is a Kind of petrified Reſin, or the Exudation of ſome reſinous Tree, concreted into a Subſtance ſo much ſeemingly different from it, that moſt of the Authors that have treated of it, or deſcribed it, have been quite of a different Opinion. Nay, even the laſt, and, I think, much the beſt, that is, *Philippus Jacobus Hartman*, who has publiſh'd a Tract, Intituled, *Succini Prußici Hiſtoria Phyſica & Civilis.* For, after he has diſproved, as he conceives, all the Opinions of thoſe who have writ of *Amber*, and, amongſt the reſt, thoſe of ſuch as have inclin'd to think it originally ſome vegetable Subſtance, &c. he thus concludes, *p.* 16. of his 2d Book. *Subterraneum utique ſuccinum apud omnes in confeſſo eſſe, idque ex hiſtoria ſatis probari; cum vero, id nec duci nec fundi poſſit, metallis non accenſendum eſſe, neque ex reliquis foſſilium generibus terris, ſulphuri aut bitumini anumerandum, quod ſoliditas ſuccino major quam quæ ejuſmodi foſſilibus ineſt: Lapidem igitur reliquum eſſe, ut dicamus, & quidem non ex ſaxorum aut marmorum, ſed nec ex lapidum peculiariter ita dictorum genere, ſed gemmam, per quam apte reſponderi poſſit ad quæſtionem, quid ſit cum naturam ejus recte exprimat.*

exprimat. Now, how much the wiser we are, as to the Knowledge of its Nature and original Substance, I leave to others to judge; to me, I confess, it seems more obscure, than if he had said that *Amber* is *Amber*; for, what he understands by *Gemma*, to me seems more obscure. He has, indeed, many pertinent Relations, and Observations, which have much assisted me in my Inquiry; but the Uses and Inferences, he draws from them, are quite contrary to those which I have remarked them for. 'Tis not my Design to contradict his Opinion, or to make Objections to his Doctrines: I think it fairer to propound my own, and leave the Choice to the Judgment of such, as shall consider impartially the one and the other Deduction from the Phænomena, which I take to be what concerns his own Observations truly delivered by him. He relates then, (in his Preface) that he has three or four times visited the *Sudavean* Coast, which is the principal Place of *Prussia*, where the *Amber* is found in the greatest Plenty: And that he there did not only inform himself by what he saw, but by Discoursing and Examining the Searcher, or Fishers, for it, and the Overseers and Governors that took Care of the whole Affair, for the Prince's Interest, that he collected, and carried away with him, not only Pieces of *Amber*, but several Sands, Clays, and other Materials found with them, that he might be inform'd by Judgment of others to whom he shew'd them, &c. This Coast faces the West, and lies about 20 Leagues N. E. by E. of the Town of *Dantzick*. He adds, that it has been found also in many Inland Parts of *Prussia*, as well as upon other Shores of the *Baltick* Sea; but thinks it to have been carried by the Sea to such Places from this Shore. He mentions a Piece found at *Gilyenburg*, 20 *German* Miles from the Sea, which was found in

making

making a Well, which proved to yield falt, not frefh Water. Alfo at *Bortenftein*, a Fountain breaking out brought with it much *Amber*. And he mentions another two Miles from *Bartenftein*, which in 1666. broke out in the fame Manner, and vomited, with the *Amber*, a great Quantity of Sea-Sand, which much damaged the Fields; and it hollowed the Mountain fo much, that the Top funk in, and left a foundlefs Abyfs, or Vorago. The *Amber* thrown off was of divers Colours, and Bigneffes; and there were various Pieces of Wood alfo mix'd with the Sand: This Efflux, at laft, ceafed; and it has now left a Lake, and prodigous Caverns. He fays further, that digging a Well at *Afchenburg*, they found *Amber* in a Bed of Sand, like Wood; but he thinks the Wood to be Clay, fhaped like Wood. He mentions alfo many Inland Lakes where it is fometimes found, far diftant from the Sea. He mentions it found in making other Wells; one in the Year 1641, another in 1663, at whofe Bottom *Amber* was found in Beds of Sea-Sand. In other Places Trees were found alfo in the fame Sand. He relates many other Places of *Pruffia*, where, after the fame Manner, it has been found; and he could have inftanced alfo in abundance more. This I find upon the Whole, that it is almoft all over *Pruffia*; that it is generally found in a Bed of Sand; and, that other Subftances, as Wood, Iron, &c. are often found in that Sand alfo. Thefe Subftances, Sand, Amber, Wood, Trees, &c. he believes (*p.* 36.) to be the Product of the Sea; but to be convey'd thither by fubterraneous Paffages: And this efpecially, for that Planks, Iron, and other Parts of Veffels, are found in the fame Sand of the Inland Lakes, and Wells, where he thinks it impoffible that there fhould ever have been any Ships or Veffels. He mentions it to be found in *Pomerania*, but

but in fmall Quantities, and that only to have come from *Pruſſia* : Defcribing further the Places of the Coafts, where 'tis found in moſt Plenty, he fays, the Rocks and Shore have many petrifaꞔted Stones, and that the Clifts, or Banks, are full of Vitriol, or Marquifite Stones ; and Plenty of Vitriol, Niter, and other petrifying Salts, are found mixed with the fame Sand, in which the Pieces of *Amber* are found. *(p. 51.)*   Quantities alfo of Thunderbolt Stones, and *Pruſſian* Diamonds, or Chryftals, are alfo found with it.   He proceeds in his 3d *Chapter* to defcribe the proper Vein, or Mine, of *Amber* ; and this, he fays, no one has truly defcribed befides himfelf. He fays, there are three Kinds of it ; namely, a clayiſh, a woodiſh, and a fandy Mineral ; in one of which it is always found : The clayiſh is a Sort of blue Clay ; the woodiſh confifts of foffil Wood, not vegetable, (as he thinks) but form'd out of the clayiſh one ; fome, he fays, rejeꞔt the clayiſh and fandy Minerals, and think them to be the only true *Minera* of *Amber :* But he, by many Arguments, endeavours to confute their Opinion ; efpecially, that of thofe who rejeꞔt the fandy, becaufe they could not conceive, how the Sea-Sand fhould be carried fo far from the Sea ; which, he conceives, might be done by the univerfal Deluge, or by the Breaking out of Fountains, like that which happened 1666. beforementioned ; or, which he fticks to, that it has been convey'd from the Sea by fubterraneous Caverns, which he thinks are now, and have been in Time, all fill'd up by it, and fo comes to be found all over *Pruſſia*.

B u t the other Authors think the woody *Minera*, to be the only and the true *Minera* of *Amber* ; yet *Wigandus* thinks, that the Places, where it is found, have been formerly covered and overflowed by the Sea *(p. 45.)* He grants, that the

Frifch

Frisch Nerwing has been so overflown, and is now firm Land; but is not satisfied concerning other Places, (*p. 46.*) The woody Vein at *Kraxtepellen* has much Vitriol mix'd with the *Amber*; and there is much Niter also with the Vitriol, (*p. 49.*) and that almost every where, where *Amber* is found, there is found much Niter, as the Miners do assert. He adds, That the Sea does petrify Substances into black Stones, as he himself observed at the Places where *Amber* is found in most Plenty, (*p. 51.*) The Diamonds are found in such petrified Stones, when broken, like those I have formerly described in the hollow Flints, (*p. 52.*) A woody Vein at *Grofs Havenig* he survey'd, and found the Hill to be all sandy, but the middle Part was Wood, like rotten Trees, very black; they seem'd a Kind of Fir-Trees, others thought them Oaks; but he seems to slight what Trees they may be like; for he will have them to be only Clay, or Earth, so shaped, (*p. 6.*) But that at *Kraxtepellen*, he grants, was yet more plainly like Wood, having nothing of Earthiness mix'd with it. That which when moist was very black, when dried discovered more plainly its Parts, and became of a reddish Colour, (*p. 61.*) In the Cavities of these Trees he found them fill'd with *Amber*, and inclosed in the Wood; yet he thinks the Wood never was from Trees; tho' yet he grants, that several of his Friends and Patrons assert them to be true Wood. He adds, (*p. 65.*) that they found them burn clear without Mineral-stinking; but, he says, what he had found, stunk of Niter mix'd with Vitriol and Sulphur: But this Stink the Alga burnt also yields, and stinks somewhat like Garlick. He has much more about the *Minera* of *Amber*, &c. which I shall not trouble you with the Epitome of, at present. I shall only acquaint you with what I collected by my Observation of the whole,

and

and that is, that all thofe Parts, where the *Amber* is found, as in Beds, has been fometimes under the Sea, and fo has been raifed from under it, as I have heretofore made it probable that *England* has been ; that it has been often tumbled with Earthquakes, as *England*, has been where the Foffil Trees are found ; that the Trees have formerly grown where the Banks are now found ; that the Gums of thefe, and fuch like Trees, having dropped from them, have been, by Rains, wafh'd down into the River, and, by their Streams, carried into the Sea ; that greater Quantities have remain'd where the Trees grew ; and when, and where, they came to be thrown down, there they have remained, and fince been petrified into *Amber*, by the nitrous, vitriolate, and other faline Subftances, the Products of faline Eruptions ; and that has been the true Caufe of the Phænomena.

T H A T fuch an Exudation may be from Trees, and that it may be fo carried into the Sea, I could produce many Obfervations ; but I fhall only inftance in one, at prefent, and that is, at *Bencoula*, on *Javaghen*, the *Englifh* have a Fort, and Factory, all their Pitch, or Rofin, is collected out of the Stream of the River, or gathered on the Banks and Shores of the Sea : And *Dampier*, in his Voyages, tells us, That the *Cochin China* Men fetch their Pitch from *Pulo Candore*, where, by cutting a Notch in the Bottom of the Tree, it will run, every Day, more than a Quart of Rofin each Tree. As to the Probability of petrifying of fuch Rofins, I fhould fay more, if I had Affiftance for making Experiments, which at prefent is wanting. But I do not in the leaft doubt, but that the fame thing may be perform'd by Art, which is in this by Nature. I could add many other Arguments for this Conjecture, from the Smells of *Amber*, from the Things inclofed in it ; as alfo fome Obfervations

about

about *Ambergreese,* and some other Petrifactions;
but for these I shall take another Opportunity.

Dr. H o o k's *second Discourse of* Amber.

I ACQUAINTED you, the last Day, with what
my Author thought the most general and
common *Minera* of *Amber,* which he conceives to
be extended over all *Prussia,* as well in all the Inland Parts as in the Sea, on the Shores, and in the
Caverns of the Clifts and Hills out of which it is
dug; and this he has confirm'd by many particular Instances, at some of which he had been a Wit-
ness, and of others he has had very pertinent Informations. This is the *Minera arenaria,* a certain Layer, or Bed of Sand, which, by the Substances found in it, does to me seem plainly to
have been the Bottom of some Sea that has formerly covered all that Country; which Country
has, in Process of Time, been rais'd above the Level of the Surface of the present Sea; but yet, at
a certain Depth, all that sandy Bottom yet remains, containing such Substances as were there
deposited whilst it was in that Estate and Condition; at least such of them as have not by Length
of Time rotted and consumed. These more durable Substances, I say, as the Pyrites and petrified Shells, which he calls Thunderbolts and Wood,
Bones, and Amber, together with the saline Bodies of Vitriol, Alum, Niter, and Sea Salt, are
found to have been, to this Day, preserved in it,
and to be found unconsumed by the general De-
vourer of all Things, *Time.* So that, when they
have Occasion of Digging into this Bed of Sand
for Wells, or the like, or upon the accidental
Eruption of Springs, Lumps and Pieces of *Amber*
are often found in it, together also with divers of
the

the other permanent Subſtances found commonly on the Shores of the Sea.

N o w, that this is not ſo impoſſible or unuſual a Phænomenon, as ſhould ſtartle any one's Aſſent, or Belief of the Truth of it, I did, 33 Years ſince, prove, by Multitudes of Obſervations (divers made my ſelf, and many more by others) that all *England* is a moſt evident Inſtance and Teſtimony of the like Phænomena here ; that is, that not only the Vales, and lower Parts of the Land, have been ſometimes the Bottom of a Sea, but even the Tops of the Hills and Mountains, (ſuch as we have) do plainly, and undeniably, confirm it. How, and when, theſe Alterations have been effected, I have long ſince given my Conjectures ; but, if God reſtore my Health, I hope I ſhall be able to give a more particular, convincing, and ſatisfactory Account ; not only founded upon the Obſervations and Phænomena I then had for my Directors, but many Hundreds of others, which I have ſince that Time collected ; which have not only confirmed, in the general, what I then pitched upon, but has enabled me to be more particular in the Mode, Time, and Method of them.

Now, if this Phænomenon be thus ſolved, by granting that all *Pruſſia* has been formerly under the Sea, and that this *Minera arenaria* is a plain Teſtimony of it ; 'twill not be difficult to conceive how the *Amber* comes to be found in it, ſince the greateſt Part of what is now taken by thoſe, whoſe Buſineſs it is to find it, is by Digging, and Fiſhing it up out of the Sand of the Shore, or of thoſe Parts that are pretty near contiguous to it, and lie not very deep under the Water ; and theſe Pieces of *Amber* are not found on the Top of this Sand, but buried in, and covered by it, a pretty Depth ; not but that, queſtionleſs, the deeper Parts of the Bottom of the Sea, if it were in the ſame Manner digged

digged and examined, would yield as great Plenty of it ; but I perceive they have not a Method of making fuch Experiments, and content themfelves to fifh for it only in the fhallower Parts, and on the Shore. But ftill the Queftion is, How, and from whence came it, and by what Means to be there placed? That then is the next Enquiry.

A N D here, for the anfwering of this, we muft *audire alteram partem,* that is, the Judgment of thofe which he acknowledges to have been the Principal who have treated of this Subject, and thofe from whom (befides his own Obfervations) he hath collected the chiefeſt of his Informations, whom he calls *Triga eruditorum Pruffiæ,* i. e. *Aurifaber, Gobelius, & Wigandus, viri de Succini notitia optime meriti :* But, tho' he praifeth thefe, yet he quotes, and makes Ufe of the Relations and Teftimonies of many others alfo. But yet, as to the true *Minera,* or Vein, or proper Scent of it, he rejects the Opinion of them all, and endeavours, by his whole Difcourfe, to confirm his own Opinion ; which he calls his own, becaufe, fays he, (*p. 55.*) *Hic locus quidem (quantum fcio) diferte a nemine explicatus.* And yet, (he adds) *Proprias autem venas ut aliorum mineralium ita etiam & fuccinorum extare, tam certum mihi quam quod certiffimum.* (We muft allow him fome Grains for his Fondnefs of his own Opinion) *Neque folum id confirmat, quod pecaliaris fignorum cognitio in foffo-ribus requiratur, ut quæ propter fingularis curæ venas indagandi & obfervandi cuipiam in angulo ad* Grofs Hubenig *ubi præprimis foditur, eft demanda-ta, fed quod hujus ætatis eruditi Phyfici Chymici qui illa loca adierunt, aut terras inde allatas fuerunt accuratius contemplati, itidem venas mecum ftatuant, fed & fequentia affertum noftrum manifeftum reddent, ubi etiam per totam Pruffiam fi qua altius ex terra effoffa, figna venarum adfuiffe con-ftiterit.*

ſtiterit. I ſhall not trouble you with the Relation of theſe *Sequentia,* but ſhall only ſay, that the Hypotheſis I have mentioned, of the whole Country's having been ſometimes overflowed by the Sea, does give a full Solution, and Explication, of them all ; and, indeed, they are, moſt of them, very confirming Proofs of that Doƈtrine, if they be duly conſidered ; as I could ſhew, if it were not too tedious : For, how ſhould the broken Pieces of the pitched Plank of a Ship otherwiſe come to be found in his *Minera arenaria,* or *Minera lutea,* at ſo great a Diſtance from the Sea.   He grants, indeed, that the *Amber* found on the Shore, in the Sand, is not there in its proper *Minera ;* but is by Accident thrown up by the Working of the Sea, and, by the ſame Cauſe, covered and buried in the Sand : But, when it is found in the Inland Parts, then he thinks it to be in its proper *Minera.   Alii arenoſam & luteam negant, & caſu vel forte immiſta ſuccina aſſerunt, unam ligneam genuinam venam autumantes.   Verum arenoſam ut ut illa probarent loca quæ ex Pruſſia & Pomerania dedimus, quæ ſcilicet arenis obtutum ſuccinum dedere.* Again, *De collibus vero & montibus arenoſis idem aſſerendum)* (that is, that the *Amber* has been accidentally, or by the Working of the Sea, mixed and buried in the Sand), *difficilior eſt ratio, imprimis quod a mari ſatis ſint remoti.   Quis vero caſus his vel fingi poteſt ſuccina & quidem non contemnenda copia credidiſſe ?   An ad inundationes terrarum recurrendum ? Sed illas nondum ubique hiſtorica fides ſatis adſtruxit.   Potius, ut quæ mea ſit ſententia exprimam, meatibus ſubterraneis eadem deberi contendo, & cum ſcaturigine aliquando ejeƈta fuiſſe non aliter quam ad* Bartenſtein *Anno* 1666. *contigiſſe recenſuimus, & hic multum arenæ ſimul egeſtum, & credo ſub illa etiamnum latere ſuccina.* So that we ſee he is forced, tho' unwillingly, to yield, that
'tis

'tis poffible the *Minera arenaria* may be a Product
of the Sea; tho', becaufe he finds no Hiftory when
the Country was overflowed by the Sea, he would
evade that Way, and introduce his Notion of fubterraneous Paffages; which is, as if a Mariner difcovering an Ifland in fome great Ocean, and finding
fome Houfe on it, but no Inhabitants, fhould conclude that this Houfe had there grown of itfelf, or
elfe had been brought thither thro' the Air by fome
violent Hurricane, and there fet down, (for I fancy a Hurricane might as eafily carry a Houfe threefcore Miles thro' the Air, as fubterraneous Paffages convey the Sand, and *Amber*, of the Sea-
Shore, to a Mountain threefcore Miles in the
Land,) and he fhould make this Conclufion, becaufe he wanted a Hiftory of the Habitation of
this Ifland by fome Men. But (as I faid before)
we muft allow the Author fome Grains for his
Kindnefs to his own Off-fpring. But, as we have
hitherto made him fome Grains of Allowance for
his Partiality for his Hypothefis of the *Minera arenaria*, and the fubterraneous Conveyances, where
he is forced to yield it may be Sea-Sand; fo we
muft now allow him fome Drachms, or rather
Ounces, where he would evade the *Minera lignea*
of *Amber*; for this *Minera* feems to fpoil his *arenaria*: For, tho' almoft all the other Authors do
make this to be the chiefeft, and moft natural *Minera*, which affords, by much, the greateft Quantity,
and the biggeft, and moft entire Pieces; and tho'
he agrees with them, by his own Experience and
Obfervation, yet, fince it would depofe his *Minera
arenaria* from the firft Dignity, by one Salvo he
evades all the Strefs of it againft his *Minera arenaria*, by making it but one Species of the *Minera
lutea*: For, he would have the Wood that is
found, not ever to have been Trees, but only Clay
fo fhaped, by he knows not what Caufe. For he
fays,

fays, (*p*. 61.) after he has told the feveral Opinions of divers Authors concerning the Species of the Trees that compofe the *Minera lignea, Verum parum intereſt ſcire cujus ligni præferat faciem, cum genuinum lignum non eſſe in Phyſicis demonſtretur ſatis.* And fo again, after he has more particularly examined the Words and Affertions of the moft celebrated Authors, concerning this Opinion, and oppofed them, as much as he was able, (which Anfwers, to me, I confefs, feem very infignificant, and, at beft, but Evafions) he fays, (*p*. 182.) *Quare cum nec hiſtoria nec rationibus ſolide probari poſſit, ſuccinum arborum eſſe ſuccum, parum intereſt diſcrimen lachrymæ, gummi, & reſinæ, hoc loco annotare & diſquirere quo nomine convenientius ſuccinum fuerit appellatum.* But, notwithftanding this, what he himfelf has obferved and related, concerning this *Vena lignea,* feems as great an Argument againft his own Opinion, as any can be brought. He fays then of his own Obfervations, (*p*. 59.) *Diverſimode contemplari contigit ad Craxtepellem totum montis jugum contectum quaſi corticibus griſei coloris vidi. Superiorem enim faciem Soli expoſitam ita calor exſiccaret : remota vero hac parte extima piceæ nigredinis terra magnis quaſi & levibus nitidiſque cruſtis perſitus concreta conſpiciebatur ; atque ſi cultro diſſecabatur, quaſi multos molliſſimos cortices diſſecuiſſes, ſecta præ ſe ferebat ; introrſum verſus vero ſoliditas compacta terra difficilem ſectionem reddebat.*

This is his firft Obfervation, which, how plainly it defcribes Trees, I leave any one, unprejudiced, to judge. Firſt, They were found at the Top of the Mountain, where, in Probability, they had grown; and where, by the way, 'tis not very likely that there fhould arife a Fountain of *Bitumen,* or that the *Amber* fhould be conveyed thither by fubterraneous Paffages; and yet Plenty is there
dug

dug out : And whilft he was there, he fays, there was taken up *unum ghetalum fuccini* : What *ghetalum* fignifies I know not. Next, How proper his Opening, or Diffecting of this Ground, as he calls it, does reprefent a rotten Tree, you may eafily judge: For, firft, he defcribes the Subftance of the Bark, or Rind ; next, the fappy Parts of the rotted Tree ; and, laftly, the Heart, or folid woody Part of it.

H i s fecond Obfervation is this, (*p.* 60.) *Aliter ad Grofs Hubenig venam ligneam cum foſſorum operis confpicere datum fuit. Mons erat arenofus, plane intermedium erat genus ligni quod putredo emolliviſſe videbatur, ut facillimo negotio bipalio inſtar molliſſimæ terræ a foſſoribus radi poſſet, nigro quaſi carbonis colore infectum ; fpecie abirgno non abſimile, imprimis cum in ejufmodi cortices circulares veluti deglubi poterat, & alias ejufmodi interſegmenta, five lineamenta oftendebat : Alii querno comparant, ficuti & fruſto ligni* Spork, *fcilicet fragmenta quæ cum fuccino ejici poſtea dicendum erit, ejufdem generis credunt.* Read the Book, *Pag.* 61, 62, 63, 64, 65, *and* 66, *to* §. VIII.

T h a t fome of thefe Pieces of Gum have been found not quite petrified, but only fo far as to have fome Degree of it, yet to be mouldable like Wax ; Further, that the Country has been fometime overflow'd, and that the Remainders of the Sea have been left in feveral Parts of the Country : But, befides the Sea-Water, it feems to me, by feveral Paffages in this Book, that I could quote, that the Land of *Pruſſia* abounds with thefe Kinds of petrifying Subftances, rather than that that Country was the only Place where thofe Kind of Trees grew ; and, that it feems by the Differences of *Ambers*, found in very diftant Parts of the Earth, that other Sorts of refinous Gums may be turn'd into *Amber*, if the petrifying Subftances be

afforded,

afforded, where fuch Gums do drop from their proper Trees : Now what is the true petrifying Subftance of *Amber*, I have not Obfervations enough to determine, nor have I wherewith to defray the Charge of Experiments for that Purpofe. Some Conjectures I have, concerning other Kinds of Petrifactions, for there are many Kinds of that Operation, which I may, fome other Time, difcourfe of, and, if I have Conveniency, fhew fome Experiments about it : 'Tis a Subject that deferves to be cultivated, for it will afford very much of Information in phyfical Productions, and 'tis, I conceive, much differing from the Sentiments of Authors I have hitherto met with, who have treated of it. But I fear, I have been too tedious on this Subject, and therefore fhall fay no more at the prefent, only I fhall fhew a Specimen or two of another Sort of Petrifaction, and thofe are of Chalk, which though from its Plenty, it be more vile, yet, for that very Caufe, it feems to me to be well worthy a more ferious and diligent Enquiry, to find out from what Subftance that Body had its firft Original, for by the Inftances that I fhall fhew, it appears plainly, that it was a fluid Body before it became a folid ; and by other Inftances alfo, it appears, that Flints were likewife fo before they were petrified into Flints, and fo feveral other ftony Concretions, of which Subjects, little is to be found in natural Hiftorians.

A third Difcourfe of Dr. Hook's *concerning* Amber, *on* May 19, 1697.

SINCE I read fome Difcourfes here the laft Vacation, concerning my Conjecture about the Original of *Amber*, in which I endeavour'd, by many Arguments, to prove it to be a Petrifaction

faction of a vegetable Juice, or the refinous Gum of fome Tree, I had Occafion to fearch into the *Acta Hafnienfia* of *Thomas Bartholine*, for another Enquiry, and fo accidentally met with fome curious Obfervations of that learned Man, concerning this Subject of *Amber* ; fome of which I conceive, if not all of them, do much contribute to eftablifh the Doctrine, or Opinion, which I endeavour'd to maintain.

T H A T which I principally took Notice of is, the 57th Head or Section of the firft Volume, for the Years 1671, and 1672. Publifhed at *Copenhagen*, in the Year 1673. It contains an Account of Obfervations and Experiments about *Amber* ; where, firft, he relates, that the Diggers of the new Ditch, about the City of *Copenhagen*, met with Pieces of *Amber* of feveral Bigneffes ; and, which was very remarkable, the Diggers took Notice, that wherever they found thefe Pieces, they found them mix'd with the *Minera* of *Amber*, namely, the Bark or Rinds of Oak-Trees, with which it was not only mixed, but ftuck, or glued faft to it, as is to be feen, fays he, in the feveral Pieces which the Diggers have fold to divers curious Perfons. There was alfo another Mineral, which was a black Wood, as if burnt, to which the *Amber* alfo ftuck. I fhould, fays he, have believed it to be fome Sort of Bitumen, or black *Amber*, if the Smell of it had not made me of another Mind ; for the ill Smell of it, when burnt, made me judge it to be the Remainders of fome Pieces of Oak. And yet *Camden*, fays he, in his Defcription of *Whitby*, mentions fuch a black *Amber*, or Jet, to be found in *England*. The Paffage in *Camden* is this; fpeaking of the Parts near *Whitby*, in the North-Riding of *Yorkfhire*, he fays, *Juxta hunc locum & alibi in hoc littore repertum eft Succinum nigrum five Geate, Gagatum aliqui effe exiftimant, quem inter rariores*

*rariores lapides gemmasq; habuerunt veteres.
Enascitur vero inter cautes ubi rimis debiscunt;
& priusquam expoliatur, colore est subrufo, & æru-
ginoso. Expolitum autem vere est, ut inquit* Solinus,
*Nigro-Gemmeus, de quo Rhemnius Palemon è Dio-
nysio.*

> ―――― *Præfulget nigro splendore Gagates*
> *Hic lapis ardescens austro perfusus aquarum,*
> *Ast oleo perdens flammas, mirabile visu,*
> *Attritus rapit hic teneras, ceu succina frondes,*

Et *Marbodæus* in suo de Gemmis Libello;

> *Nascitur in Lycia lapis & prope Gemma Gagates,*
> *Sed genus eximium fœcunda* Britannia *mittit;*
> *Lucidus & niger, est levis & lævissimus idem:*
> *Vicinas paleas trahit attritu calefactus,*
> *Ardet aqua lotus, restinguitur unctus olivo.*

Audi etiam Solinum. *Gagates in* Britannia *plurimus
optimusq; lapis; si colorem requiras, Nigro-gemmeus;
si qualitatem, nullius fere ponderis; si naturam, aqua
ardet, restinguitur oleo; si potestatem, attritu cale-
factus, applicita detinet.* Thus far *Camden*; from
all which to me it seems very probable, that the
true Jet is a Kind of *Amber*, and differs from the
common yellow *Amber* only in its Colour, which
is very black; but 'tis found, as the other *Amber*
generally is, only in small Pieces, most commonly
in the Clefts of Stones, and which is further remarkable, where there are also found several other
Substances, preserved by Petrifaction; for just before this Passage, about black *Amber*, in the same
Page 485 of my Edition, he, mentioning other Re-
markables of the same Place, says, *Lapides hic in-
veniuntur, serpentium in spiram revolutorum effi-
gie, naturæ ludentis miracula, (quæ ut inquit ille)*
(he means *Bede) natura cum veris & seriis negotiis*
quasi

quafi fatigata ludendo efformat. Serpentes olim fuiffe crederes quos lapideus cortex intexiffet. Hildæ autem precibus adfcribit credulitas, tanquam illa commutaffet, &c.   I fuppofe he means the *Cornu-Ammonis* Stone, of which Kind, many are found in *Yorkfhire* by feveral, but more particularly by Sir *Jonas More,* who affured me, he had feen one, and knew where to fetch it, which was full as big as the fore-Wheel of a Coach, which he promifed to get, and convey to *London,* whenever he went into that Country ; and that there was great Plenty of others of fomewhat fmaller Sizes, yet of the bigger Kind ; divers of which Kind are in the Repofitory, though found in other Parts, as particularly in the Quarries of *Portland,* and at *Keynfham* in *Somerfetfhire,* by Mr. *Waller ;* nor are thefe Kind of Petrifactions in *Yorkfhire* only about *Whitby,* but Multitudes alfo are found in *Richmondfhire,* as the fame Author, Mr. *Camden,* teftifies, *(pag. mihi* 489.) *Incifis rupibus & montofa collium eminentia hæc regio fere tota eminet quorum convexa funt alicubi, funt fatis herbida,* &c. *Montes plumbo, carbone foffili, necnon ære gravidi,* &c.—— *Quod in eorum autem fummitatibus ut etiam alibi, lapides nonnunquam fuerint reperti, cochleas marinas & alia aquatilia referentes, fi non fint naturæ miracula : refufi in omnem terram fub Noe diluvii certa effe indicia cum Orofio Chriftiano hiftorico judicabo. Sic enim ille fcribit,* &c.   But to pafs this by, which I have only taken Notice of, to fhew, that about thofe Parts there are fufficient Indications of Petrifactions of other Subftances alfo ; and thence we have the more Reafon to conclude, that *Amber* alfo, both White, Yellow, and Black, are Petrifactions alfo, and that the Colour may proceed, either from the particular Nature of the original Gum, or elfe from the differing Sorts of the petrifying
mineral

mineral Salts; for 'tis fufficiently known, that Oak turns to Black with a vitriolate Mineral, and to Red with an aluminous; or that the Black may have been produced by the Effects of a fubterraneous Fire there having broke forth, as Pitch and Tar are ftrain'd by the Power of Fire, in the artificial Making them, by burning of the Wood, out of which they are forced; and as the vitrious Jet, of which we had formerly fome Specimens here, prefented by Sir *Robert Moray*, which were brought from the burning Mount *Hecla* in *Ifland*; which black Subftance was a perfect Glafs, and, by melting of it in the ftrong Flame of a Lamp, I reduced feveral Pieces of it to clear tranfparent Glafs, the Thicknefs thereof vanifhing, by the keeping it for fome Time melted in the hot Flame of the Lamp: But however this black *Amber*, or Jet, comes to receive fuch a Tincture, it feems plainly to me, to be of the fame Nature with yellow *Amber*, and both of them very different from thofe Subftances that are originally mineral, as *Afphaltum*, or other bituminous Subftances, efpecially by their Lightnefs and Finenefs of Texture, as their artificial Polifh does plainly manifeft. And *Bartholine* feems plainly to be convinced of the Truth of this Hypothefis by many Paffages, related in this 57th Obfervation; as particularly, that it has been left where it was found at *Copenhagen*, by the Sea; and that all that Country has fometimes, formerly, been overflowed by the Sea. Next, That all *Amber* has been firft foft, and, by Procefs of Time, indurated; that, when foft, it was the Gum of fome Tree; and, while fo, thofe feveral Subftances were immers'd in it, which afterwards became cafed up, and inclofed in the fame Subftance hardened, or petrified; as, *Joh. Gobelius* had a green Frog fo inclofed, and *Frederick* III. King of *Denmark*, had a Lizard after the fame Manner: And Monf. *Picart*

was

was prefented, by *Scholerus*, with the Cone of a Fir-Tree inclofed in the fame Manner. *Non igi-* *tur dubitamus,* fays *Bartholine, liquidam fuiffe Re-* *finam vel Lachrymam ex arbore profluentem, & vel* *fale, vel temporis diuturnitate in maris littoribus* *concrefcere & indurari : Quanquam probabili ra-* *tione quoque, alii ex pingui bitumine in iftam foli-* *ditatem compingi fufpicentur.* As to his other Trials about the Diffolution of *Amber,* mentioned in this 57th *Section,* I omit them, as affording lit-tle of Information pertinent to the Solution of this Query, Whether it owes its Original to a vegeta-ble or mineral Subftance? And pafs on to the 122d Obfervation of his fecond Tome ; where, upon the Occafion of fome Objections made a-gainft his Suppofition, by *Joh. Dan. Major,* Pro-feffor in the Univerfity of *Kilee,* he has enumera-ted all the Obfervations which he conceives to be pertinent to the determining this Controverfy.

1. THE Cone of a Fir-Tree included in *Amber,* my Friend *Sextus Scholerus,* Conful of *Copenha-gen,* had.

2. I faw, at Mr. *Henry Monachen*'s, my honour-ed Kinfman's, a Piece of *Amber,* compofed of white, yellow, and green Parts, in which was in-cluded a Gnat, and fome of the Mofs of a Tree.

3. *Wigandus,* in his Hiftory of the *Pruffia Am-ber,* relates, that he faw a green Frog, which is ufed to fit on the green Leaves of Trees, included in a Mafs of *Pruffia Amber.*

4. THE Sticking of Gumlac to its Sticks gives a Sufpicion that *Amber* may ftick in the fame Manner ; tho', being liquid at firft, it may not ftick to the Twigs, but drop down from them.

5. THAT moft Gums, which flow out of Trees, do not carry with them the Impreffion of thofe Trees.

6. That

6. T H A T thofe fmall Creatures, as Flies and Gnats, which are found in *Amber*, do pitch on fuch Parts of Trees where the Gum trickles down, and fo are as likely there intangled in it, as in the Earth; where they do not only abfcond, during the Winter Months, benummed as 'twere, and half dead.

7. I F you believe *Tacitus*, Birds alfo have been found in *Amber*, whofe Words, in the Book of the Manners of the *Germans*, are remarkable, and not disbeliev'd by any. *Succum tamen arborum efle intelligas* (fays he) *quia terrena quædam atque etiam volucria animalia plerumque interlucent, quæ implicita humore mox durefcente materia eluduntur. Fœcundiora igitur nemora lucofque, ficut Orientis fecretis, ubi Thura Balfamaque fudantur, ita Occidentis infulis terrifque inefle crediderim, quæ vicini folis radiis exprefla atque liquentia in proximum mare labuntur, ac vi tempeftatum in adverfa littora exundant. Si naturam fuccini admoto igne tentes, in modum tedæ accenditur, alitque flammam pinguem & olentem; mox ut in picem refinamve lentefcit*: Thus far *Tacitus.* Now, fays *Bartholine*, If this Account be true, why fhould we doubt the former Arguments; efpecially, fince the natural Hiftorians, *Solinus*, and others, agree with him: Nor is the Fidelity of *Olaus Magnus* to be wholly rejected, tho' he had dreamt in fome Things.

8. T H E Barks of Trees are always found mingled with the *Amber*, where-ever it has been dug up with us.

9. T H E Feathers of Birds have not been obferved in *Amber*; becaufe the Bird fits on the Branches, and not againft the Body of the Tree, where the Gnats, Flies, and other fmall Infects do creep.

10. I N *Norway*, where the Pines, and other refinous Trees abound, there are found Lumps of Gums emulating *Amber*. The Inhabitants call it
a Stone,

a Stone, and my honoured Kinsman, *Joh. Finchius*, brought hither one of those Lumps, which was a Kind of *Amber*; for it seem'd to be a light Stone, or a black Sort of Horn, which would kindle, and burn with Flame; but it stunk much: Otherwise, it seemed a Kind of *Lignum fossile*; yet it did neither burn so readily, nor stink so much, as black *Amber.*

11. As to *Ambergreese*, which is brought from *Florida*, tho' it be doubted by me, whether it be made of the Sperm of a Whale, or the Semen of an Elephant, as *Ctesias* is said, by *Aristotle* in the 2d Book of the *Generation of Animals*, to assert; or of the Dung of certain Birds of the *Maldives*, which feed on odoriferous Plants, as *Ferdinando Lopez* conceives; or a Composition of Lignum Storax, Aloes, Civet, and Laudanum, as *Fuchsius* supposes; or a Kind of Bitumen ouzing out of the Bottom of the Ocean, as *Guliel. Du Val*, in his *Phytologia* asserts it; yet, I dare affirm, that it has the same Original as yellow *Amber*: For, there has been lately found some of it in *Prussia*; and, I cannot doubt, that there may be Trees found in the New World, yielding odoriferous Gums. Thus far *Bartholine.*

To whose Arguments I have only six of my own to add, which seem to me as convincing, if not more, than all these. And those are,

1. T h a t it appears, by all the Relations we have of the finding of the yellow, black, or gray *Amber*, that they are never found in any very large Pieces; but only in such Lumps or Pieces, as may very well be supposed the Exudation of a Gum out of one or two Vents of the same Tree. Whereas, were they mineral, I see no Reason why they should not be found in as great Masses as Asphaltum, Canall, Scots-Coal, or Bitumen, are usually found.

2. T h a t

2. That all Kinds of *Amber*, of whatever Colour, whether white, yellow, green, or black, are very light, and almoſt of the ſame Weight with Water, being but ⅟₇₃ Part heavier; ſo that it will but juſt ſink: Whereas thoſe other Subſtances, as Canal, or Scots-Coal, are very heavy generally, and more than double the Weight of Water.

3. None of theſe Subſtances do ſeem to have any peculiar Figure, as to be formed into plated or priſmatical Bodies, as thoſe Subſtances I laſt mentioned have, eſpecially ſuch as have Tranſparency, as Talk, Selenites, Chryſtals, &c. and the Uniformity, or Continuity, of the Maſs, plainly proves, that it was perfectly united, whilſt yet fluid, and not form'd by Chryſtallization, or Concretion, as Salts out of Brines, or Sugar-Candy out of Syrups; or petrified Spars, or Chryſtal, out of Sea-Water.

4. That Turpentine, by being buried in the Earth, for ſome conſiderable Time, will yield, upon Diſtillation, an Oil perfectly reſembling Oil of *Amber*, for Colour, and Smell, as was above 30 Years ſince proved by Dr. *Daniel Cox*.

5. That there is no other mineral Subſtance that is ſo light and rarified as this, which will take and receive ſo curious a Gloſs, and Poliſh, as this will receive; whereas, of vegetable Subſtances, we have Inſtances enough in hardened Gums, &c.

6. That there are Inſtances enough to be found of the Petrifaction of vegetable Subſtances; and ſo this cannot be look'd on as a Singularity in the Parts.

These, I confeſs, to me ſeem to be *Experimenta Crucis*, as the Lord *Verulam* ſays; and I very much doubt, whether there can be any one Argument as convincing, as each of theſe, for the contrary Opinion. However, I leave every one to judge of both as he ſhall ſee moſt reaſonable,

and

and propound theſe Arguments only, as thoſe which have inclin'd me to be of this Opinion.

THE Weight of a Piece of *Amber* in the Air is, —————— ——— — —— 2443 *grs.*
AND in Water ——— —— 202.
AND is to Water near as $1\frac{82}{1000}$ or $\frac{8}{100}$ Parts.
Amber to Water is as 12 to 11.

2443 ($1\frac{202}{2241}$.
202

—————

2241

———————————————

Obſervations concerning the Refractions of the Atmoſphere.

THESE *Obſervations, I conceive, were the Reverend Mr.* Lowthorp's, *being written in his Hand. They bear Date* February 14, 1698-9, *and precede the Experiment he made at the Requeſt of the* Royal Society *the Month following,* March 28, 1699. *Of which an Account is given in* Phil. Tranſact. N° 257.

W. DERHAM.

THE Doctrine of Refractions does ſo ſenſibly affect almoſt all Aſtronomical Obſervations, that, till that be well eſtabliſh'd, theſe will be too weak to ſupport the Concluſions which are generally inferr'd from them. At preſent, this Doctrine is involv'd in this one great Uncertainty, *viz.* The Air being no uniform Fluid, the Rays of Light are not refracted in any one terminated Superficies, but continually into a Curve ; and it is not eaſy (if poſſible) to determine the Nature of that Curve, till we know the Proportion of the

Powers

Powers of Refraction in the several Densities of the Atmosphere.

That the Attempts, hitherto made by Astronomers, are not satisfactory, I think, will be allow'd, when it is consider'd, that, if (according to the receiv'd Opinion) the Distance of the Moon be about 60 or 61 Semidiameters of the Earth, and the Horizontal Refraction above 30, the Moon at an Eclipse passes thro' the Focus of the Atmosphere, or very near it; and that every distinct Point of the Moon's Hemisphere is illuminated (even in the Middle of a Central Eclipse) by Rays flowing from every Point of the Sun's Hemisphere, which is directly contrary to the Nature of an Eclipse. We seem, therefore, under a Necessity, either to remove the Moon in the Planery System above 20 Semidiameters nearer to the Earth, that it may fall into that Part of its Shadow, which the Duration of Central Eclipses require; or to form a new Theory of the Refractions of the Atmosphere. I am sure the first would so far confound our receiv'd Astronomy, that he would be a very bold Man who durst venture to maintain such a Paradox: But I hope the Proposal of the following Experiment, relating to the latter, will be excused; because it may, perhaps, be of Use towards the removing this great Doubt.

Upon an Air Pump place a small Receiver of Copper, having, on each Side, an even, well-polish'd, flat Glass, and moderately thin : Let their Angle of Inclination to each other be about 65 Degrees, *viz.* with a Telescope, thro' these Glasses, whilst the Receiver is full of Air, a Thread placed at least 40 Foot from them ; and, as the Pump reduces the Air to several Degrees of Rarity, (which may be measured by a Barometer inserted into an End of the Receiver) remove the Thread, till it appear

in

in the ſame Place in the Focus of the Teleſcope, as at firſt. By this Means the Angles of Refracti- on, and Incidence, may be eaſily found, and more certainly determin'd, than any other Way yet pub- lick. And if this Experiment be repeated in ſe- veral Temperatures of the Air, I doubt not but ſuch a Theory of Refractions may be eſtabliſh'd, as may be depended upon, to confirm, or reform, Aſtronomy.

PERHAPS this Experiment may be made, more conveniently, by filling the Receiver with Quickſilver, and pumping it out ; which will leave the Receiver abſolutely void of Air.

THIS Experiment muſt be made with great Nicety and Exactneſs : For, according to the com- mon Tables of Refraction, this Inclination of the Glaſſes to each other (one of them being at Right Angles to the Axis of the Teleſcope, and may be its Object Glaſs) will not produce, for the Angle of Refraction, above 4.

THE Charge will not be above two Guineas, or two and an Half, if made with Quickſilver; and the Materials will be worth moſt Part of that Money again, whenever diſpoſed of : But, if the Air Pump can be ſo fix'd, as not to ſhake, or change its Situation with Working, the Charge will be very little

LET *a b c d* be the Superficies of the Earth, and *e f g h i k* of the Air, having a common Cen- ter *C*. Then ſuppoſe *e g i m* to be a Cylinder of Light flowing from a ſmall Part of the Sun, equal to the Earth, and the extreme Rays *l g* and *m i* refracted (by their Immerſion into the Air) to- wards the Perpendiculars γ *C* and ☽ *C* becoming thereby Horizontal at *b* and *d*; and by their Emer- ſion out of Air, from the Perpendiculars ϙ *C* and ϰ *C*, and to interſect   and the Axis of the Cylinder of Light at the Focus F.   Let the
Angles

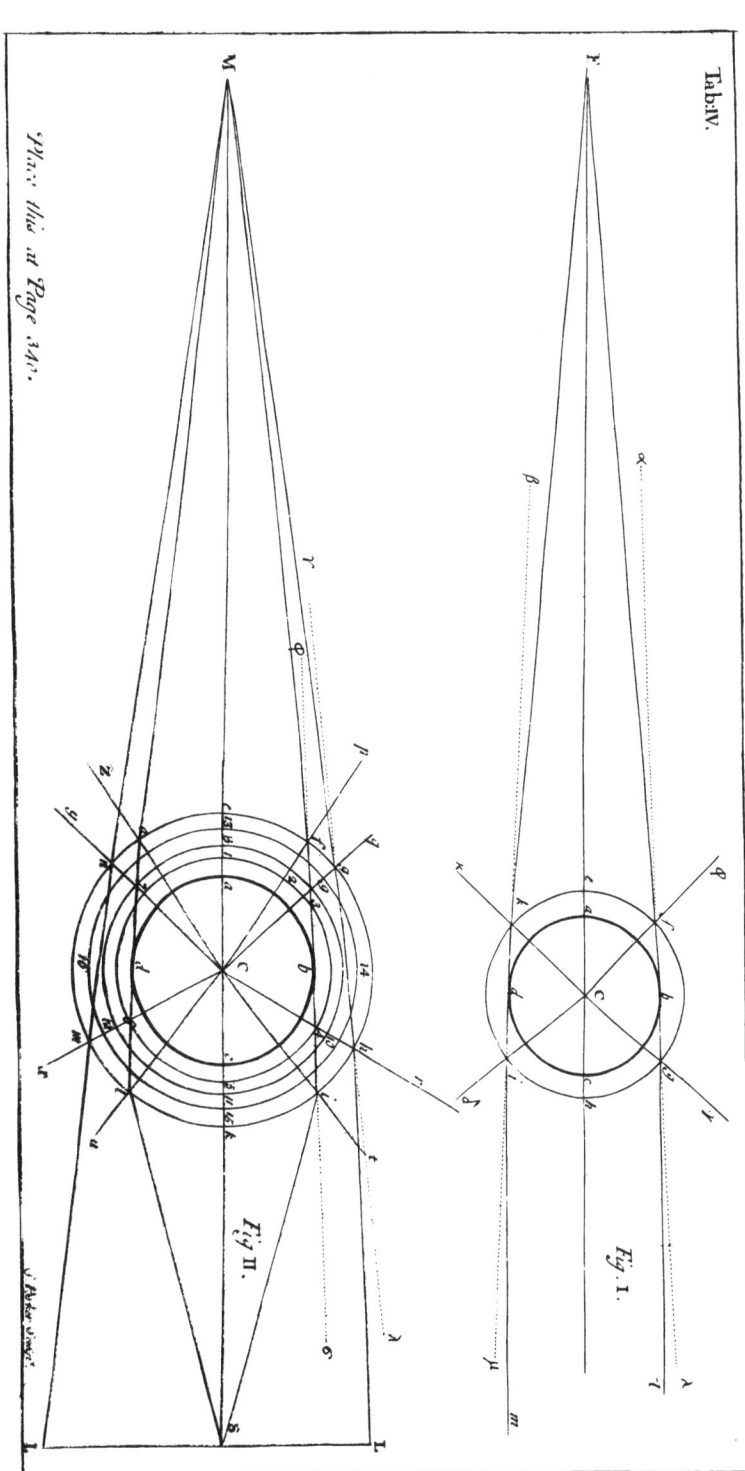

Place this at Page 340.

Fig. I.

Fig. II.

Angles of Horizontal Refraction, towards the Perpendicular $\lambda\, g\, l$, and $\mu\, i\, m$ be 30′, then the Angles of Horizontal Refraction, towards the Perpendiculars $\alpha\, f\, F$, and $\beta\, k\, F$, will be also 30′, and then the Angles $f\, F\, C$ and $k\, F\, C$, will be 60′, or 1°. And then the Semidiameter of the Earth, $b\, C$, subtending that Angle (and near   ) is near $\frac{1}{30}$ Part of the Distance of the Focus $F\, C$, and therefore the Moon's Place at an Eclipse.

Fig. II. L e t $a\, b\, c\, d$ be the Superficies of the Earth, and let the Concentric Circles 1, 2, 3, 4, 5, 6, 7, and 8, 9, 10, 11, 12, and 13, 14, 15, 16, and $e\, f\, g\, h\, i\, k\, l\, m\, n\, o$, be supposed so many distinct Densities of the Air, and $L\, L$ the Diameter of the Sun. If the Rays $S\, i$ and $S\, l$, flowing from the Center of the Sun S, be refracted by their Immersion into Air, towards the Perpendiculars $t\, C$ and $u\, C$, and become Horizontal at b and d; and be refracted by their Emersion from the Perpendiculars $p\, C$ and $z\, C$, and intersect each other at M; then the Parallax of the Sun $i\, S\, C$ (suppose) 48″, being substracted from the Horizontal Refraction, by Immersion $\sigma\, i\, S = 30′$, and the remaining 29′ 12″ being added to the Horizontal Refraction, by Emersion $\varphi\, f\, M = 30′$, the Sum is the Angle $f\, M\, C = 59′\ 12″$; therefore the Semidiameter of the Earth, $b\, C$, subtending this Angle (and near at Right Angles to $M\, C$) is more than $\frac{1}{31}$ Part of the Distance of this Intersection, and therefore not far from the Moon's Place in the Middle of a Central Eclipse.

F r o m hence 'tis obvious, that, if the rarer Air have a less Power of Refraction, the Rays $L\, R$ and $L\, m$, flowing from any Parts of the Sun's Hemisphere, (as L and L) may fall upon some Part of the Atmosphere, where the Angles of Refraction $\lambda\, b\, L$, and $\gamma\, g\, M$ (being less $\sigma\, i\, S$) will direct it to the same Point M. Therefore M is illuminated

by

by Rays flowing from every Part of the Sun's Hemifphere: And therefore, if the Moon be at this Diftance, every diftinct Part of its Hemifphere will be illuminated by Rays flowing from every Part of the Sun's Hemifphere.

I am in too much Hafte to be exact, either in the Exceptions, or Reafonings, but I hope thus much will fufficiently appear, that it is very difficult to account for the Phafes of the Moon upon the receiv'd Hypothefis, and that further Satisfaction is to be wifh'd, which is all the Ufe I defign'd to make of them.

The

The Sun's Eclipse, as it was observed at Canterbury, *in the Year* 1699, September *the* 13*th, in the* Forenoon, *by Mr.* STEPHEN GRAY.

Phaſes.	Digits Eclipſed.	Time by the Min. Watch Correct.		What more was worthy Obſervation.
		h		
1	0	8	12	The Eclipſe began.
2	1	8	18	
3	2⅛	8	24	
4	3	8	29	
5	4	8	34	
6	5	8	39	
7	6	8	47	The Center of the Sun is eclipſed.
8	7¼	8	55	
9	8	9	0	
10	9	9	8	
11	10	9	17	
12	10½	9	24	The greateſt Obſcuration.
13	10	9	30	
14	9	9	38	
15	8½	9	41	
16	8	9	45	
17	7½	9	49	
18	7	9	53	
19	6½	9	57	
20	6	9	59	The Center of the Sun is emerged.
21	5	10	7	
22		10	9	Cloudy.
23		10	20	Cloudy yet.
24	3	10	23	
25	2	10	29	
26	1	10	34	The End of the Eclipſe.
27	0	10	42	Latitude of *Canterbury* 51° 15′

By

By comparing *thefe Obfervations of Mr.* G R A Y *with others of the fame Eclipfe,* I *find the Difference between the Times of Mr.* G R A Y'*s Obfervations at* Canterbury, *and thofe of other* Places *lying* E. *and* W. *of* Canterbury, *to be as in this Table.*

Places where Obfervations were made.	By the Beginning.		By the Middle.		By the End.		Bearing.
	Min.	Sec.	Min.	Sec.	Min.	Sec.	
Oxford, ———			17	00	17	51	W.
Paris,———	3	00	3	00	3	00	E.
Greenwich,——					8	30	E.
Hervelfing,——	43	00			49	00	E.
Nurenburg,——	45	14	53	54	51	56	E.
Ciza, ——					53	00	E.
Leipfick, ———	59	00			56	30	E.

Obfer-

Observations on the Noſtock ; *proving it to be a real Plant. By Monſ.* Geoffroy, *Jun. From the Memoirs of the* Academie Royale des Sciences, *June the* 6th, 1708. *Mem. Edit.* Amſt. *p.* 293.

THE *Noſtock* of *Paracelſus*, which he alſo ſometimes calls *Cærefolium*, and which other Writers name *Cœli Flos*, *Cœli Folium*, *Flos Terræ*, looks like a Kind of Jelly, ſometimes clear and tranſparent, ſometimes greeniſh, trembling when freſh. It is found often in the Summer Months between the vernal and autumnal Æquinox, before Sun-Riſing, in Fields, and on dry ſandy Grounds, after a Shower of Rain. After the Sun is up, the Heat of his Rays dries it up, ſo that there remain only the Skins, or Membranes, of it, of a brown Colour.

THERE is a Doubt, as to the Production of this Subſtance : Some would have it, that it falls from above, like a Dew ; and that it is the Excrement of the Stars : Others look upon it as a Product of the Earth, and a Sort of Plant.

MONS. *Magnol*, in his *Botanicum Monſpelienſe*, names it, *Muſcus fugax, membranaceus, pinguis.* Monſ. *Tournefort*, in his *Treatiſe of the Plants about* Paris, calls it *Noſtock Ciniflorum.* I take theſe two to be the only Botaniſts who have taken Notice of it.

Mr. Ray *ſaith nothing of the* Noſtock, *as a Plant : But if it be the ſame with that Jelly which we call* Star-fall *(which many imagine to be the Subſtance of thoſe Meteors, that go by the Name of* Falling-ſtars, *that ſhoot croſs ſome Part of the Heavens, or ſeem to fall down in the Night)*
if,

if, I ſay, the Noſtock *be this* Jelly, *then it is what Mr.* Ray *takes to be vomited up by Crows, that have overgorged themſelves with Frogs.*

W. Derham.

I thought good to ſhew it to the Academy, in its different Ages, that they might be the better aſſured, that it is a Subſtance produced by the Earth; and that it is joined to it, or communicates with it, by many ſmall Roots, or Fibres.

The Embrion of this Plant, at firſt, looks only like a little, ſoft, fleſhy Tubercle, full of little Inequalities, like thoſe on ſome Strawberries. It is of a green-brown Colour; it grows brighter, or lighter-coloured, as the Skin grows larger; and, at laſt, this Membrane ſeems quite opened, and ſpread upon the Ground, on which it ſometimes leaves the Marks impreſs'd by it.

When the Plant is arrived to this State, it keeps ſo, as long as the Seaſon remains wet; nor does it dry up, or wither, till the Sun and Wind dries and parches the Earth; and, by Conſequence, deprives it of Nouriſhment.

I have obſerved it, in its natural State, to turn up, and bend, uſually lengthways; and, it ſeemed to me, that the two Ends, coming afterwards to meet and join, made a Kind of membranous Bag, or Packet.

In the Year 1667, Mr. *Duclos* brought to the Academy a clear inſipid Water, diſtill'd from the *Noſtock*, which turned white, with a Solution of corroſive Sublimate.

In 1678, Mr. *Bourdelin* made a more exact Analyſis of it; which afforded him a great deal of Phlegm, a conſiderable Quantity of a volatile Salt, either concrete, or diſſolved in the Liquor, and a fœtid Oil.

The

THE Analyfis, I have made of it, agrees very well with that of thefe Gentlemen ; for, at firft, I drew from it a clear, taftelefs Water, which turned white with the Sublimate, and turned Syrup of Violets green. The other Liquors, I drew from it, confirmed what I have remaked of the firft.   Laftly, I gained from it a fine, volatile, concrete Salt, chryftallized on the Sides of the Recipient ; a volatile, urinous Spirit ; and a fœtid Oil.   A Lixivium, being made of the Caput Mortuum calcined, afforded but a very little fix'd Salt, and that mix'd with an earthy Matter : It turned a Solution of corrofive Sublimate a little yellowifh, and made Syrup of Violets green.

IF this Plant be put to ferment of itfelf, in a clofe Veffel, it rots and diffolves into a very ftinking Liquor ; which, at the End of 20 Days, looks red ; and, at 10 Days more, blue.

I HAVE obferved, that thefe two Sorts of Liquors, even after a confiderable Time, were the one acid, the other alcaline.   The red Liquor had no Effect at all on the Solution of Sublimate, and reddened Syrup of Violets but a very little : The blue Liquor turned white, with a Solution of Sublimate Corrofive, and made the Syrup of Violets green.

GREAT Power and Vertue is atrributed to this *Noftock*.   The Common People of *Germany* ufe it to make the Hair grow thick.   It is thought to be an excellent Remedy for Cancers and Fiftula's. A *Swifs* Phyfician, having powder'd it, gave two or three Grains of it to eafe inward Pains : The fame made Ufe of it externally for Ulcers.

IT is a Part of the Compofition called *Sperniolum compofitum Cnoeffelii pro Principe van Eggenberg* ; the Defcription of which may be feen in the *German Ephemerides* for the Year 1676, amongft the Secrets of *Cnoeffelius*.

THE

THE *Chymiſts* imagine that the *Noſtock* contains the univerſal Spirit.   They draw from it a ſoft Spirit, *(Eſprit doux)* to which they attribute great Vertues ; and this they believe to be the radical Diſſolver, or *Menſtruum* of Gold.

THEY diſtill the Water off by the Heat of the Sun only, or of a gentle Fire, otherwiſe it riſes very faſt.   This Water is reckoned a very gentle, mild Diſſolver.

IT is reported that it eaſes Pains admirably, and cures the moſt ſtubborn Ulcers.

Concerning the Burning-Glaſſes *of the Ancients, from the Hiſtory of the* Academie Royale des Sciences, *for the Year* 1708. *With ſome Remarks. By* RIC. WALLER, *Eſq. &c.*

ALTHOUGH the *Academy* does not propoſe to make Enquiries after Antiquities, and is rather employed in Diſcoveries of Matters, as they are at this preſent, than to know what was formerly thought of them, or what Additions may be ſtill made to the Arts, than what has been already practiſed ; yet in it there was a conſiderable Regard made to Monſ. *de la Hire's* Remarks, That *Burning-Glaſſes were not unknown to the Ancients.*

THAT they knew the Uſe of Burning Mirrors, or *Specula Uſtoria* by Reflexion, is unqueſtionable ; ſince ſome Hiſtorians have related, that *Archimedes* made Uſe of theſe for ſetting on Fire the Enemies Ships, in the Siege of *Syracuſa* ; and though they attribute a Power impracticable to them, yet it proves, that at leaſt they were known to them.

BUT

Bᴜᴛ it is certain, thefe Mirrors were of Metal, and concave, and had their Focus by Reflexion: And it is a common Opinion that the Ancients knew nothing of Burning by Refraction, by convex Glaffes.

Monf. *de la Hire* has found this Invention in a Paffage in the *Clouds* of *Ariftophanes,* not ftrain'd, or far-fetch'd. *Strepfiades,* a ftupid old Fellow, tells *Socrates,* That he had found out an excellent Invention not to pay his Debts: The Words are in *Act* the IId. *Scene* the ıft. towards the End.

Tʜᴇ *French* Author having omitted to give them in the *Greek,* I fhall fupply that Omiffion; and the rather, becaufe I am of Opinion there is a Miftake in the *French* Tranflation, which I fhall obferve by and by.

Strepfiades. Ἤδη παρὰ τοῖσι φαρμακοπώλαις τὴν λίθον ταύτην ἑώρας τὴν καλὴν τὴν διαφανῆ ἀφ' ἧς τὸ πῦρ ἅπτυσι; Socrates. Τὴν ὕαλον λέγεις; Strepf. Ἔγωγε. Socr. Φέρε, τί δῆτ' ἄν; Strepf. Εἰ ταύτην λαβὼν, ὁπότε γράφοιτο ἦ δίκην ὁ γραμματεὺς, Ἀποτέρω τὰς ὧδε πρὸς τὸν ἥλιον, ται γράμματα ἐκτήξαιμι τῆς ἐμῆς Δίκης. Socr. Σοφῶς γε, νὴ τὰς Χάριτας. Strepf. Οἴμ' ὡς ἥδομαι, ὅτι πεντετάλαντG διαγέγραπ]αι μοι δίκη.

I fhall give the *Latin* Tranflation of this Paffage, by *Nicodemus Frifchilinus,* to which I fhall add the *French,* and laftly my own.

Strepf. *Vidiftin' apud Unguentarios & Aliptas, lapidem illum pulchrum & pellucidum, unde Ignem accendunt?* Socr. *Num Vitrum dicis?* Strepf. *Utique.* Socr. *Quid cum illo ages?* Strepf. *Si fcriba mihi fcribat dicam, Ego procul ftans, ad hunc modum ad folem, vitro delevero Literas intentæ mihi Dicæ.* Socr. *Sapienter, ita me Gratiæ ament!* Strepf. *O geftio. Dicam quinque Talentorum effe expunctam mihi.*

Fr.

Fr. Str. *As-tu vu chez les Droguiſtes cette belle Pierre tranſparente, avec quoi on allume du feu?* Socr. *N'eſt-ce pas du ver que tu veux dire?* Str. *Juſtement.* Socr. *Et bien, qu'eſt-ce que tu en feras?* Str. *Quand on me donnera une Aſſignation, Je prendrai cette Pierre là, & me mettant au ſoleil, Je ferai fondre* de loin *toute l'Ecriture de l'Aſſignation.*

I ſhall render the *Greek* Words thus:

Strepſ. *Haſt thou ſeen at the* Apothecaries *that fine tranſparent Stone, with which they kindle Fire?* Socr. *Doeſt thou ſpeak of the Glaſs?* Str. *Yes:* Socr. *Bring it: What then?* Str. *When the Attorney hath written an Action againſt me, I will take this Glaſs, and ſtanding at a Diſtance, in this Manner, againſt the Sun, I will efface the Letters of my Action.* Soc. *Cunningly done, by the Graces.* Str. *O! How I rejoice, that the five Talent Action againſt me is defaced.*

I ſhall here only obſerve, that this was indeed to be performed by the Rays refracted through a Glaſs Body, in which I agree, with this Gentleman: Yet, I am of Opinion, it does not come up to a full Proof, that the Ancients knew any more than the Uſe of Spheres, for collecting the Rays, and not the Way by Lenſes, which I take to be a modern Invention; but of this more hereafter. To proceed then with the Tranſlation.

It appears plainly, by this Paſſage, that the Writing was graved in the Wax, which covered a more ſolid Body. That the Glaſs, which did light the Fire, and melted the Wax, was not a Concave; for altho' ſuch a Figure would have its Focus by Reflexion, yet, that being neceſſarily made upwards, would have rendered its Uſe very improper, and unfit for the common Uſe of lighting the
Fire;

Fire; and it would have been neceſſary to have had the Deed held up in the Air to have effaced the Writing; which would be an unnatural Suppoſition, whereas, with a convex Glaſs, which throws the Rays downwards, they may be directed, where one pleaſes.

T H E Scholiaſt, upon this Place of *Ariſtophanes*, ſays, It was a round, thick Glaſs, made on purpoſe for this Uſe. This they rubb'd with Oil, and heated it, to which they fitted, or brought near, a Match, (for the *Greek* Word here is equivocal) and after this Manner the Fire was lighted.

I do not well underſtand what the Oil was for, unleſs it were to poliſh the Glaſs; but, be that as it will, what is ſufficient here, he conceiv'd this Glaſs to be convex, and that in his Time, much later than *Ariſtophanes*, they uſed ſuch Glaſes to kindle a Fire.

I have no Deſign here to make a learned Diſſertation, in which it were a Shame to let any Paſſage of Literature eſcape. I ſhall only remark that *Pliny*, in his 36th and 37th Books, ſpeaks of Balls of Glaſs, and Balls of Chryſtal, which, expoſed to the Sun, burn'd the Cloaths and the Fleſh of ſick Perſons, which needed Cauterizing. And *Lactantius*, in his Treatiſe *de Ira Dei*, ſays, That a Ball of Glaſs, fill'd with Water, and held in the Sun, would kindle the Fire, even in the greateſt cold Weather. Here then we ſee the Effects of convex Glaſſes undoubtedly proved.

B U T if they knew that they would burn, how came it to paſs that they did not alſo know they would magnifie Objects? For it is hard to be imagined, that an Invent on ſo entertaining and uſeful, and withal, ſo ſimple and eaſy, ſhould ever have been loſt, even in the greateſt Barbarouſneſs of any Age; and all Hiſtory fixes the Origin of magnifying Glaſſes about the End of the 13th Age,

Age, when the Uſe of Spectacles began to be diſcovered. If the *Greek* or *Latin* Philoſophers had known this Augmentation of Objects, would they not have made Mention of it very frequently in their Writings, and ſeveral Metaphors, and Alluſions to it, would have been brought into their Language. It is true, there are two or three Paſſages in *Plautus*, which ſeem to hint at the Uſe of Spectacles; which yet, more nearly conſidered, do not at all prove it. We will not inſiſt upon them to avoid a Literature, to which I am a Stranger.

WHENCE came it then, that the Antients were ignorant of the chief Uſe of Burning-Glaſſes? Firſt, The falſe Ideas, the Philoſophers had of Viſion, might contribute to it. They thought, that Viſion was either cauſed by an Emanation of I know not what Sort of Subſtance, which came from our Eyes, and went in Queſt of the Objects; or, by little Repreſentations of the Objects, in Miniature, which came from them, and ſought out our Eyes: All their Difficulty lay, in which of theſe two to chooſe, both equally falſe; they had no Suſpicion of Pencils, of the Rays, nor of our Focus's; and, by conſequence, they could ſee no Agreement between a Burning-Glaſs and the Manner of Viſion, ſo that the one of theſe could not lead them to the other. Beſides, it ſeems, that it was with Balls of Glaſs, ſolid, or fill'd with Water, that they burnt any thing; and Dioptricks demonſtrate, that the Focus of a Sphere of Glaſs is at the Diſtance of half the Radius; ſo that if theſe Balls, or Spheres, had been ſix Inches Diameter, which is the moſt they could be, the Object to be magnified muſt have been placed at one Inch and half to be perceived to be magnified; and it is natural, and almoſt neceſſary, that when any one had look'd thro' theſe Glaſſes, he would have
look'd

look'd only at diftant Objects, which, inftead of appearing bigger, wou'd only have looked con-fus'd. A defined and diftinct Augmentation of diftinct Objects requires either very large Spheres, (which is impracticable, nor ever put in Ufe, or of Portions of large Spheres, as is now practifed with great Succefs) which cou'd fcarce ever be found out by Chance, nor eafy to be invented by Rea-foning.

BESIDES, they muft have known how to have wrought, and ground their Glaffes as we do; and, in all likelihood, the Ancients knew only how to blow their Glafs, to make Veffels of it. It is no ftrange Matter, therefore, that their Know-ledge of *Burning-Glaffes* carried them no farther: It is more ftrange, that from the Ufe of Spectacles, to the Invention of Telefcopes, there fhould be an Interval of 300 Years. Every Thing goes on flowly with us; and, 'tis poffible, we are at this Time on the Brink of fome important Difcovery, which may be furprifing, one Day, that we did not find it out.

THUS ends this ingenious Gentleman's Dif-courfe, to which I fhall beg the Freedom to add fome few Remarks on the fame Subject, or nearly related to it, partly in Confirmation, and partly, as I take it, in clearing the Matter, and fetting it in a true Light, without, in the leaft, pretending to Literature or Criticifm.

IT feems then to me, in the firft Place, that Monf. *de la Hire* would infinuate, that the Anci-ents knew not only Spheres burning at a 4th of the Diameter, but fuch Burning-Glaffes as would have their Effect at a confiderable Diftance; fince he tranflates the *Greek* Word Ἀποτηρω, *De loin, je ferai fondre de loin*; as likewife the *Latin* Word is *Procul.* This Word I rather *Englifh, at a Diftance*; which Senfe, I take it, the Word will more truly

bear

bear in this Place. So that by this Paſſage, it is not neceſſary the Glaſs ſhould burn at any remarkable Diſtance from the Writing on the Obligation, provided it did not touch it; which I take the Meaning to be here by Ἀποτίρω, *Longius, at a Diſtance,* or *farther off.* So that a Sphere of Glaſs might do all that was requiſite in this Caſe. Beſides, if it ſhould be urg'd that this could not be done, when the Scrivener was preſent, without his taking Notice of it. I reply Neither was it: And if it be remarked, that the whole Deſign of the *Nubes* of *Ariſtophanes* being only to ridicule *Socrates,* it was proper enough to bring in an old Coxcomb boaſting of an Invention for doing what indeed it would not perform; ſo the old Man, having ſeen a Fire kindled with a Globe of Glaſs, never conſidering the Diſtance requiſite, might fooliſhly think it would do ſo at any Diſtance.

T H A T the Ancients had ſeveral Ways of kindling the combuſtible Matter placed on their Altars, without making uſe of common Fire for that Purpoſe, might eaſily be ſhewn, were it requiſite, or to the preſent Purpoſe; which it is probable the Prieſts made uſe of, to raiſe the greater Admiration and Devotion in the ignorant and ſuperſtitious Beholders. The moſt ſolemn was that of the re-kindling the Veſtal Fire, when it happen'd to go out.

T H A T the ſacred *Veſtal* Fire was continually kept burning, with great Care, by the *Veſtal Virgins,* is certain; and if at any Time it happened to be extinct, the Virgin, to whoſe Care it was at that Time committed, was ſeverely whipp'd, (*Flagris cæſa Veſtalis*) by the *Pontifex Maximus*; which Damage and Loſs was not to be repaired, by making uſe of any common or culinary Fire to re-kindle it. A particular Account how this was done, *Dionyſius* ſomewhere ſays, he wrote himſelf;

felf; which, as *Juſtus Lipſius* obſerves, muſt have been in ſome of thoſe Books of that Author, that are loſt, *(Lipſ. de Veſta & Veſtal. Syntag. cap. 8.)* The ſame *Lipſius*, out of *Feſtus*, cites this Paſſage, *Tabulam fœlicis materiæ tam diu terebrare mos erat, quouſque acceptum ignem cribro æneo Virgo in ædem ferret.* Which Method of ſetting Wood on Fire is more clearly expreſs'd by *Ariſtotle*, Lib. 3. *De Cœlo. Ignem e lignis excutiunt, alterum lignorum tanquam terebram, in altero circumvertentes :* Which Way of ſetting Wood on Fire, by boring it with another pointed Piece of Wood, *Lipſius* ſays, is ſtill in Uſe among the Natives of the *Weſt Indies.* This *Terebra*, or Borer, *Theophraſtus* ſays, was often made of Laurel, and the other Piece to be bored of Oak.

Plutarch mentions another Way of re-lighting this Fire. *Si quando extinctus ibi ſacer Ignis: negant eum fas eſſe ex alio igne accendi, ſed novum parandum eliciendumque ex ipſo ſole : quod faciunt ſcaphiis ſive vaſculis, quæ parantur ex latere trigoni rectanguli, quod duo latera æqualia habeat, divergunt autem ex circumferentia in unum Centrum. Cum igitur ſoli opponuntur, ut radii ejus in ipſum centrum cogantur & implicentur, aere attenuato, fomenta leviſſima & ſicciſſima apponunt, quæ facillime per renixum & reflexionem concipiant accenſum ignem.* This Paſſage of *Plutarch's Lipſius* attempts to illuſtrate, by a Sort of Funnel, whoſe Sides meet at Right Angles in the Bottom, which he calls the Center of it, and repreſents the Contrivance by a Figure. (See *Fig.* 1.) But either *Plutarch* did not well underſtand the Matter himſelf, or *Lipſius* has miſtaken his Meaning: For a Veſſel, ſo made, will never throw the Rays of the Sun into a Center, or Point; it will indeed reflect the parallel Rays into a Line, (as in *Fig.* 2.) **where**

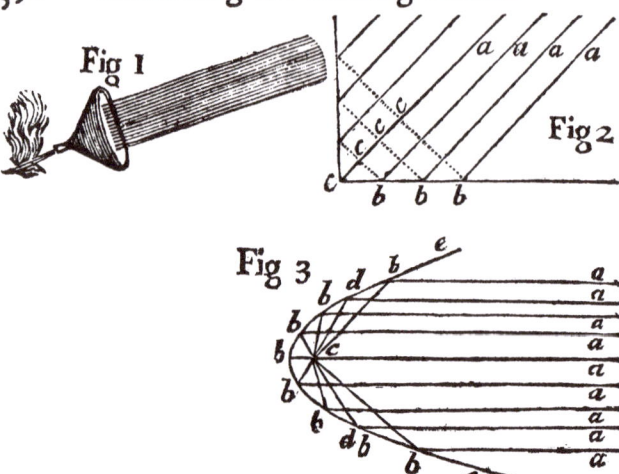

where the parallel Rays of the Sun are reflected,
into the Line *a* C, in the ſeveral Points *c c c*, &c.

Possibly this Veſſel (if any ſuch was uſed
by the Ancients) was of a parabolic Figure, as
(*Fig.* 3.)

Which reflects all the parallel Rays *a b*, *a b*,
into the Focus at *c*, and the parabolic Sides of the
Veſſel appearing, for a great Part, *viz*. from *d* to
e, very like ſtrait Lines, *Plutarch* might take
them for ſuch, which made him call it a Rectan-
gular Triangle ; whereas it was really a parabolic
concave Veſſel, made by the Section of a Rectan-
gular Cone.

As to this *Veſtal* Fire, there is a Paſſage related
by *Dionyſius*, *(Lib.* 2.) concerning the Veſtal
Æmilia ſomething obſervable. Scilt. *Hæc dicens,*
& e veſte linea faſciam abſtrahens, qua cincta erat,
dicunt illam poſt orationem jactaſſe in aram, eque
frigido cinere, quod longe antea fuit abſque ſcintilla,
magnum per linum exiiſſe flammum, &c. This Paſſage ſhews plainly, that the ancient Prieſts knew
how to raiſe a Fire, nay, Flame, out of appear-
ingly

ingly cold Afhes, fince there had been not fo much
as a Spark of Fire on the Altar for a long Time;
whence, it feems, they then knew the *Phofphorus*,
or fomething of the fame Nature, with which the
Linnen Fillet was rubb'd over, and which the un-
difcern'd Warmth of the Afhes fet on Fire, and
poffibly fome other Chymical Preparation might
be concern'd in producing this Flame; which, being kept as a Secret among the Priefts, might well
raife Wonder in the Beholders. They had likewife feveral earthen Veffels, for the Fire, on the
fame Altar; *Et æternos Veftæ focos fictilibus etiamnum vafis contentos*, &c. Valer. lib. 5. cap. 4.

Onuphrius Panvin. de civitate Romana, cap. 29.
(Gr. Vol. II. p. 228.) fpeaking of the Veftal Fire,
has thefe Words; *Ignem amiffum nequaquam dece-
bat ex altero igne fumere; fed maximis precibus
Numen Veftæ placantes, multis facrificiis novum ex
Solis radiis eliciebant, pura & immaculata flamma,
ex lagena aqua plena folis fplendori oppofita, accenfa.* *Onuphrius* does not quote his Authority for
this Way, by a *Lagena*, as he calls it, placed, as
he expreffes it, oppofite to the Sun's Rays, *Ex
lagena aqua plena folis fplendori oppofita, accenfa*;
fired by Means of a *Lagena*, fill'd with Water, and
placed oppofite to the bright Rays. As to what
the Figure of the *Lagena* of the Ancients was, I
fhall give my Thoughts prefently; only firft I
muft obferve, that *Onuph. Panvin.* had no clear
Notion of what he has here written; he tells us,
that this Veffel (whatever its Shape was) was fill'd
with Water; then its Ufe muft have been to refract, and unite the Rays tranfmitted thro' it; for
which Reafon it was not *oppofita fplendori folis*,
but *interpofita*, in the fame Manner as Burning-
Glaffes by Refraction.

As

As to the Figure of the *Lagena*, I find, by a Paſſage in *Pliny*, that it had a Neck, *Lib.* 28. *cap.* 11. where treating of Remedies for a Pain in the Ear, amongſt others, he mentions Goat's U-rine, &c. heated in this Veſſel, the Steam being conveyed hot, thro' its Neck, to the Ear of the Patient, *Urina capri vel tauri aut fullonia vetus calefacta, vapore per lagenæ collum ſubeunte.* Whence 'tis plain, this Veſſel had a Neck, and a large Belly, poſſibly not unlike our Glaſs Bottles, only its Belly was more nearly of a ſpherical Figure, which rendered it more proper for this Purpoſe of refracting the Rays into a common Focus. That the Ancients had the Uſe of Glaſs, is undoubted; a Veſſel therefore, made of Glaſs, of a ſpherical Figure, ſuch as are now commonly ſold at the Glaſs-Shops, under the Name of Jewellers Glaſſes, performs this whole Matter, of ſetting the dry combuſtible Fuel on Fire, by the Sun's Rays. Some ſuch Veſſels are repreſented in the ancient Baſs-Relieves of Compotations or Banquets.

So that this Paſſage out of *Panvin* might be added as a farther Proof of what this ingenious Gentleman ſays, of the Ancients knowing the Uſe of Burning-Glaſſes.

H E R E I ſhall obſerve, that the Ancients made Looking-Glaſſes, *Specula*, not only of Metal, but of Glaſs: *Pliny*, Lib. 36. cap. 26. having before treated of the Art of Glaſs-making, adds, *Authores ſunt, in India e cryſtallo fracta fieri, & ob id nullum comparari Indico. Levibus autem aridiſque lignis coquitur, addito Cyprio, ac Nitro, maxime Ophirio. Continuis fornacibus, ut æs, liquatur: maſſæque fi-ant colore pingui nigricantes, &c. Ex maſſis rur-ſas funditur in officinis, tingiturque: & aliud flatu figuratur, aliud torno teritur, aliud argenti modo celatur, Sidone quondam iis officinis nobili: ſiqui-dem etiam SPECULA excogitaverint.* From which

which Paffage it may be gathered, that the *Sido-nians* made reflecting Looking-Glaffes of this black femiopaque Glafs, and that they knew the Way of grinding Glaffes alfo ; as feems to be plainly hinted, by *Torno teritur.* The Glafs was ground, or work'd off, by a Turner's Engine, or by turning it into the Figure defired: This, without any Foil, being made of black, or femiopaque Glafs, will make a Sort of Burning-Glafs, tho' not fo good as thofe foiled on the Back with Quickfilver.

B u t to come to the Paffage cited by this Gentleman, concerning Burning-Glaffes, out of *Pliny, Lib.* 36. *cap.* 26. the Words are thefe : *Eft autem caloris impatiens (fcilt. vitrum) ni præcedat frigidus liquor : cum addita aqua vitreæ pilæ fole adverfo, in tantum excandefcant, ut veftes exurant.* Whence, by the Way, 'tis apparent that the Ancients did not know the true Reafon why the Balls of Glafs, filled with Water, which they made Ufe of, fet on Fire the combuftible Matter placed in its Focus, fince they fuppofed that the Glafs itfelf was confiderably heated ; whereas it is no more heated for this Tranfmiffion of the Rays, than if it were of a Plane, or any other Figure.

T h e next Place, quoted out of *Pliny, Lib.* 37. *cap.* 2. by this Gentleman, is this ; *Invenio Medicos, quæ funt urenda corporum, non aliter utilius id fieri putare, quam Cryftallina pila adverfis pofita folis radiis.* This Ufe of Cryftal Globes, for Cauterizing, is alfo mentioned by *Matthiolus Diofcorides.* Diofc. *Lib.* 5. *cap.* 116.

B u t altho' it is evident, from all thefe Paffages, that the Ancients knew that the Rays of the Sun, tranfmitted either thro' a Sphere of Cryftal, or a round Ball of Glafs filled with Water, would fet any combuftible Matter on Fire, at a fhort Diftance, *viz.*   Yet, I prefume, this can be no Proof of their ever knowing
the

the Way of making either plano-convex, or double convex Glaſs-Lenſes, ſuch as are now made, either by their turning or grinding them of two ſpherical Segments, or by faſtening two Portions of ſpherical-blown Glaſs Globes, and then filling the Space between them with Water.

So that I am of Opinion that their Knowledge did not reach to the Contrivance of Lenſes, as now made ; therefore their Burning-Glaſſes might rather be termed Burning-Spheres, ſince they were wholly ignorant of the true Cauſe of Viſion, and of the Union of the Rays, by Refraction, into one Focus. Much leſs were they capable of contriving either Microſcopes or Teleſcopes.

THERE is indeed a Paſſage quoted by *Petrus Borellus,* out of *Porta,* in his Treatiſe of Natural Magick, *(De vero Teleſc. Invent. cap.* 1. *Porta Lib.* 17. *cap.* 5.) concerning *Ptolomæus,* that he could diſcern Ships approaching, to the Diſtance of 600 Miles ; *(ſexcenta millia)* but, as *Borelli* obſerves, this Contrivance, if true, remains *inter arcana;* and indeed the Convexity of the Earth contradicts ſuch a Diſtance ; for allowing, the moſt that can be, to the Refraction by the Air, the greateſt Diſtance a Ship can be ſeen at, by Teleſcopes, now amounts but to   ; ſo that *Ptolomæus,* in all Probability, had his Intelligence by the Means of *Specula,* or *Watch Towers,* placed at ſeveral intermediate Diſtances, which, by ſome Signs, gave Notice to each other, of the Ships and their Number. From all which, and much more that might be alledged, I think it is evident, that Burning-Glaſſes, of two ſpherical Segments, were not known to the Ancients. But it is not my Deſign to enquire into the firſt Inventor of Teleſcopes, in this Place, or whether *Roger Bacon,* as it is very probable, knew the perfect Reaſon of Viſion, and the Conſtruction of Teleſcopes, long

long before either *Metius, Galileo,* or *Drebell,* or rather *Joannes Lipperſein* of *Middelburgh* in *Zealand,* as *Borelli (Cap.* 11.) ſhews in the forementioned Treatiſe, about the Year 1609; or his Father, *Zacharias Joannides,* about the Year 1590; but of this enough.

Upon the Whole, I am of Opinion, that the Ancients were wholly ignorant of refracting Burning-Glaſſes, except Spheres, and therefore agree with this Gentleman, that it is no ſtrange thing that they had neither Teleſcopes nor Microſcopes; both which noble Inventions have diſcovered new Worlds to the laſt and preſent Age.

But theſe Remarks I ſubmit to the more learned Judgment, and Cenſure, of this illuſtrious Society.

Mr. WALLER'*s Account of a Book, intituled,* Trattato dell' Apopleſſia, &c.   Dal Dottor Domenico Miſtichelli da Fermo. *In Roma* 1709.

THIS Treatiſe is divided into two Books, and each Book into three Sections, which are ſubdivided into Chapters.

In a ſhort Preface to the Reader, the Author informs him, that the unuſual Number of Perſons, who died ſuddenly at *Rome,* in the Years 1705 and 1706, was the firſt Motive to his writing this Diſcourſe; to which, a ſecond was his Deſire of making known a new Remedy, which his Experience had confirmed very advantageous in this Diſtemper; to which Publication alſo, the Sollicitation of his Friends concurred.

THE

THE firſt Book treats of *Matters relating to the Theory of this Diſtemper:* In which the firſt Section gives an Account of the Body anatomically conſidered, with Reſpect, more particularly, to *Apoplexies.*

THE five firſt Chapters treat of the Head, with its Coverings, and Contents, the *Brain, Cerebellum,* and *Origin of the Nerves*; in which there is little Difference from other Anatomical Treatiſes of the like Nature. I ſhall take Notice of ſome: He obſerves, that the *Dura Mater* is furniſh'd with very numerous Branches of the Blood-Veſſels of the *Carotids* and *Jugulars*; that tho' it ſeems to be a ſimple Membrane, yet it may rather be called a Tendinous *Muſcle, ſui generis*; ſince, as he ſays, it has the Force, and performs the Office of a Muſcle. It has a Motion of Depreſſion, and Elevation, from the Pulſe in the Arteries, which drives the volatile Spirits of the Blood into the ſmall Pipes of the Brain, and thoſe Parts which are the immediate Roots of the Nerves; which volatile Eſſence, conveyed farther on, and being mix'd with the Blood, are called the *Animal Spirits,* and irradiate the ſenſitive and motive Parts of the Body.

IN the 6th Chapter, of the *Medulla Oblongata,* he ſets down ſome Particularities obſerved by himſelf. The *Medulla Oblongata,* he ſays, is a Continuation of the callous Subſtance of the *Brain* and *Cerebellum,* derived from four Heads, which join into one Stalk, of a conical Figure, about three Inches long: This, ſtripp'd of its Membrane, differs not from the Subſtance of the Brain. On the lower Part, it has a ſtrait Furrow running up the Middle, on the Outſide; it has that cineritious Subſtance on the Inſide, which makes the cortical Part of the Brain. He ſays, he could never find, in Brutes, or Men, dying of a violent Death, any tubulous,

tubulous, fiftular, or fibrous Parts; but, rather, a foft, mucous, tomentofe Body: That he had obferved it raw, boiled, and infufed, for feveral Days, in Brandy, Vinegar, and Oil, and always found the fame Confiftence, only a little dark, or livid, Alteration of the Colour of the outward Part: That being cut tranfverfe, and gently pref- fed on the Outfide, a tomentofe, medullary Sub- ftance, ouzed out in little Grains, as it were, from fo many *Tubuli*; whence he fufpected, that the Fi- bres, as well thofe of the *Pia Mater*, which pene- trate the Windings, as thofe of the Membranes, which cover the Ventricles, infinuate into the Sub- ftance of the Brain; and, being prolonged to the *Medulla oblongata*, and *fpinal*, form fo many *Tu- buli* to contain the forementioned tomentofe Sub- ftance. Again, what is very particular, is the Fi- bres of the Membranes which encompafs the *Me- dulla* round: Having kept it covered, with its Membranes, 8 or 10 Days in Vinegar, in which they were thicken'd to about the Breadth of the Back of a Knife, he diligently feparated the Blood- Veffels, which form a Kind of a Net-work; and, then, taking off the external Fibres, and coming to the laft Covering, he obferved, that the Stalk look'd like a braided Trefs of Hair. Many fmall Bundles, or Collections, of ftrait Fibres, are brought over many tranfverfe; and many oblique, again, wove over the tranfverfe, and ftrait ones; fo that, following one Collection of Fibres, you will find it fometimes uppermoft, and fometimes undermoft in the Brede, till they pafs out on the Sides, to form the fpinal Nerves on each Side. This, he fays, is more obfervable on that Part, or *Caudex*, of the *Medulla*, which is inwards, or the Fore-Part, than on the Back-Part, next to the *Nucha*, where only fome ob- lique Fibres run over the ftrait and tranfverfe ones, which feem to come from the Center, to make up, with

with the others, the fpinal Nerves. 2*dly*, This Texture is only obfervable in the fuperficial Part, not wholly ftripp'd of its membranous Coat; therefore in that membranous Coat itfelf: For, when that is quite taken off, there remains only the fimple, tomentofe, Subftance of the Brain; in which, with Signior *Campani*'s Microfcopes, he could difcover nothing obfervable. 3*dly*, Thefe Fibres which thus concur in Bundles, to form the fpinal, lateral, Nerves, at the Place where they pafs thro' the Holes of the *Vertebræ*, are bound round, as it were, with a fmall Ring. 4*thly*, He fays, this Obfervation of the Texture may rather be applied to the membranous Fibres which encompafs the *Medulla*, than to the *Medulla* itfelf, as is done by Dr. *Willis*, in his *Anatome Cerebri*.

THE 7th Chapter treats of the Nerves proceeding from the *Medulla oblongata*, and fpinal Marrow. In the Enumeration of thefe, he follows the Order of Dr. *Willis*, reckoning up 10 Pair of Nerves proceeding from the *Medulla oblongata*: Thofe of the fpinal Marrow he diftributes according to the Parts they proceed from.

IN the 8th and 9th Chapters he fpeaks of the Ufe of the Brain. Here he difagrees from *Willis* and *Malpighi*, that it is a great Gland, for feveral Reafons; and that it cannot be the Place for the Generation of the animal Spirits: Which he confirms by an Obfervation of a Child born without a Brain, only it had the *Meninges* filled with a ferous Liquor. He fays then, that the *Meninges* are an Expanfion of the Tunicles of the *Carotide* Arteries of the Neck, and jugular Veins; that this Membrane encompaffes the whole Brain, the fpinal Marrow, and the Nerves: Whence, fays he, it may, without Difficulty, be apprehended, that the Spirits, or volatile Effence, of the Blood, carried thro' thefe Veffels of the *Meninges*, and, by reafon

reafon of their Subtility, brought into the fibrous Interftices of thefe continued Membranes, (which Membranes alfo encompafs every Fibre of the Mufcles in the Body and the Spirits) are, by the Motion of the faid Membranes, forced forwards to all the fenfitive and moving Parts of the Body. In fine, our Author maintains, that the animal Spirits are the more fubtile Parts of the Blood feparated from the capillary *Carotide* Arteries in the *Pia Mater*, chiefly by means of the Interftices of the Fibres, of which the Membrane confifts; that alfo along the fpinal Marrow, which is but a Continuation of the fame Membrane, the fame Separations are made; and, that to have a greater Plenty of Spirits in Readinefs for Ufe, for all the animal and voluntary Functions; and that the chief Ufe of the Mafs of the Brain is, by its Softnefs, Coldnefs, and Bulk, to diftend, and bear up the *Meninx*, and help it in the Separation of this fubtile Spirit from the Blood; which he endeavours to confirm by feveral Reafons and Obfervations.

THE 10th and 11th Chapters fhew, *how Senfation and animal Motion is performed.* As to the firft, he fays, three things are to be confidered; the *Objefts*, or fenfible Bodies, the *Organs*, and the *Soul*: The *Objeffs*, by their Materiality, or extended Quantity, muft either immediately touch the Senfory, or mediately imprefs upon it their Motions, which the Schools call *Species*: Whence all Senfation may be reduced to Touching. This he exemplifies in the Hearing, Seeing, and the reft of the Senfes. As for the *Organs*, tho' *Donato Rofetti* makes them 11, yet he is contented with five; to all of which the Nerves arifing from the *Medulla oblongata*, conveying the animal Spirits, are continued. That thefe animal Spirits are corporeal, is evident from a certain Modification which happens either in the foft nervous,

Fila-

Filaments, or in the animal Liquids contained in them, which they communicate *partem poſt partem* to the *Meninges*, from whence the animal Spirits are derived in the greateſt Plenty. The *Anima*, or *Soul*, being immaterial, is not ſubject to Modifications, or of receiving Impreſſions from the Spirits or Nerves, which are material. But as *Hippocrates* ſays, *Qualiacunque patitur corpus, talia videt anima*, which has a Power of comprehending theſe Impreſſions, and diſtends its Powers, and raiſes the Paſſions, which it does, or ought to regulate with its Approbation, or Diſapprobation. That indeed, without this *Soul*, the Impreſſions would be made, but they would neither be diſcern'd, nor any Uſe made of them.

As to the next Thing, how animal Motions are performed, he ſays, The Muſcles are a Collection of fleſhy Fibres; that it is to be obſerv'd, they are all envelop'd with a Membrane denſe, ſtrong and nervous; in which external Membrane all the Nerves terminate with their numerous Ramifications, that it is impoſſible to ſeparate this Membrane from the contained fleſhy Fibres, without breaking innumerable Filaments of the Nerves; ſo that the animal Spirits paſs by theſe Filaments into all the fleſhy Fibres that make up the Belly of the Muſcle. By Means of which, the animal Spirits, which are fluid Bodies, enter into, ſwell, and ſo contract the Muſcle, by drawing the tendinous Parts, that are at each End, nearer together: This he illuſtrates by a Cable, which, being wetted, is thereby ſhortened. That when, according to the Empire of the Soul, there is ordered more of theſe Spirits to one Part, than to another, of the *Dura* and *Pia Mater*, either within the Scull, or along the Canal of the ſpinal Marrow; then the correſponding Branches of it, on that Side, are acted upon, and the correſponding

Muſcle

Mufcle fhortened, and the Member moved accordingly; that this is done independently on the Will, fometimes, and thefe are called involuntary, or natural, Motions.

An *Apoplexy* often happening from a Defect in the Heart, in the 12th, and four following Chapters, he confiders the *Thorax*, *Pericardium*, and the *Heart*, with its Auricles and Ventricles, its Subftance, Ufe and Motion; in all which, there is nothing different from other Anatomifts; only as to the Heart he agrees with Sig. *Giacomo Sircibaldi*, in his *Apollo Bifrons*, that the Subftance about the Heart, commonly taken for Fat, is made of the ferous *Lympha* contained in the *Pericardium*, brought to that Confiftency by the Heat of the Heart, like a Sort of Glue, hardened, and fticking clofe to the Heart, fince it is not melted by Heat, like Fat, and crackles in the Flame of a Candle. As to the Motion of the Heart, he fays, indeed, that its *Syftole* is caufed by the Spirits, conveyed by the Branches of the *par Vagum* to the Membrane that covers it; *but it were to be wifhed he had more particularly explain'd, how this Contraction is fo regularly and alternately caufed and continued.* He endeavours to explain it by the alternate Vibrations of the Balance of a Watch, which the circular Motion of the Balance Wheel continues backwards and forwards, by the different Pofition of the Pallets; fo the circular, yet alternate, rufhing in of the Blood and Spirits, caufe the alternate Motions of the Heart and Pulfe. As to the Obfervation of a Frog's, and fome other Creatures Hearts beating, after they are taken out of the Body, he compares that to a Steel Spring, which, being bent one Way, will continue its Vibrations backwards and forwards for fome Time, after the firft bending Force is removed. *In this, I think, he is fhort.*

THE

THE 17th and 18th Chapters of this Section, describe the Veins and Arteries, with their several Coats and Structures, together with the Nature, Motion, and Use of the Blood. As to which, he says, many Principles of it are discovered, *viz.* Certain subtile, airy, volatile Particles, discernible by the Plenty of Vapours that arise from it, while it remains hot, when fresh taken out of the Body. 2*dly*, Salts of divers Figures, observed in the *Serum*, by the Microscope. 3*dly*, Several fibrous *Stamina*, or Fibres, observ'd in the thick, or grumous, Part, when wash'd in warm Water. 4*thly*, Some small red Globules, made of little oval, plane Corpuscles, which, separated, are transparent, but, being joined, appear more or less of a purple Colour. 5*thly*, Several Particles of Sulphur, which Chymistry procures out of the thick Part, of a yellow, or red Colour. 6*thly*, Several little *Moleculæ*, derived from the various Combinations of the fore-named Principles. 7*thly*, A great Proportion of a watery Fluid, serving as a Vehicle to the rest. 8*thly*, a great Quantity of Chyle, not yet converted into Blood. To this Fluid, or Blood, he gives a threefold Motion; an Agitative, from the different specifick Gravities of the Contents; a Fermentative, and a Circular, from the Action or Pulse of the Heart: All which Motions he applies to the Increase, Nutrition, and Preservation of the Individual.

THE second Section relates to the Theory of an Apoplexy, and is divided into 14 Chapters. I shall only take Notice of what I think most observable. He says, that the Apoplexy, as was remarkable in that at *Rome*, so frequent from the Autumn of 1705, throughout the whole Winter, and Spring following, being a sudden Deprivation of Sense and Motion, it must be granted, that the Parts affected, are either the animal Spirits, or

the

the Nerves, or both; and fince this Stroke is fo inftantaneous through the whole Body, 'tis reafonable to believe that the Mifchief is imprefs'd on the Principle of all the Nerves that is on the *Meninges*, tho', fince there is a continual Circulation, he allows that the Part immediatly affected, in an Appoplectick Fit, may be in the *Thorax*, the Heart itfelf failing to fend a requifite Quantity of Blood to the Brain.

HAVING thus mention'd the Parts affected, he proceeds to confider the Signs of it. Thefe Signs he diftinguifhes into, Thofe which fhew Perfons fubject to it; An impendent Evil, or Fit; A real prefent Fit; and, Thofe which diftinguifh this from other Ailments: For which I muft refer to the Author; taking Notice only of fome Remarks: As, that fometimes in an Apoplectick Fit, the Pulfe is full and ftrong, and without any Fever; and this accompanied with a Snorting in Breathing, and a Relaxation of the Sphincters of the *Anus* and *Urethra*. In the next Place, amongft external, or remote Caufes, he reckons Evacuations either fuddenly ftopp'd, or unufually large, of what Kind foever.

IN the 5th Chapter of internal Caufes, he enumerates feveral, fome relating to the Brain itfelf, and others to the Heart.

IN the next Chapter, he mentions Apoplexies caufed by a Blow on the Head, or Stomach; the firft caufing an Extravafation of Blood in the capillary Veffels: And here he gives fome Inftances of fudden Death from a Blow on the Head, efpecially near the Temples, with the Reafon of fuch fudden Deaths; fuch as the Loofening the Contact of the Brain from the *Meninges*, Extravafation of the contained Fluids, &c. which muft neceffarily interrupt the Courfe of the Spirits.

AS to Blows on the Pit of the Stomach proving mortal, he cites a Cafe in *Hippocrates*, of a Boy
kick'd

kick'd by a Mule, and agrees with *Willis*, that the outward Coat of the Ventricle, being all nervous, and the Nerves of the *Par Vagum*, brought thither, form, near its Orifice, remarkable *Plexus's*; whence it has a wonderful Communication with the Brain and Heart, and so Convulsions, Syncope's, and the like mortal Syptoms, happen upon a Hurt there.

T H E 7th and 8th Chapters, treating of Apoplexies from Hurts on the *Pericranium*, and Fractures of the Skull, have little remarkable, more than is generally known.

I N the 9th Chapter, he observes, that Hurts on one Side of the Head cause a paralytic Affection on the contrary Side. As to this, having observed, that *Hippocrates* has taken Notice of this Case, he explains it by what he had before related of the Nerves, in the *Meninges of he Medulla oblongata*, that they are interwoven and braided, so as those which proceed, at first, from the Left Side Fibres of the *Meninges*, have their Branching-out to the Limbs, or other Parts, on the Right Side. He says farther, That the little Rings, which bind round the Nerves, at their Parting from the *Vertebræ*, may be-convulsed, and so, stopping the Nerves, cause a Palsey.

T H E remaining Chapters of this Section treat of the internal Causes of Apoplexies, the Vitiousness of the solid Parts, *viz.* Nerves, Membranes, Tunicles, musculous or tendinous Fibres, and the like, which, he says, proceed, either from their too strong Tension, or from their too great Flaccidity, or Feebleness.

T H E first of these may cause a Strangulation, or Stoppage, of the *Canaliculi*, of the Nerves, and instantaneously stop the Heart. This he farther explains in the *Meninges*, and in the *Lymphaticks*, within the Head.

O N

O N the contrary, too great a Relaxation is as mifchievous, from the Parts in that Cafe failing to fend a fufficient Supply of Spirits, to the feveral Organs of the Body. And, as this Palfey is frequent in the outward Part of the Body, fo it may, and does fometimes, feize the Heart, or *Meninges*. This Weaknefs of the Parts fometimes happens to the Arteries, which he makes the Caufe of Aneurifms, the Varices, &c. This Cafe, generally, is preceded by very long Indifpofitions, or lingering Diftempers.

T H E 12th Chapter is concerning *Apoplexies* caufed by the Denfity of the fluid Parts, the Chyle, Blood, Lympha, and *Succus Nervofus*, of all which he treats briefly.

A N D, as all thefe Fluids are, fometimes, too thick, fo, on the contrary, they are alfo, at other Times, too fluid, which is the Subject of his next Chapter. This, he fays, he has frequently obferved in the Cavities of the Body, efpecially in the Heads of dead Perfons, they being filled with a bloody *Serum*.

T H E laft Chapter is of *Apoplexies* from *Narcotic Steams*. Speaking here of *Opium*, (which by the Way he feems not to have a good Account of) he makes the fulphureous and vifcous Quality of it to bind, and, as it were, glue up, and fo ftop the Paffages of the Spirits : Whence Sleep, and, if taken too largely, Death follows. He makes the Suffocation by Charcoal, to be from the fame Caufe, in which he is, without doubt, miftaken, their Effects, aud Manner of acting on the Body being quite different.

H E R E he obferves, that the Wines of *Rome*, when mix'd with Water, will not depurate, unlefs helped with Flower of Brimftone, which their Vinteners call *Ciambella* (a *Simnel* ;) but if they put too much into it, as they are apt to do in refining

fining either too grofs, or thick, Wines, the Narcotic Sulphur, thereby mix'd with the Wine,
proves very mifchievous.

The Third SECTION.

Of the particular Caufes producing the frequent
Apoplexies *at* Rome *in* 1705-6.

IN order to explain this more fatisfactorily, our
Author premifes feveral *Lemmata*.

Lemma I. OF Refpiration and its Neceffity.

THE Blood-Veffels in the Lungs, being deftitute
of the flefhy Fibres that accompany all the Arteries
of the reft of the Body, are fupplied, in this Refpect,
by the Spring of the Air admitted into the *Veficu-
læ* of the Lungs, on which the capillary Blood-
Veffels are ramified; which not only helps forward
its Motion, but carries off, when exfpired, the
noxious Humours from the Blood. Here he mentions feveral other Ufes of Refpiration: And, in

THE 2d *Lemma*, treats of the principal Ufe of
Refpiration, *The Introduction of an aerial Nitre
into the Blood.*

Here he mentions this Experiment: If you omit
to tie up, very clofe, the pulmonary Vein, and
Artery, and blow up, by the Windpipe, the Lungs
of any Animal, and then tie up the *Afpera Arteria*;
yet the Air will find a Way out, and the Lungs
fink: Which, on the contrary, will not happen,
if the Extremities of the pulmonary Vein, and
Artery, are well tied up: Whence he argues a
Communication of the Air with the Blood. He
alfo obferves the Difference of Colour in the Blood,
before, and after, its paffing thro' the Lungs: Obferving farther, that our Atmofphere is impregnated with this nitrous Spirit; he adds, that if fome
few Drops of the Chymical Spirit of Nitre be
dropp'd on black, cold, and coagulated, grumous,
Blood,

Blood, it will not only render it fluid, but florid, and like arterial Blood.

T o this he fubjoins, that Nitre, having an expulfive and elaftic Power, communicates to the Blood, by Means of Refpiration, that which caufes its Fermentation, and continued internal Motion; citing *Galen (Lib. de Refp.) Aer non ad refrigerandam, fed ad nutriendam vitalem flammam, animalibus ineft.*

T H E 3d *Lemma* is to fhew, that this nitrous aerial Spirit, mix'd with fome other Principles in the Blood, compounds, in the Veins and Arteries, a Subftance very like the Air which encompaffes us.

A s to this Point, having obferved that our Atmofphere is a Compound of all Sorts of Particles exhaled from Earths, Minerals, Vegetables, Animals, &c. he fays, that what pure Part foever may be received into the Blood, yet, in that Blood, it meets with the like Particles conveyed in the Chyle, from the feveral Foods eaten; whereby, when mix'd therewith, it becomes like the encompaffing Air.

B E S I D E S, finding thofe Perfons, that dwell in marfhy Places, fubject to ill Habits of Body, he argues, that the Air of fuch unhealthy Places, fome Way or other, gets into the Blood, and that, by the Breath, feems the moft likely.

T H E 4th *Lemma* is, that the Air, mix'd with the Blood, agrees with, and participates of the Condenfation and Rarefaction of the Ambient.

H A V I N G mention'd the feveral States of the Air, in refpect to Condenfation and Rarefaction, and compared its component elaftic Particles, to incurvated Steel Springs, always endeavouring to dilate themfelves; and obferved, that it is the Particles of Air, in Spirit of Wine, in Thermometers, which dilate, or contract, by Heat and Cold: He urges,

urges, that, for the fame Reafon, the Air, contained and intermixed with the feveral Fluids of the Body, muft alfo participate with the Alterations of the Ambient.

H a v i n g premifed thefe *Lemma's*, in the fifth Chapter, he treats of *Apoplexies* caufed by the Rarity, or Denfity of the Air, external and internal.

T h i s Alteration of the Temperament of the Air, when to Excefs, hinders that due Separation of the Humours, and more fpiritous and ufeful Parts from the Chyle and Blood, in the Harmony of which, Health and Strength confifts : Too great a Condenfation, clogging, and thereby hindring this due Separation ; and the Contrary, forcing off unfit Particles, efpecially to the Brain and Meninges, where the Separation of the animal Spirits is made.

F r o m thefe *Lemmata*, our Author, as fo many Corollaries, deduces the Caufes of Faintings, or a Sort of *Apoplexies*, in the too excefiive Heats of the Summer, from a too great Rarefaction. As on the Contrary, the Fixation of the Fluids, by excefiive condenfing Cold. The falling of Fruits from the Trees, at both thefe Extreams, *&c.* confirming it.

T h e 6th Chapter, being his 5th *Lemma*, is to fhew that the *animal Spirits are compounded of a two-fold volatile Effence*, viz. *a fulphureous from the Blood, and a nitrous from the Air.*

S i n c e, it muft be granted, there is in the Blood a continual Motion and Fermentation of the feveral different compounding Principles, it may eafily be allow'd, that there is feparated in the Brain a more fine and fubtile Effence, which, communicated to the Nerves, is what may be called the animal Spirits, the animal Liquid, or *Succus Nervofus.*

H e

HE fays, as from Wine fermented, an ardent fulphureous Spirit is extracted ; fo Blood, after its frequent Motions and Fermentations, affords the like fulphureous Spirit to the Brain or Nerves, mix'd with the nitrous Spirit taken out of the Air.

THE 7th Chapter of *Apoplexies*, from the Condenfation of the nitrous Spirit, relates this Experiment.

IF, near an unftopp'd Bottle of frefh-drawn Spirit of Nitre, another open Bottle of Spirit of Urine, or Sal Armoniac, be placed, the Steams from the Nitre will be thereby condenfed, like a white Smoak, which, inftead of evaporating into the Air, falls down on the Table, or Place, where the Bottles ftand. Whence he argues, that whenever an urinous Spirit abounds in the Blood, it produces the fame Effect in the nitrous of the animal Spirits, and fo caufes an *Apoplexy*.

The 8th Chapter of *Apoplexies*, from the Condenfation of the fulphureous Part of the animal Spirits.

THIS he explains by rectified Spirit of Wine, coagulating with a fmall Quantity of the urinous, or Sal Armoniac, Spirit ; and whereas he had before afferted Wine and Blood to confift nearly of the fame Principles, he hence deduces another Caufe of *Apoplexies*.

As to the Objection, that Spirit of Sal Armoniac, Hartfhorn, and the like, is given, with Succefs, in Apoplectic Fits ; he fays, if fuch Spirits were immediatly mixed with the animal Juice, the Mifchief would foon appear ; but after paffing thro' fo many Alterations, as they fuffer in the *Vifcera*, they do neither Good nor Hurt ; and if in a Fit, as it is poffible, they do any Good, it is by their violent irritating the Nerves of the Palate and Tongue, and likewife thofe of the Stomach, which,

which, as one Nail drives out another, so it may shake, and open the present Obstruction.

THE 9th Chapter, being the 6th *Lemma*, shews how, from these Principles, new, or second, Principles may be generated in the Blood, and other Humours, which may prove morbific and mischievous.

WHETHER the Blood be composed of Galenic, Chymic, or Democratic Principles, yet it must be granted, that it may, and does receive such Alterations, both in its more fluid and solid Parts, as to cause great Disorders in the Body. Thus, by the Circulation, some Parts are brought together and stopp'd, where they ought not to be ; and, by Fermentation, some are raised up, and rendered conspicuous, in Places where they should not.

THIS he exemplifies in Wine, which, according to its Fermentations, receives great Alteration from the Winds, Storms, Thunders, &c. so as to become turbid, and quite alter'd in the Texture of its compounding Parts. So tho' the Blood has not, in it, any visible, fix'd, or tartareous Salts, yet such are often brought together in strumous and schirrous Affections; which, tho' invisible in the Blood, yet are, by the Mechanism of the Body, united and stopp'd, in the Glands most commonly.

THE same may be said of the Bile, the pancreatic Juice, and other Humours ; all which, when vitiated, prove noxious to the Body.

THE 10th Chapter of *Apoplexies*, arising from morbid Principles produced within the Body, and there condensed in the solid and fluid Parts.

OUR Author begins this Chapter, with the Experiment of calcined Tartar condensing the Air in damp Places ; whence its Oil, improperly so called, *per deliquium* : Alcalizate-Nitre, the white Magnesia, I suppose he means the Pyrites, do the same, &c. The same may happen in the Humours of the

Body,

Body, by condenfing the more aerial Parts of the Blood into Water, or fixing, into a Sort of Salt, the nitrous Spirit. The alcalizate, acrid, fix'd Particles, he believes, to be what *Hippocrates* called the *Atra Bilis.*

H E remarks alfo, that as Spirit of Nitre, fix'd by Oil of Tartar into a nitrous Salt, diffolves, in warm Water, or damp Air; fo the volatile Effence of the animal Spirits, either fix'd into, or condens'd into a Kind of Salt, by fome Alchaly either produced, or introduced into the Blood, and eafily after diffolved by the warm *Serum,* breaks the fibrous Texture, and thereby difpirits the Blood; fo that it no longer furnifhes that ætherial Spirit to the *Genus Nervofum,* which is the Original of all Motion and Senfation.

I N the next Chapter he applies what he has before mention'd, to the Cafe of *Apoplexies.* Thefe Condenfations, *&c.* either fometimes proceeding flowly, in chronical Diftempers, or fometimes very quick; and, as it were, in a Moment, the forementioned Alchaly being communicated from one Part of continued Veffels, to another; fo that quickly, the Whole becomes broken, difordered, and fpoil'd.

T H I S he endeavours to explain, by thefe Sort of Dews on Shrubs, and the Grafs in Autumn, which look like the fineft Spider's Webs; but, upon the leaft Touch of the Finger, on their Center, they fly away into a fingle Drop of Dew : So, by a fmall Touch, as it were, of this noxious Matter, the whole Order and Texture of the animal Spirits become broken, from Head to Foot; and from fine, rare, and delicate; become a thick, grofs, and unactive Juice, and the whole animal Machine ftopp'd in a Moment.

H E

HE adds, that it is not always neceſſary that a lixivial Alchaly ſhould deſtroy this volatile Eſſence; ſince without any Error, or external Cauſe, *Apoplexies* may happen, ſince, as *Galen* ſays, *Etiam in ſanguine poteſt generari venenum*: But this uſually happens, when the Conſtitution of the Air contributes to ſuch Diſtempers.

THE 12th Chapter contains his Conjectures, as to the Cauſes of the frequent *Apoplexies* at *Rome*, in 1705, and the Beginning of 1706.

OUR Author ſays, that he makes no Doubt, but that in the many ſudden Deaths happening at *Rome*, in the fore-mention'd Time, ſeveral might proceed from the Cauſes ſet down, in the ſeveral Chapters of the ſecond Section; ſo that all of them cannot properly be called *Apoplexies*: Yet they being ſo unuſually frequent, he judges what he has laid down, in the preſent Section, had a great Share in producing this Evil.

HE propoſes therefore to conſider of three Matters, in ſo many Chapters.

The Thirteenth CHAPTER.

Why the forementioned Cauſes were capable of producing Apoplexies *at* Rome, *more than in other Places.*

HERE he takes Notice of the Situation of *Rome*, in the 42d Degree of Latitude, in a large low Plane, divided by the *Tiber*, where the Air being little moved by the Winds, and impregnated with mineral Exhalations, but chiefly with putrid Impurities from the neighbouring ſtagnant Waters, cannot but be prejudicial to the Health of the Body.

THIS Air, being overcharged with Impurities, becomes thick, ſo, as at a Diſtance, to look like a hovering Cloud; wherefore, being ſo denſe, it muſt preſs down, or load, more than it ſhould, its

elaſtic

elaftic Principle; fo that from the 4th and 5th Chapters, it may caufe fuch *Apoplexies* as proceed from a thick Air.

B e s i d e s, *Rome* lying expofed to the South Winds, is too often mifchievoufly affected by them. Since it is known, by common Obfervation, that when thefe Winds prevail, there is a fenfible Languifhing of the Strength and Spirits; which our Author attributes to the rarifying Heat of the Air, and, by its Dampnefs, a Diffolution of the Salts; fo that there being conveyed to the Nerves an oppreffive Quantity of Humidity, it renders them unactive.

T h e Tramontane, or North Winds, are alfo, at fome Times, very violent at *Rome*, and in its Diftrict, efpecially in the Winter; thefe, coming often unexpectedly, alter, of a fudden, the ambient Air, which, communicated to the Air within the Body, renders the Veffels unable to carry the fpiritous Effence up to the Brain and Meninges; whence Apoplectic Affections may arife.

Laftly, The mineral Impurities from Vitriol, Alum, and Sulphur, which abound in the Diftrict of *Rome*, either taken in with the Air, or Nutriment, vegetable and animal, infinuating into the Humours, may either produce in them an urinous or lixivial, alkalizate Effence, either of which may condenfe the volatile Effence of the animal Spirits. Whence the Inhabitants of *Rome* are more fubject to thefe fudden Deaths, than thofe of other Countries.

The Fourteenth C h a p t e r.

Whence Rome *was, at that Time, more than ufually fubject to* Apoplexies.

I n the Summer and Autumn of 1705, the moift hot South Winds blew almoft continually, at which Time *Apoplexies* began to be frequent.

I n

IN the following Spring a very cold Seafon fuc-
ceeded, with ftrong North Winds, with a confiderable Froft ; each of which ftop, or retard, the
Motions of the Spirits, which he confirms by two
Aphorifms of *Hippocrates*.

THE Fruits of the Year 1705, were unripe,
and the Wines poor, four, and auftere, which fince,
Ex iifdem conftamus, quibus nutrimur, muft lay
the Seeds of future Mifchiefs in the *Vifcera,* efpecially in the *Serum,* and other Fluids in the Body.
Thefe Salts being, by a continued Fermentation,
raifed into an urinous Nature, and, by the wet South
Winds, diffolv'd, and carried thro' the Body, even
to the Head and Meninges ; and afterwards, by
the cold North Winds, fix'd, in the feveral Humours, might, by an Excefs in either Cafe, caufe a
Failure, or Stoppage, of the animal Spirits.

HE believes alfo, that continual Fermentations
may turn thefe immature Salts into a Kind of lixivial Salts.

HE obferved old Men to be more fubject to this
Diftemper, than young, as he fuppofes from this
Reafon : The young Men abounding more in a
fulphureous Effence, which, when the North Winds
bring the nitrous Particles, there being a fufficient
Quantity of other, to mix therewith, increafes the
Spirits ; whereas, for Want of that Sulphur in the
old, the Blood, by the Nitre, is ftagnated, and
the few Spirits, they have, ftopp'd.

The Fifteenth CHAPTER.

Wherefore fince, in Rome, *the Caufes of this Diftemper were univerfal, yet the Diftemper was not
fo ?*

FOR the Caufes of this Difference, he gives the
different Ages, Sexes, Conftitutions, Manner of
Diet, and Way of Living : Whence, in fome,
there is fuch a juft Balance and Proportion of So-
lids

lids and Fluids, of volatile and fix'd Parts, fuch a
due Formation of the Glands, and other excretory
Veffels, that there arifes a due and regular Fermentation and Circulation of the Blood, and other
Fluids in the Body ; all which contribute to
Health. Whereas, when any of thefe are faulty, the Evil more readily feizes on the Patient;
and efpecially, if they lay up the Seeds of it, by
eating immature Fruits, or drinking four, auftere,
Wines.

As a Corollary, he adds, that the ill Temperature and Difpofition of the Air and Winds, in
thofe Years, was the occafional Caufe ; and, as a
more remote Caufe, he reckons up the unwhole-
fome Food, and bad Wine, then generally taken.

The Sixteenth CHAPTER,

*Gives feveral Remarks on the malignant Fevers,
which, at* Rome, *frequently terminate in* Apoplectic *Symptoms.*

HERE he obferves firft, that every Summer
and Autumn, at *Rome*, and in the neighbouring
Campaign, there is an univerfal malignant Fever,
commonly call'd, Fevers from the Air. This In-
fection is very fatal to Strangers and Travellers,
nay, to the Inhabitants themfelves, if they come
at that Time from a more healthy Place; or, if
leaving the City, they go to other more healthy
Places, and ftay there, or fleep there, and then
return Home.

THESE Fevers, he fays, when it is little expected, end in a fatal *Apoplexy*.

To account for this, he fays, that the Air of
different Climates has different Effects, and that it
requires fome Time, before the internal Air in the
Body can be reduced to the Conftitution of the
ambient ; which, while doing, caufes Alterations
in the feveral Fermentations. Whence the Fermentation

mentation, at that Time, is either too violent, or too remiſs. Again, Sleeping, in a different Air from what we are uſed to, cauſes thoſe Separations which are uſually made in Sleep, to be differently performed from what they uſed to be.

THE Cauſe of theſe, happening chiefly in Summer and Autumn, is from the Heats then reigning, which cauſe too great a Rarefaction of the Humours and Fluids, whence they may more eaſily be altered by the noxious Exhalations; all which entering into the Body by the Breath, or Food, produce thoſe diſorderly Rarefactions, or Fixations, of the Animal Spirits before treated of; which happening either at the Beginning, or Declenſion of the Fever, may cauſe *Apoplectic* Symtoms.

THE laſt Chapter treats of ſeveral Phænomena accompanying *Apoplexies.*

AMONG theſe he reckons up, Failure of Motion, Senſe, and Speech; Falling down; the Breathing hindered, or violent, and diſorderly; a froathy Foaming at the Mouth; a full Pulſe, vibrating, and ſometimes natural; a Relaxation of the *Anus* and *Urethra*; the Intellect and Faculties of the Mind (which, without the Nerves, cannot act) failing, &c. all which he explains, and concludes his firſt Book.

The ſecond Book is alſo divided into three Sections: The firſt of Chirurgical; the ſecond of Medicinal Methods uſed in the Cure of this Diſtemper; and the third concerning the Diet: Of all which I ſhall be but ſhort, having been alrealy too prolix in the former Part.

THE firſt Chapter concerns Chirurgical Operations in general; and the three next of the Cure of Blows, or Wounds, on the Head, Fractures of the Skull, and the like ; with the Prefcriptions of feveral Ointments, Plaifters, Salves, &c.

I n the 5th Chapter he treats of Blood-letting in *Apoplexies.* This he recommends as beneficial, and, in many Cafes, neceffary, with the Lancet, in the Arm, or Jugulars, and fometimes has been practifed in the Forehead ; but with due Refpect to the Age of the Patient, and other Circum-ftances.

I n the 6th Chapter, treating of hot Irons, he mentions hot Pans held over the Head, Stupes in Brandy fired upon the fhaved Crown of the Head, with other Cauteries applied to the Neck, Arms, Pit of the Stomach, and other Places. But above all, as the moft efficacious Remedy, he advifes the Application, to the Soles of the Feet, of an Iron heated, lefs, or more, according to the Exigence of the Patient ; of which Iron, and Manner of applying it, he gives a Figure ; affirming it the moft certain Remedy, which rarely failed of Suc-cefs.   He produces feveral Authorities for this Practice : And,

I n the next Chapter, he fhews the Method of Curing the Burn, after it has rouzed the *Apoplectic* Patient.

T h e 8th Chapter concerns Veficatories, Sina-pifmes, and leffer Cauteries, &c.

T h e laft of this Section mentions Frictions, Li-gatures, and Cupping.

T h e fecond Section relates to the Part of the Phyfician, in this Diftemper, which he handles in 14 diftinct Chapters, giving particular Directions and Recipe's, as the Cafe requires.

T h e third Section refpects the Diet, both of Perfons cured, and fubject to it ; with his Advice as to Prefervatives ; in all which there is little ex-traordinary.

T h e

THE Author concludes his whole Work with fome remarkable Cafes of Perfons, chiefly in the Hofpitals at *Rome*, either dying, with fome Obfervations on their Diffections, or happily cured, and that, moftly, by the hot Iron applied to the Bottoms of their Feet.

IN the Diffections mentioned by our Author, I find, he opened only the Heads of the dead Perfons; taking that Part to be, chiefly, if not only, affected in *Apoplexies*; which, poffibly, may be true, as to Diftempers properly fo called : Tho', on the other Hand, fudden Deaths may proceed from an immediate Stop on the Heart ; and, indeed, he obferves fome had a good and natural Pulfe, when at the fame Time they lay in an *Apopleftic* Fit.

IN all thofe who died of Hurts in the Head, he found extravafated Blood, or Matter, or both, on the *Dura Mater*, or between the two *Meninges*, with a copious *Serum*, fometimes in the Ventricles of the Brain.

IN thofe dying *Apopleftic*, after malignant Fevers, the Blood-Veffels of the *Meninges* were turgid, with a black Blood.

The Pores of the Senfitive Plant.

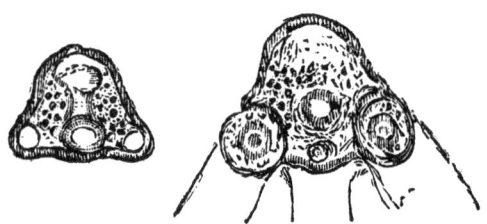

Altho' I do not find any verbal Account of the Senfitive Plant, *that thofe Figures relate unto, yet I think fit to infert them, becaufe they may probably be of Ufe to Perfons that are minded to enquire into the Mechanifm of that uncouth Vegetable.*

W. DERHAM.

The Mechanical Way of Drawing Conical Figures.

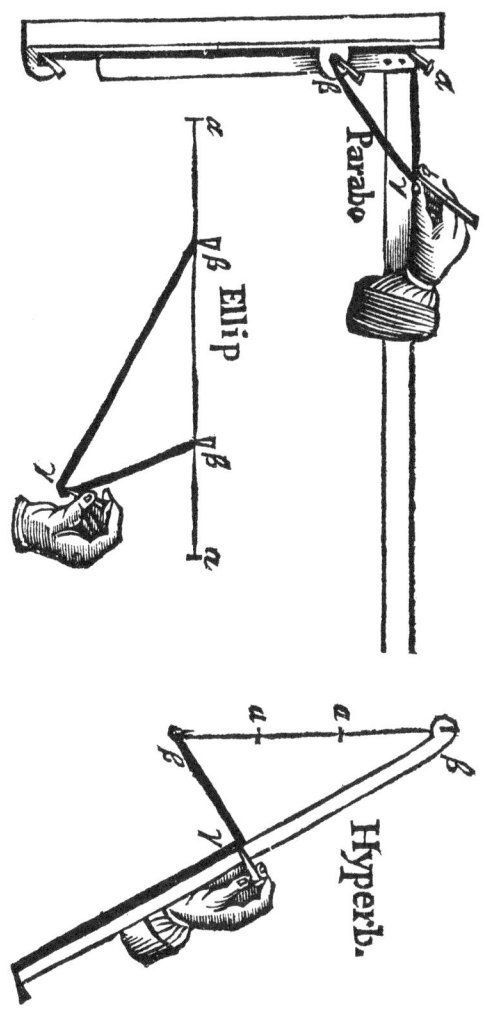

α α *Vertices,* ⎫
β β *Foci,* ⎬ *Sectionis.*
γ *Punctum,* ⎭

An

An Extract of a Letter to R. WALLER, *Efq; from* Dr. Cotton Mather, *dated* Decemb. 1, 1713, *at* Bofton *in* New-England, *of a Woolly Subftance falling in a Shower of* Snow.

THo' I have unhappily miflaid the large and well-attefted Account of what follows, yet, however, my Memory fufficiently ferves me, to affert fo much as may afford you a tolerable Satisfaction: Which is, That at a Town in one of our Colonies, called *Fairfield*, in the Depth of the Winter, there fell a *Snow*, as at other Times; but there was a large frozen Spot, (of I have now forgot juft how many Acres) which, inftead of the *Snow* that lay covered in other Places, was covered with a very confiderable Quantity of that *Wooll*, whereof I now tender a Specimen to your Acceptance.

Mr. WALLER'*s Relation of petrified Bodies of Men,* &c.

November 12, 1713. *Mr.* Baker, *who had been Conful at* Tripoli, *&c. gave me this Relation.*

ABOUT 40 Days Journey, S. E. from *Tripoli*, and about feven Days from the neareft Sea-Coaft, there is a Place called *Ougila*, in which there are found the Bodies of Men, Women, Children, Beafts, and Plants, all petrified, of a hard Stone like Marble: That about 1654, or 5, the *Corfairs* having taken feveral of the *Englifh* Ships, Admiral *Blake* was fent with a Squadron of Men of
War

War to *Tripoli* ; from which Place and *Tunis*, he had all the Captives delivered without Ranfom ; at which Time, the Report of this Difcovery of the above-mentioned was new, fo that he obliged the Alkade to procure a whole Figure for him, which he promifed.   But *Blake* not ftaying long enough there, but failing to *Leghorne*, he fent a fmall Frigat to *Tripoli* to fetch it a-board ; in which Frigat one Mr. *Hebden* (then a young Gentleman) went, who told Mr. *Baker*, that he himfelf faw a Figure of a Man petrified, which was conveyed to *Leghorne*, and thence to *England*, and that it was carried to Secretary *Thurlow*.

T H E fame Mr. *Baker* told me, That when he was at *Tripoli*, he fpoke with feveral *Turks*, who affirmed themfelves to have been Eye-Witneffes of the faid Petrifactions : That, particularly, an Officer that commanded a Garrifon of 200 Men, on a Frontier Place, called *Derney*, not many Days Journey from the Place, had promifed him to procure a Figure thence ; the fame affirmed the Relation ; that, accordingly, he fent fome Spies to find the Place, which, at that Time, they could not, as he fent him Word, it being wholly buried in the Sands, which in that Country are carried in great Clouds ; that a ftrong North Wind blows the Sands off, and by that Means difcovers the Place ; which, at other Times, is covered by thefe Sands.

H E farther told me, That this Mr. *Hebden* died about two Years fince, a Prifoner in the Fleet, tho' he had been formerly fent to *Mofcow* by King *Charles* II.   He faid, he had procured the Arm of a *Fig-Tree*, as big as his Arm, petrified ; whereon the Bark and Wood were plainly vifible ; the Bark grey, the Wood yellowifh, of the true Colour of the Plant ; that in the Bark was a Grove, in which were feveral fmall Infects like the Lady-
Cow

Cow petrified; that he had preſented this Piece of petrified Wood to my Lord *Torrington*, in whoſe Poſſeſſion he believes it now is.

Dr. Hook's *Anſwer to ſome particular Claims of Monſ.* Caſſini's, *in his Original and Progreſs of Aſtronomy.*

HAVING lately peruſed a Diſcourſe of Monſ. *Caſſini*, concerning the Original and Progreſs of Aſtronomy, and of its Uſe in Geography, and Navigation, I could not chuſe but take Notice of ſeveral Paſſages of it, which ſeem more particularly to concern this *Honourable Society*; and the rather, becauſe I do not find that it hath been mentioned by any hitherto, but ſuffered to paſs into the World for Authentick, and will be ſo concluded by the future learned World, if it be not otherwiſe informed of the Errors, or Miſtakes, therein contained.

THE firſt is, concerning the Beginning, and Original, of the *Royal Society* : Concerning which he might have been much better informed, if he had taken Notice of what is ſaid concerning it in Dr. *Sprat*'s Hiſtory thereof; but that, it ſeems, did not ſo well ſuit to his Deſign of making the *French* to be the firſt. He makes, then, Mr. *Oldenburg* to have been the Inſtrument, who inſpired the *Engliſh* with a Deſire to imitate the *French*, in having Philoſophical Clubs or Meetings; and that this was the Occaſion of founding the *Royal Society*, and making the *French* the firſt. I will not ſay, that Mr. *Oldenburg* did rather inſpire the *French* to follow the *Engliſh*, or, at leaſt, did help them, and hinder us. But 'tis well known who were the principal Men that began and promoted that
De-

Defign, both in this City, and in *Oxford*; and
that a long while before Mr. *Oldenburg* came into
England. And not only thefe Philofophick Meetings, were before Mr. *Oldenburg* came from *Paris*;
but the Society itfelf was begun, before he came
hither; and thofe, who then knew Mr. *Oldenburg*,
underftood well enough, how little he himfelf knew
of Philofophick Matters.

T h e next Thing, I take Notice of, is his afferting the *Royal Academy*, at *Paris*, to be the
Inventors of many Inventions, and Improvements,
of Aftronomical Helps, which were invented, and
improved here, by fome of this *Society*, before
that at *Pari* was founded.

T h e firft Thing, he inftances in, is the Pendulum Clock, which, he fays, was invented by one
of the Members of that *Academy*. I fuppofe he
means Monf. *Chr. Huygens*, becaufe he mentions
the Regulation of them by the Cycloid: Now,
'tis well known, that this Perfon was a Member of
the *Royal Society* four or five Years before the
Royal Academy was founded, which was not till
the Year 1666: The *Royal Society* has, therefore,
more Right of Claim to that Improvement, than
the *Royal Academy*; but, indeed, the Invention
was precedent to both, and was made in *Holland*,
and from thence fent into *England* about the Year
1659, or 1660.

T h e next Thing, he lays Claim to, is the Regulation of Watches, by a Spring applied to the
Balance; but that is fomewhat more injurious
than the former: For, it was not pretended to by
Monf. *Zulichem*, till about the Year 1675; whereas it was here invented, before the Year 1660; in
which Year, I, and three other Members of this
Society, had a Grant of a Patent for the Ufe thereof; and fome Years after, when Monf. *Zulichem*

came

came to be informed of it, he wrote a Letter a-
gainſt it as a Thing not practicable.

T h e 3d Thing is about the finding a Standard
for an univerſal Meaſure by the Length of a Pendulum vibrating a certain Time. This, I believe, was
firſt invented, and tried, by Sir *Chriſtopher Wren*,
ſome Years before the Beginning of the *Society*.

B u t that this Length would not be the ſame,
all over the World, was diſcovered by me to this
Society, 32 or 33 Years ſince, as will appear by the
Regiſters of this *Society*.

T h e 4th Thing, he inſtances in, is the Improvement of Teleſcopes, both for Length and
Goodneſs, which was firſt performed here by Sir
Paul Neile, Sir *Chriſtopher Wren*, and Dr. *Goddard*,
who inſtructed and employed Mr. *Reives* in the
manual Operation ; and, by that Means, it was carried to the Perfection of making Object-Glaſſes of
60 and 70 Foot long, very good, before any Mention
was made of ſuch being made in *France*. Some ſuch
Attempts, indeed, had been made in *Italy*, by *Divini* and *Campani* : But upon comparing one of
the beſt of them, brought hither by Mr. *Monconys*, I found, that a Teleſcope I had then by me,
of Mr. *Reives's* making, of the ſame Length with
the *Italian*, was full as good, if not better ; which
Mr. *Monconys* acknowledged.

A 5th Thing, he inſtances in, was a Way of uſing
theſe Object-Glaſſes without Tubes. This I practiſed here long before any Mention was made of
its being known beyond Sea, where, I ſuppoſe, it
was firſt uſed by Mr. *Huygens*, who hath printed
a little Diſcourſe concerning it ; but that was above
20 Years after I had uſed it here in *England*.

A 6th Thing is the Application of Clock-Work,
to keep the Glaſs directed to the Object ; but who
contrived this Application, will appear by my Ani-
madverſions on the *Machina Cæleſtis* of *Hevelius*.

A 7th

A 7th Thing, he inftances in, is the Application of Telefcope Sights to Inftruments, which was invented and perfected here long before any fuch were to be found, or heard of, in *France.* And Mr. *Bullialdus,* and feveral other of the *French* Aftronomers, as well as *Hevelius* in *Dantzick,* and Dr. *Wallis* here, did difapprove of them, after I had publifhed the Ufe and great Benefit of them, for Sights of Inftruments, in my Micrography, in my Attempt to prove the Parallax of the Earth's Orbit, and in my Animadverfions; and by the Letters publifhed by *Olhof* for *Hevelius,* it will appear how much the World was then of another Mind.

A N 8th Thing is the Ufe of a Micrometer, &c.

Concerning which I fhall refer to our Philof. Tranfact. N° 352, *where I have given a fufficient Anfwer to his Claim of the* French *Gentlemen, by afferting that and other Inventions to Mr.* Gafcoigne.

W. DERHAM.

It would be too tedious to mention all the Particulars, which he intitles the *Royal Academy* to the Honour of the Invention of, to which, in Truth, they have no juft Pretence of Claim. However, I conceive, it might not be improper for fome Perfon to vindicate the right and juft Claim of this *Society,* that may ftop the Mouths of fome malicious Men, who will needs fay, that this *Society* hath invented or improved nothing of real Ufe.

F I N I S.

THE
INDEX.

INDEX.

INDEX.

INDEX.

Metals

INDEX.

INDEX.

INDEX.